# Handbook of Engineering Heat Transfer

# Handbook of Engineering Heat Transfer

Editor: Edgar Miller

NY Research Press
New York

Published by NY Research Press
118-35 Queens Blvd., Suite 400,
Forest Hills, NY 11375, USA
www.nyresearchpress.com

Handbook of Engineering Heat Transfer
Edited by Edgar Miller



International Standard Book Number: 978-1-64725-386-8 (Hardback)



**Cataloging-in-publication Data**

Handbook of engineering heat transfer / edited by Edgar Miller.
p. cm.
Includes bibliographical references and index.
ISBN 978-1-64725-386-8
1. Heat--Transmission. 2. Heat exchangers. 3. Energy transfer. 4. Thermodynamics. I. Miller, Edgar.

QC320 .H36 2023
621.402 2--dc23

# Contents

# Preface

Heat transfer refers to a branch of thermal engineering which deals with the conversion, generation, exchange and utilization of thermal energy within physical systems. There are several mechanisms of heat transfer including thermal radiation, heat conduction and heat convection. There are various examples of heat transfer, such as greenhouse effect, refrigeration and maintenance of optimum temperature within the human body. Heat-transfer principles can be employed for maintaining, decreasing or increasing temperature in a variety of situations. It has numerous applications in industries such as milk pasteurization, temperature control for purified water, laser cooling, night vision cameras, and magnetic cooling. Many disciplines utilize heat transfer methods including materials processing, automotive engineering, power station engineering, insulation, chemical engineering and climate control. This book outlines the processes and applications of heat transfer in detail. It will provide comprehensive knowledge to the readers.

This book is a comprehensive compilation of works of different researchers from varied parts of the world. It includes valuable experiences of the researchers with the sole objective of providing the readers (learners) with a proper knowledge of the concerned field. This book will be beneficial in evoking inspiration and enhancing the knowledge of the interested readers.

In the end, I would like to extend my heartiest thanks to the authors who worked with great determination on their chapters. I also appreciate the publisher's support in the course of the book. I would also like to deeply acknowledge my family who stood by me as a source of inspiration during the project.

**Editor**

# 1

# Fouling in Heat Exchangers

*Sergio García and Alfredo Trueba*


## Abstract

A major problem in industries that use heat exchanger equipment cooled with water in their industrial processes is biofouling. In the design and operation of heat exchangers cooled with water, a coefficient of biological must be considered, which affects the efficiency of the equipment. For this reason, it is necessary to apply appropriate antifouling treatments to the design of each heat exchanger. In order to minimize the undesirable phenomenon of biofouling, various mitigation methods have been developed over the last 30 years, both online and offline, of a physical, chemical, or biological nature. Most of these methods are well contrasted and are applied in the regular operation of the facilities, although some methodology approaches are in the research and development phase. However, the application of most of these methods requires interrupting the production, periodically, in order to clean the biofouling, seriously damaging the performance and operation of the installation. The "online" methods to biofouling control are chemical (oxidizing, and non-oxidants), biological and physical treatments. Nowadays, other meth-odologies of biofouling mitigation that do not affect the environment are being investigated, although, until now, none have been found that are substitutes for chemical agents and that have the same or with higher efficiency.


**Keywords:** biofouling, biofilm, fouling, antifouling, energy, efficiency

## 1. Introduction

A major problem in industries that use heat exchange equipment cooled with seawater in their industrial processes is biological fouling (biofouling) (**Figure 1**). Biofouling acquires greater relevance when it develops in the marine environment, due to the high biological activity of seawater [1]. In the design and operation of heat exchangers cooled with seawater, a biological fouling coefficient must be considered, which affects the efficiency of the equipment [2]. For this reason, it is necessary to apply appropriate antifouling treatments to the design of each heat exchanger [3].

In order to minimize the undesirable phenomenon of biofouling, various mitiga-tion methods have been developed over the last 30 years, both online and offline, of a physical, chemical, or biological nature. Most of these methods are well contrasted and are applied in the regular operation of the facilities, although some methodol-ogy approaches are in the research and development phase. However, the applica-tion of most of these methods requires interrupting production, periodically, in order to proceed with biofouling cleaning, as this seriously impairs the performance and operation of the facility (**Figure 2**).

Oxidizing chemical methods are available [4] and non-oxidants [5], biological [6–8] for "on-line biofouling control. Nowadays, other methodologies of biofouling mitigation that do not affect the marine environment are being investigated [3],

**Figure 1.**
*Presence of mussels in the water box of a condenser (Source: By courtesy of Nuclenor).*

**Figure 2.**
*Tubular heat exchanger in maintenance due to the presence of biofouling in its tubular bundle (Source: By courtesy of Merus-the water conditioner company).*

although, until now, none have been found that are substitutes for chemical agents and that have the same or with higher efficiency.

Other "offline" methods, such as antifouling coatings (AF) or mechanical cleaning, are widely used in heat exchangers. AF coatings have the drawback of using chemicals that can harm the environment. Nowadays, new coatings are being researched to avoid the growth of biofouling, without harming the environment [9, 10].

## 2. Classification of the fouling

For a better understanding of the biofouling problem, this chapter begin of the description general term "fouling" and its types, among which is biofouling. This term is used to define the accumulation of biotic and/or abiotic deposits on a surface in contact with a fluid. This type of soiling can occur in nature in many different ways and is usually classified in the following categories [11]:

- Biological fouling
- Fouling by chemical reaction
- Fouling by corrosion

- Ice-cold fouling
- Fouling by precipitation
- Fouling by particles

## 2.1 Biological fouling

The biological fouling or biofouling in the Anglo-Saxon literature depends on the material on which the accumulation of macrofouling or macrobial biofouling occurs, which originates from microfouling or microbial biofouling.

## 2.2 Fouling by chemical reaction

Deposits caused by chemical reaction can be initiated by the contact of two chemicals in a heat transfer process. In some cases, the metal surface favors precipi-tation by acting as a catalyst.

This type of fouling is frequently observed in the petroleum industry, in which the refining processes induce the polymerization of unsaturated hydrocarbons (**Figure 3**).

## 2.3 Fouling by corrosion

This type of fouling occurs as a result of electrochemical or microbial corrosion of a surface in contact with a fluid (**Figure 4**). The composition of corrosion products is normally a function of the material from which they originate. According to [12] cor-rosion must be taken into account in the design of heat exchangers in order to gener-ate more information on the thermal resistance associated with fouling by corrosion.

## 2.4 Freezing fouling

It is the accumulation of solid deposits on a refrigerated surface due to the solidification of a fluid or one of its constituents. The accumulation of this type of fouling depends on the temperature of the surface and the shearing effect produced by the fluid on the surface as a function of its velocity of circulation [13, 14]. This phenomenon is seen in the solidification of the wax, whose melting point is high, in a hydrocarbon solution on a heat exchange surface (**Figure 5**).

**Figure 3.**
*Residue of hydrocarbon in the tubes of a heat exchanger (Source: By courtesy of H&C Heat Transfer Solutions Inc. 2013).*

**Figure 4.**
*Oxides of corrosion in the tubes of a heat exchanger (H&C Heat Transfer Solutions Inc. 2013).*

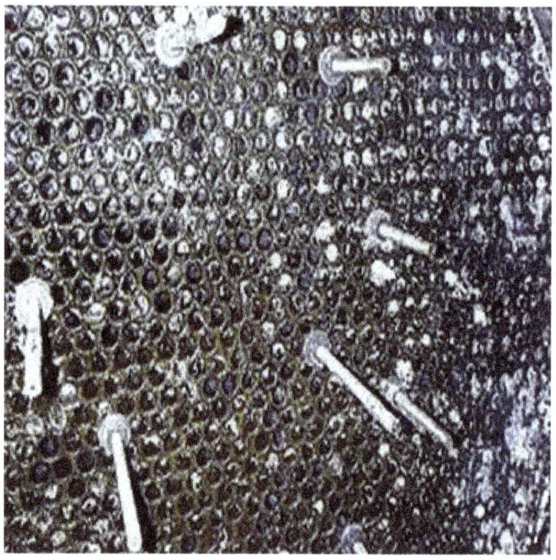

**Figure 5.**
*Icy icing of polymer products in a heat exchanger (Source: By courtesy of H&C Heat Transfer Solutions In.).*

## 2.5 Fouling by precipitation

The origin of this type of fouling is the precipitation or scaling of dissolved substances on the surfaces. Precipitation usually occurs from aqueous solutions based on their temperature and the solubility of dissolved substances, such as calcium carbonate ($CaCO_3$), whose precipitation of the solution occurs on a superheated surface when the water contained in the water evaporates (**Figure 6**).

The precipitation of the salts dissolved in the water originates when coming into contact with the wall of the hot tube and supersaturation occurs. The deposits adhered to the surface may have different hardness depending on their composition. Thus, when the deposit is only composed of salts, it is hard and very adherent. On the other hand, if the deposit is a mixture of salts and suspended solids, its hardness is low, and they adhere weakly [15].

The precipitates usually observed in industrial facilities that use aqueous solutions are the following [16]:

- Calcium carbonate
- Calcium sulfate
- Calcium oxalate
- Barium sulfate
- Magnesium hydroxide
- Silicates
- Aluminum oxides
- Aluminum silicate
- Copper
- Phosphates
- Magnetite

These precipitates are very common in boilers and heat exchangers that operate with water of high hardness.

### 2.6 Fouling by particles

It is the accumulation of solid particles that are suspended in a fluid (liquid or gas). The particles are deposited on surfaces by means of various mechanisms, some of which are closely linked to properties of particles as specific as coagulation. The amount of deposit and the rate of accumulation on the surfaces depend on the size and nature of the particles, as well as the environmental conditions. When gravity is the main mechanism of deposition of suspended particles on a surface, fouling is called sedimentation (**Figure 7**) [17].

**Figure 6.**
*Precipitated salts in the tubes of a heat exchanger (Source: By courtesy of H&C Heat Transfer Solutions Inc).*

**Figure 7.**
*Accumulation of solid particles in the plate of a tubular heat exchanger (Source: By courtesy of AH&C Heat Transfer Solutions Inc.).*

## 3. Biofouling

The term biofouling is clearly defined and accepted by the scientific community, although in the various definitions proposed by different authors some aspects of others are highlighted, depending on the importance given by each author to the different facets of the phenomenon. Below are several definitions of this term, published over the past years by different authors:

- Biofouling is simply the union of an organism or organisms to a surface in contact with water over a period of time [18].
- Biofouling is the colonization of submerged surfaces, produced by unwanted organisms such as bacteria, algae, and barnacles. Biofouling has detrimental effects on transport and leisure vessels, heat exchangers, oceanographic sensors, and aquaculture systems [19].
- Biofouling can be defined as the union and growth of a community of plants and animals [20].
- Biofouling is defined as the fixation and growth of organisms on submerged surfaces of an artificial nature. The accumulation of biofouling includes barnacles and other sessile marine invertebrates [21].
- Biofouling is the unwanted accumulation of microorganisms, algae, arthropods, or mollusks on a surface [18].

The definition of the term biofouling accepted is the following: "biofouling can be defined as the undesirable phenomenon of adhesion and accumulation of biotic deposits on an artificial surface submerged or in contact with seawater." This accumulation or embedding consists of an organic film composed of microorganisms embedded in a polymer matrix created by themselves (biofilm), where they can reach and retain inorganic particles (salts and/or corrosion products) as a result of other types of fouling developed in the process. This biofilm composed of microorganisms (microbial biofouling or microfouling) can lead to the accumulation of macroorganisms (macrobial biofouling or macrofouling) [22].

### 3.1 Evolution of the research of biofouling

Zobell [23] examined a natural marine population through a direct microscope and observed that the number of bacteria adhered to the surfaces was much higher than that found in the medium. In addition, these bacteria were characterized by showing an activity and high growth rates, concluding that the bacteria were attracted to the surfaces to which they adhered to form sessile populations.

The first detailed examination of the nature and composition of the biofilm had to wait for the appearance of the electron microscope, which provided a greater extension with respect to optical microscopy and, in its scanning and transmis-sion modalities, was able to show the variety of microorganisms that made up the biofilm in a treatment plant [24]. Staining the biofilm with ruthenium red and fixing it with osmium tetroxide allowed to show that the material that surrounds and encloses the cells that compose it is composed mainly of polysaccharides. Costerton et al. [25], taking as a starting point the observations of sessile communities in the mountain streams, elaborated a theory that explains the mechanisms through which microorganisms adhere to living or inert materials, as well as the benefits obtained for their ecological niche. From this moment, numerous studies of the biofilm were developed in both industrial and ecological scenarios [26].

The possible effects of the biofilm in industrial processes are derived from the beneficial or harmful reactions that can be carried out by the microorganisms that compose it and that depend on the environmental conditions of the environment. These conditions have a great influence on the growth and metabolic activity of the biofilm [27].

The bacteria that make up the biofouling can be up to a thousand times more resistant to antibiotics than the same bacteria grown in a controlled liquid medium. The mechanisms responsible for this resistance include (1) the physical and chemical diffusion barrier that constitutes the matrix of the biofilm to the penetration of antimicrobials, (2) the slowed growth of biofilm bacteria due to nutrient limitation, (3) the existence of microenvironments that antagonize the action of the antibiotic, and (4) the activation of stress responses, which cause changes in the physiology of the bacteria and the appearance of a specific biofilm phenotype that actively combats the negative effects of antimicrobial substances [28]. Due to this resistance, antifouling substances must be highly effective and incorporated at considerably high concentrations, which can lead to harmful effects on the environment.

### 3.2 Microfouling

The biofilm (microfouling) is composed mainly of water, in intervals that are between 87 and 99% of the total content. The rest of the components is very variable depending on the medium in which it is studied [17]. The matrix of the biofilm is a complex formed mainly by exopolysaccharides and bacterial cells. Other compounds that can be found are macromolecules such as proteins, DNA, and vari-ous products derived from the destruction of the cell wall of bacteria [29].

#### *3.2.1 Development of the biofilm*

The biofilm begins to be generated when an individual cell initially joins a surface. The ability of this cell to carry out this binding and initiate cell growth depends on factors such as temperature and pH of the medium, genetic factors that encode its motor functions and environmental sensitivity, and those that generate adhesins and other proteins [30–33]. The factors that directly affect its development depend mainly on the microbial species, although a part of these characteristics is common to numerous bacteria. Once initial biofouling adhesion is produced, cell growth and expansion begin on the surface, forming monolayer microcolonies. At the same time, the cells modify their activity and begin the complex process of structure forma-tion of the biofilm. The most obvious of these changes is the production of the exopolymer matrix (EPS), which will unite the whole [34]. If the conditions of the medium allow it, the biofilm will grow and spread to non-colonized areas releasing cells that will be distributed through the water in search of new sur-faces to colonize.

The formation of the biofilm is a systematic process of predictable evolution, in which five phases are differentiated [35]: (1) the reversible adsorption of the bacterium to the surface, (2) the irreversible union, (3) the first maturation phase with growth and division, (4) the growth phase with production of the exopolymer, and (5) the final development of the colony with dispersion of colonizing cells (**Figure 8**).

### *3.2.2 Conditioning of the surface*

The ability to bind to different materials depends on the specific proteins of its coat and the bacterial motor appendages. According to Pedersen [36], stainless steel can be as susceptible to the formation of biofilms as plastic. This is because the organic matter present in the water previously comes in contact with the surface, depositing an organic layer in the water/surface interface that changes the chemical and physical properties of the surface, improving the possibilities of fixing the bacteria.

### *3.2.3 Adsorption of molecules*

The adsorption of ions and other dissolved substances (sugars, amino acids, proteins, fatty acids, etc.) begins when the material submerges under seawater

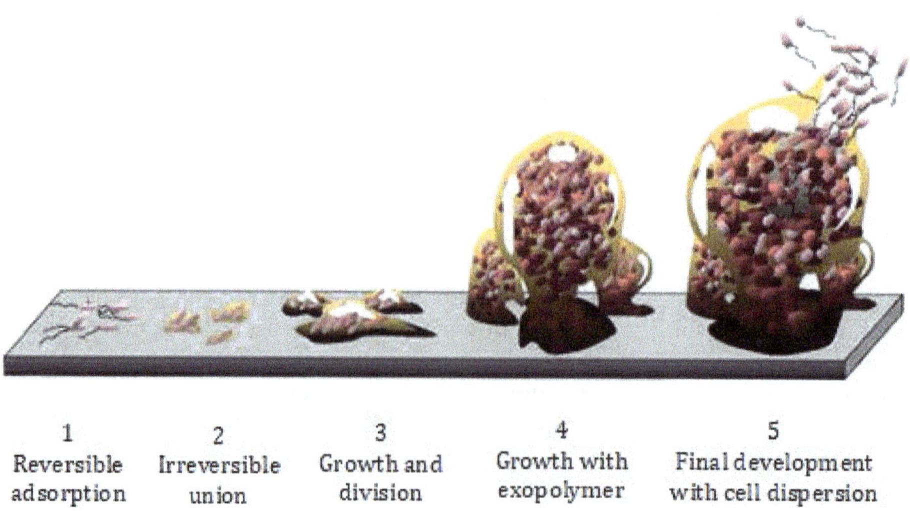


**Figure 8.**
*Phases of biofilm formation (Source: By courtesy of Center Biofouling Engineering).*

and accelerates when the fluid begins to circulate through the heat exchanger. This adsorption occurs rapidly on the surface, saturating the concentration of substances on the surface in a few minutes [37].

### *3.2.4 Adhesion of microorganisms*

The adhesion of microorganisms to a substrate can be active (by flagella, fimbrias, adhesins, capsules, and surface charges) or passive (by gravity, diffusion, and fluid dynamics). In the absence of these mechanisms, the bacterial cells would be repelled by the surface when presenting electric charges of the same sign [38].

In a few minutes, the free bacteria form a reversible EPS matrix with the "conditioned" surface (**Figure 9**) [39], whose characteristics depend on the electrical charges of the bacteria. These attractive forces have their origin in hydrogen bonds, cation bonds, and van der Waals forces that compete with the forces of repulsion. If this union is maintained long enough, new chemical and physical structures appear that make it permanent and irreversible [40].

In cases of high microbial population density or lack of nutrients in the water, some microorganisms are able to individually alter their cell wall to make it more hydrophobic and increase its greater affinity of adhesion toward the surfaces. When the microorganisms approach the surface, with almost no water flow, they are attracted, proving their affinity for union and fixation (**Figure 10**) [41].

During the reversible adsorption stage, the bacterial cells still show Brownian motion and are easily removed with a nonaggressive cleaning method. The irreversible union implies the anchoring of bacterial appendages and the production of exopolymers, which determines that the mechanical action necessary to detach them will be greater depending on the time that the biofilm is active.

Bacteria undergo important transformations in their structure to adapt to the environment. These transformations activate different genes that encode new struc-tural proteins and enzymes, which explains the adhesion and resistance of biofilm bacteria to antibiotics and disinfectants. In recent years, the advances made in the field of proteomics and genomics have allowed the identification of 800 proteins that modify their concentration throughout the five phases of the biofilm develop-ment and clarify the complex process of biofilm formation [42, 43].

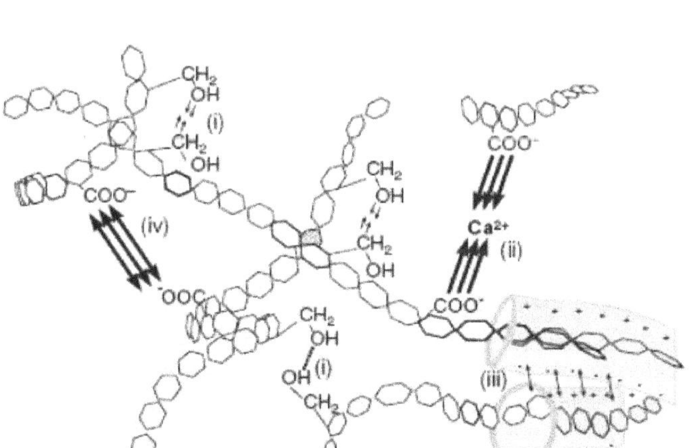


**Figure 9.**
*Binding forces in an EPS matrix: (i) hydrogen bonds, (ii) cation bonds, (iii) van der Waals forces and (iv) repulsion force [38].*

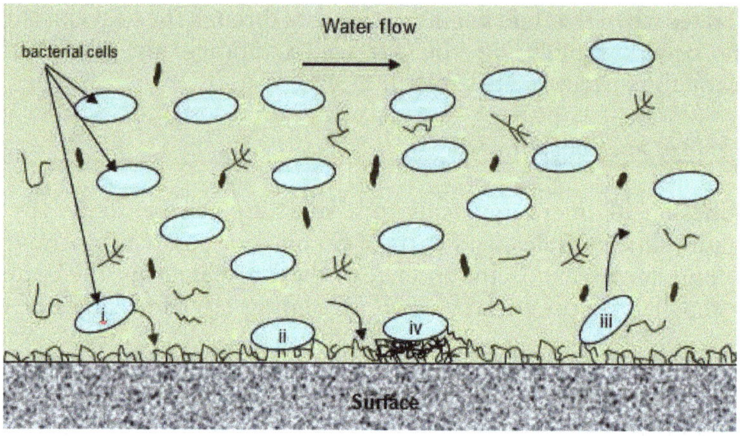


**Figure 10.**
*Water flow transport of a bacterial cell to the conditioned surface: (i) adsorption, (ii) reversible adsorption, (iii) detachment and (iv) irreversible adsorption [40].*

### *3.2.5 Maturation*

Environmental stability favors biofouling growth and multiplication of cells which allows to generate a polyanionic polymer mixture of silty and sticky consistency whose are excreted to the outside to facilitate union cells onto surface.

Although its composition is not completely known, the mixture of exopolymers is formed by polysaccharides or glycoproteins of various sugars (glucose, fructose, mannose, N-acetylglucosamine, and others) [44] and, additionally, may contain free proteins, phospholipids, and nucleic or teichoic acids [45]. These free proteins are useful for retaining nutrients and protecting bacteria from various biocides.

Extracellular exopolymeric material or glycocalyx is expelled from the bacterial cell wall and adopts a reticular structure reminiscent of a spider. This structure is formed from groups of polysaccharides both neutral and carriers of electric charges, which act as ion exchange systems, capable of capturing and concentrating the nutrients present in the medium.

The structure of the biofilm interacts in a complex way, showing a behavior similar to that of multicellular organisms. If a microorganism generates toxic waste, another will use it as food and, in this way, coordinate the biochemical resources of all the beings that inhabit the matrix of the biofilm. In addition, some bacteria clump together within the matrix with a series of enzymes that allow them to digest nutrients that no isolated species could digest. Also, these enzymes will be used to respond to the attack of various biocides.

Anaerobic biofilms can be developed under the aerobic layer whose structure is permeabilized with a mesh of furrows crossed by water, bacterial debris, enzymes, nutrients, metabolites and oxygen. The gradients of ions and molecules that are established between the different zones generate the necessary impulse to take the substances to the periphery of the biofilm where most of the cells are located. The number of these cells is reduced with the age of the biofilm, being 80% in a young matrix and 50% in an aged shade [42].

### *3.2.6 Growth and dispersion*

Biofilm formation has continuous divisions of the cellular matrix colony, which means a periodic detachment of groups of cells that deposit downstream.

These new colonies take advantage of the release of residues and nutrients from the original colonies to prepare the new surface and to feed other cells, and, as a consequence, the new colonies will grow and spread much more rapidly than in the original biofilm. This colonization is related to the long-term evolution and survival of bacteria [30].

If the conditions of the fluid allow it, the equilibrium that is established between the growth of the colony and the movement of the water releases few cells (**Figure 11**). With an intense or turbulent flow, many more can be released, even releasing whole areas of the biofilm colony.

### *3.2.7 Contribution of nutrients to the biofilm*

The main factor controlling the growth of the biofilm is the availability of dissolved nutrients and their conversion into accumulated biomass. In cooling water circulation systems, the transfer of nutrients to the biofilm tends to increase with flow velocity [41]. Also, the rough surfaces of biofilms increase the transfer of nutrients about three times in relation to smooth surfaces [30, 46]. The control of nutrients is a way to control the development of the biofilm [47–49]. Melo and Bott [50] observed, in an industrial refrigeration system, an increase of 400% in the thickness of the biofilm at a speed of 1.2 $ms^{-1}$ for an increase in the nutrient level of 4 mg $L^{-1}$ at 10 mg $L^{-1}$.

The chemical composition of the waters determines the number, diversity, metabolic state of the bacteria, and their tendency to adhere to surfaces [51]. So far there has been no study that directly relates nutrients, biofilms, and colonization. Huang et al. [52] demonstrated in the laboratory that the availability of nutrients and the synthesis of new proteins for the formation of biofilms of *Pseudoalteromonas spongiae* under static conditions and without added nutrients affected the induction and adhesion of the biofilm to the surface. The effects of organic substances in the form of amino acids on the bioactivity of the biofilm were studied by Jin and Qian [53, 54]. The results of this study showed that the incorporation of aspartic acid and glutamic acid causes a significant increase in the bacterial mass, modifies its structure, and increases the inducing effect of biofilm formation. In addition, Huang et al. [55] found that the characteristics of biofilms generated in habitats with different environmental conditions show remarkable differences in the bioactivity of the larval settlements of the barnacle, which

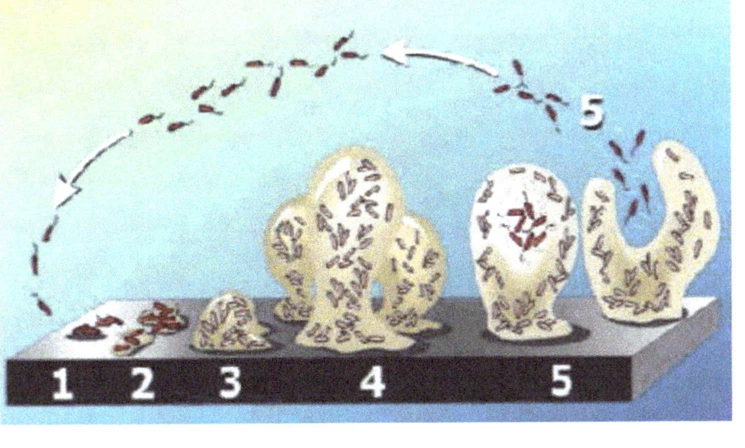


**Figure 11.**
*Microbial cell transport during the growth phases of the biofilm (Source: By courtesy of Center Biofouling Engineering).*

suggests that the nature of the nutrients is a determining factor of the biological activity in the biofilm.

The nutrients required by microorganisms for feeding are divided into two broad categories [56]:

- Macronutrients (C, H, O, N, P, S, K, Mg)
- Micronutrients or trace elements (Co, Cu, Zn, Mo, etc.)

In nature, these elements are combined as part of organic and inorganic substances. Some of the nutrients will be incorporated to build macromolecules and cellular structures; others only serve for energy production not directly incorporated as cellular material, and others can perform both functions [56].

The bacterial colony exhibits a great metabolic versatility in the use of nutrients. Autotroph bacterials obtain their carbon by reducing $CO_2$ and other elements from inorganic sources. On the other hand, heterotroph bacterials use to wide ranges of organic carbon sources. In turn, within the heterotrophs, you can find many and varied types of nutrition, from methylotrophic bacteria that only use methane or methanol as carbon and energy sources to the very versatile *Pseudomonas*, which can resort to degrade more than 100 types of carbon sources (including aliphatic and cyclic hydrocarbons). In any case, among the heterotrophs, the most widespread source of carbon is glucose [56].

### *3.2.8 Abiotic and biotic factors that influence the development of the biofilm*

Alterations in the structure of biofilms can be caused by abiotic factors such as [30]:

- Depth.
- Illumination.
- Exhibition time.
- Height of the tides.
- Water circulation regime.
- Physical alteration.
- Latitude.
- Season.
- Water chemistry.
- Supply of nutrients.
- Physical and chemical (abiotic) conditions at the seawater/biofilm interface are an important factor for the development of the biofilm. As the thickness of the biofilm increases, variations in pH, dissolved oxygen, and metabolic by-products are more important and generally directly influence its development [57].

The biotic factors that can alter the structure of the biofilm are [40]:

- Availability and physiological state of organisms, interaction between organisms, and biological disturbances [58–60]
- Properties of adhesion of microorganisms under conditions of turbulence.
- Physiological state of the organisms that have a selective advantage for the formation and development of the biofilm [61]. This capacity is reinforced by the secretion of EPS that are more resistant to high flow rates and chemi-cal agents [62]. Microorganisms in the biofilm are released when conditions become unfavorable [63].

### 3.3 Effects produced by biofouling in heat exchangers

Biofouling has a great influence on the thermal performance of heat exchangers, due to the accumulation of biotic deposits with insulating characteristics in heat exchange surfaces. This accumulation adds not only an additional thermal resistance to the flow of heat but a greater frictional resistance to the passage of the fluid. The formation of these biofilms produces important modifications, since they alter the physical–chemical conditions at the metal-solution interface and form barriers for the exchange of elements between the metallic surface and the surrounding liquid medium [64]. Among the nega-tive consequences, the decrease of the heat output of the heat exchangers and the lower durability of the construction materials of the equipment can be mentioned [65].

The main consequences produced by microfouling (biofouling) in heat exchang-ers are the following [5]:

- Production losses due to the decrease in efficiency and the scheduled and unscheduled shutdowns of the installation
- Maintenance costs, resulting from the elimination of biofouling deposits with chemical and/or mechanical devices
- Increased corrosion processes in metallic components, induced by microbio-logical corrosion (MIC) [66]
- Increased consumption of water, electricity, fossil fuels, and other sources to counteract the effects of biological fouling
- Increase in manufacturing costs of heat exchangers to consider an acceptable biological fouling without losing design power and to be able to mechanically clean them
- Increase in environmental risks due to the use of biocides and $CO_2$ emissions to the environment to counteract biological fouling

The main consequences of macrofouling in a heat exchanger are [40]:

- High costs caused by the complexity of the cleaning of the macroorganisms of the installation
- Losses of power produced by the loss of flow and thermal efficiency in the heat exchangers of a refrigeration installation

- Costs produced by the management of the macrofouling waste after its removal from the installation
- Production of faults due to the accumulation of macroorganisms in auxiliary systems, such as valve drives, grids, pumps, instrumentation, etc.

## 4. Design aspects that minimize the adhesion of biofouling

The design of the heat exchanger can represent an important factor to minimize the presence of biofouling on its surfaces. The main design variables that intervene in the biofouling growth process are [27]:

- Speed of the fluid
- Composition and roughness of the support
- Temperature of the support and cooling water
- Physicochemical properties of the fluid
- Biological variables
- Geometry of fluid passage

Although it is very difficult to relate the colonization of the surface with only one of these variables, it can be affirmed that they all have their influence on the process of developing the biofilm on the surface in contact with seawater. In this way, it is recommended to maintain minimum flow rates above 0.9 $ms^{-1}$ and use smooth surfaces and high temperatures (>90°C). Otherwise, the presence of biofouling on the heat exchange surface is unavoidable.

## 5. Chemical antifouling treatments

The chemical compounds used as antifouling agents in heat exchangers are called biocides. In addition, sequestrants, dispersants, or chelators can be used for other purposes (**Table 1**).

The biocides contain one or more active substances, whose function is to destroy, counteract, neutralize, and impede the action or exert a control of another type on any harmful organism by chemical or biological means. Properly dosed reduce drastically the number and activity of cells and biofilms in water. Due to its toxic nature, inadequate doses can be harmful to the environment and are therefore

| Agent | Methodology |
|---|---|
| Sequester | Form a chemical complex with the tank keeping it in suspension |
| Dispersant | It transmits load to the surface keeping the particles in suspension |
| Chelator | It forms a strong chemical complex with the species that form the encrustation |
| Biocide | Kills or incapacitates generating irreversible damage to the structure or vital functions of microorganisms |

**Table 1.**
*Chemical agents used in the treatment of seawater [56].*

| Biocide | Ionic character | pH | Optimal activity |
|---|---|---|---|
| $Cl_2$ gas | No | 5–7.5 | Algae and bacteria |
| NaClO | Anionic | 7–8.5 | Algae and bacterias |
| $Br_2$ | No | 5–9 | Algae and bacteria |
| $ClO_2$ | No | 6–10 | Algae, bacteria, fungi, and sulfate reducing bacteria |
| Chlorophenols | Anionic | 5–8 | Algae, bacteria, and fungi |
| Sales de $NH_4^+$ | Cationic | 7–9.5 | Algae and bacteria |
| Sulfur organ | Anionic | 5–7.5 | Bacteria and fungi |
| Brominated organ | No | 5–8 | Algae, bacteria, fungi and sulfate reducing bacteria |
| Isothiazolones | No | 5–8.5 | Algae, bacteria, fungi, and sulfate reducing bacteria |
| Glutaraldehyde | No | 5–8.5 | Algae, bacteria, fungi, and sulfate reducing bacteria |

**Table 2.**
*Specific operating conditions of different biocides [67].*

| Common name | Compound |
|---|---|
| BNP | 2-Bromo-2-nitro propane-1,3-diol |
| BNS | Beta-bromo-beta-nitrostyrene |
| Carbamates | Sodium dimethyl dithiocarbamate, ethylene bis dithiocarbamate disodium, disodium cyanodithioimidocarbamate, Potassium N-methyldithiocarbamate, potassium methyldithiocarbamate |
| Chlorine | Sodium hypochlorite, calcium hypochlorite trichloroisocyanuric acid, sodium dichloroisocyanide and potassium acid dichloroisocyanide |
| Chlorophenols | Sodium pentachlorophenate and sodium trichlorophenate |
| Dazomet | Tetrahydro-3,5-dimethyl-1-2-H-1,3,5-thiadiazine-2-thione |
| DBNPA | 2,2-Dibromo-3-nitrile-propiomide |
| DGH | Dodecylguanidine hydrochloride |
| Isothiazolone | 5-Chloro-2-methyl-4-isothiazolon-3-one |
| Glutaraldehyde | 1,5-Pentadial |
| HPMTS | Hydroxypropyl methanethiosulfonate |
| Hydantoin | 1-Bromo-3-chloro-5,5-dimethylhydantoin, 1,3 dichloro-5,5 dimethyl hydantoin, 1,3-dichloro-5-ethyl-5-ethylhydantoin |
| MBT | Methylene-bis- (thiocyanate) |
| Sodium bromide | Sodium bromide |
| Sulfone | Bis- (trichloromethyl) -sulfone |
| TBTO | Bis- (tri-n-butyl) tin oxide |
| TCMTB | 2(Tiocianatometiltio) benzotiazol |
| Triacin | Dimethyl propyl-n-ethyl-6- (methylthio) triazine diamine Chloro-4-ethylamino-6-t-butylamino-s-triazine |
| Tris-nitro | Tris (hydroxymethyl) nitro methane |
| Quaternary ammonium chloride | Didecyl dimethyl ammonium chloride, n-alkyl dimethyl benzyl ammonium chloride, dialkyl dimethyl ammonium chloride poly oxyethylene (dimethyl ammonium), ethylene (dimethylammonium) ethylene dichloride |

**Table 3.**
*Biocides used in cooling water [71].*

subject to strict regulations of use [10]. These rules are increasingly restrictive, so that the situation can be reached and that the concentration of allowed biocide is lower than the effective, which requires the search for other alternatives.

Biocides used in industrial refrigeration systems must have a broad spectrum, as they must limit the growth of a wide variety of microorganisms including bacteria, fungi, and algae. In addition, they must be effective in a wide range of specific conditions of equipment operation (**Table 2**).

There is a wide variety of chemical products that have been used as biocides to inhibit the formation of biofouling deposits in heat exchange systems that use seawater as a cooling fluid. According to their general principle of action, they are classified as oxidants or as non-oxidants.

They act by oxidizing the organic matter (cellular matter, enzymes, proteins, etc.) of the microorganism, and as a consequence they cause its death [68]. Its use is widespread, and its effectiveness as a widely demonstrated antifouling agent [69, 70].

In industrial refrigeration systems, different chemical compounds are employed that have the function of drastically reducing the number of cells present in the seawater-surface interface of heat exchange and minimizing their activity, having the capacity to limit the growth of a wide variety of microorganisms (bacteria and algae) and to be effective in a wide range of equipment operating conditions (**Table 3**).

## 6. Biological treatments antifouling

Biological antifouling biocides are active principles created by the agencies themselves (algae, batteries, fungi, etc.) as self-defense against other organisms in their natural environment. These substances are usually of a protein nature and are called enzymes.

Biological biocides are in the initial phase of research and development. Its use requires an in-depth analysis of the organisms that are intended to be eliminated and the environment in which they are found. The most recent research has shown the inhibition of marine biofouling by different types of bacteria, cyanobacteria, and marine fungi. Recent experiments have shown the antifouling properties of some organic compounds [72, 73], marine larvae of epibiotic bacteria [74–76], and various natural chemical compounds [77]. Dobretsov et al. [78] studied the microbial interference as a mechanism to control marine biofilms, focusing on the mechanisms of altering the biofilm by bacteria, algae, and various larvae.

## 7. Physical antifouling treatments

They represent an alternative to the use of biocides and an emerging research channel. They can be applied with the installation in operation (online systems) or with the installation in technical stop (offline systems) [5].

### 7.1 Online systems

#### *7.1.1 Taprogge system*

Mechanical cleaning system by sponge balls or elastomer material abrasives is a biofouling treatment which periodically introduce balls by the water flow to clean tubes in heat exchangers. The balls of suitable roughness are introduced into the water box of the condenser from where they pass into the tubes. Balls have a diameter greater than that of the tube which pass through tubes to clean inner surface avoiding the settlement of organic or inorganic matter.

### *7.1.2 Cleaning brush system housed in baskets*

At the ends of the tubes, there are baskets containing a brush that moves in both directions through the interior of the tube by reversing the flow of the cooling water. The investment of flow is made with a valve system that enables this action, not being necessary more than three daily cycles to maintain good conditions of the plant.

### *7.1.3 Heat treatment*

It consists of heating the cooling water by recirculating it in a closed circuit without renewing cold water or by installing a secondary hot water supply circuit from an internal combustion engine, for example. The water recirculates until reaching temperatures between 50 and 70°C, which produces the death of microorganisms [5]. The periodicity of this treatment is established taking into account the biological activity of the water. Its effectiveness depends mainly on the temperature of the water, the time of exposure, and the frequency of repetition [40].

### *7.1.4 Variation in cooling water flow rate*

The speed of water flow through the tubes of the exchanger is a determining factor in the growth of biofouling. In addition to influencing the contribution of nutrients that determine the growth of organic matter that is part of the biofouling film, deposits adhered to the inner surface of the tubes can be detached as a result of the greater cutting effect on the surface are exerted to higher water flow rates. The heat exchanger can be designed to work at high flow rates (from 1 to 2 $ms^{-1}$) or so that the water flow rate can be varied for a short period of time (e.g., from 1 to 3 $ms^{-1}$) at regular intervals. This increase in flow velocity is only possible if the pumping capacity of the system can be increased.

Pulsed injections can also be made by introducing a gas (compressed air or nitrogen) at very high speed (3 $ms^{-1}$) periodically at short intervals of time (5, 10, and 60 minutes). The effect is to increase the turbulent regime, creating pressure fluctuations inside the tube and higher shear stresses on the surface. As a consequence, a greater detachment of biofouling is obtained than that obtained by simply increasing the speed of water flow [5]. However, according to Mussalli and Tsou [79], the effectiveness of this system decreases as the biofouling film becomes more compact.

### *7.1.5 Streams current antifouling*

The principle of operation of this system is based on the generation of a voltage between the copper anodes (Cu) and the integrated steel cathodic plates so that when the Cu dissolves, the medium is contaminated by repelling the marine organisms. The level of Cu dissolution is regulated by a control unit adapting the system to the factors that influence biofouling growth (e.g., the season of the year) and lengthening its useful life.

### *7.1.6 Ultraviolet radiation*

The sterilization of water by ultraviolet radiation (UV) is a potentially useful method to avoid the formation of the biofouling film and as a cleaning technique. Its principle of operation is based on attacking the deoxyribonucleic acid (DNA) of the cells, causing their death. In this way, the amount of microorganisms present in the cooling water flow is minimized.

This treatment is very effective against microorganisms in the larval stage, and their environmental impact is zero [5]. On the contrary, the useful life of UV lamps is limited (approximately 8 months), and the cost of operation and maintenance is high, which is why it is not normally used in industrial installations that require a large flow of water to treat [79].
In order to reduce the treatment and broaden the spectrum of organisms attacked, it is usually combined with another chemical or physical method [5].

### *7.1.7 Ultrasound*

These are acoustic waves whose mode of action is based on the creation of vibrations to generate cavitation bubbles. The size of the bubbles can be increased by means of compression-decompression cycles until reaching a critical value at which they explode transforming their energy into heat. These microexplosions produce a disturbance in the surface that breaks the cohesion forces of the deposits to the surface causing their detachment [5]. Inside the cavitation bubbles, the temperature and pressure conditions can reach 5000°C and 500–2500 bar [80, 81], so that even water molecules decompose generating HO- and H+ radicals. The radicals formed can be recombined in the same way or react with substances present in the medium causing their degradation. Obviously, the generation of radicals is facilitated if in the medium there are molecules that break easily, such as ozone or hydrogen peroxide, which also act as usual precursors of hydroxyl radicals in the oxidation processes. The greater the effectiveness of ultrasound treatments, the higher their frequency (100 Hz and 100 MHz).

This antifouling treatment is effective against microorganisms in the larval stage and also has no environmental impact. On the contrary, the high cost of the ultra-sonic units must be added to the energy cost of the installation.

### *7.1.8 Filtering systems*

These systems are considered complementary methods for the elimination of biofouling in any industrial facility. The filtrate prevents larvae of mussels, crustaceans, and other organisms from entering the circuits of refrigeration systems and therefore develops into a biofouling film.

In cooling water conduction systems, the most commonly used filters are of the panel type for flow rates of less than 10 $m^3\ s^{-1}$ and of the drum type for higher flow rates [40]. Before the installation, the overall cost/benefit of the system must be analyzed due to the pressure drop produced in the installation [79].

### *7.1.9 Injection of fresh water in seawater systems*

The microorganisms present in the biofilm feed on the nutrients existing in the water through osmotic processes through a semipermeable membrane. When injecting fresh water, the saline concentration of the water is modified, and the pressure inside the cells increases, causing death [5]. According to Cho et al. [82], the exposure time must be high, reaching 48 hours for the mussel *Mytilus californianus* and 63 hours for the mussel *Mytilus edulis*. This method is used in piping systems of refrigeration systems that remain out of service for long periods [83].

### *7.1.10 Inserts in the tubes*

This system consists of inserting a helical element, whose vibration, distortion of flow, or rotation mitigate the growth of biofouling adhering on the inner surface of the tubes [5].

This system can present the following variants [84]:

- SPIRELF®. Metallic helical element attached to the ends of the tube and adjusted to its length. When the fluid circulates inside the tube, it vibrates axially and radially, eliminating biofouling adhering to its walls.
- FIXOTAL® or HITRAN®. They use metallic helical elements that are fixed inside the tubes, improving the turbulent effect on the wall to reduce the thick-ness of biofouling.
- TURBOTAL®. Rigid helical device that is inserted into the tubes and uses the water flow itself as the impeller of the rotating cleaning mechanism.

The insertion of metal elements inside the tubes has proven to be an effective method to prevent the formation of biofouling in heat exchangers of different industrial processes [84]. However, the use of this type of device means increasing the cost of the equipment and its maintenance, as well as a higher consumption of pumping energy due to the increase of load losses in the system.

#### *7.1.11 Circulation of polymers*

The method consists of introducing polymer fibers entrained by the fluid stream in order to erode and detach the layer of adhered deposits. According to Bott [5], its effectiveness on the process of forming the biofouling film in the tubes of a heat exchanger depends on the concentration of fibers and the speed of the flow of cooling water. The biggest drawback is that of removing the fibers from the effluent before it is discharged into the natural environment.

#### *7.1.12 Treatments with electromagnetic fields*

Electromagnetic fields have the ability to increase the interaction of the $Ca^{2+}$ and $CO3^{2}-$ ions present in seawater, increasing their nucleation and precipitation in $CaCO_3$. Below 35°C, $CaCO_3$ precipitates in the form of aragonite instead of calcite, resulting in being less adherent and less insulating [3, 85, 86]. The precipitation process of $CaCO_3$ reduces the superficial tension of the water, minimizing the adherence of microorganisms, and affects the intermolecular union of the extra-cellular polymers, weakening the matrix of the biofouling layer and diminishing its capacity of adhesion to the surface [7]. In addition, $CaCO_3$ entrainment by the seawater flow produces an erosive effect on the weakened biofouling layer that reduces its thickness [87, 88].

### 7.2 Offline systems

The use of online antifouling treatments does not prevent periodically scheduled stops in which aggressive cleaning methods are used to recover the efficiency of the tube at 100%. The frequency of scheduled stops is based on experience, and the cost/benefit ratio is derived from its application.

Brushing the surfaces of the heat exchanger in a dry or wet surface condition to eliminate adhering deposits. The cost of cleaning is high and involves access to the surfaces of the exchanger.

Although the techniques used to clean the surfaces of an industrial heat exchanger are varied, the most common methods to clean the internal surface of the pipes are air, water, and water steam.

The choice of cleaning agent depends on the degree of adhesion of the tanks and the design of the heat exchanger. According to Wilson [89], the use given to the different cleaning agents is as follows:

- Air under pressure. Eliminates deposits that are not very adherent
- Water at high pressure. Eliminates deposits and calcium deposits in tubular exchangers, condensers, and low-pressure boilers
- Water vapor. Eliminates calcareous incrustations and other substances of great hardness

To increase cleaning efficiency, detergents or abrasive materials are added to water or air under pressure. The nature of the abrasive materials depends on the hardness of the tanks and the construction material of the equipment. Sand is used as abrasive materials, pulverized mollusk shells, etc. [89].

## 8. Conclusions

Nowadays, heat exchangers play a vital role in the operational efficiency and effectiveness at every processing facility in which heat is essential, whether it is an oil refinery, combined heat and power plants, solar and biomass power plant, etc. As energy efficiency importance grows, the role of heat exchangers will become even greater and their technologies more advanced. Fouling on heat transfer surfaces of power plants are a major economic and environmental problem worldwide. Estimates have been made of fouling costs due primarily to wasted energy through excess fuel burn that are as high as 0.25% of the gross national product (GNP) of the industrialized countries. Many millions of tons of carbon emissions are the result of antifouling inefficiency.

After this study of the state of the art of fouling in heat exchanger, the main conclusion is that fouling problem is latent and unresolved. For this reason, new research projects must be developed to investigate and resolve this old problem without environmental effect.

## Author details

Sergio García* and Alfredo Trueba
Biofouling Group Research, University of Cantabria, Santander, Spain

*Address all correspondence to: sergio.garcia@unican.es

## References


[1] Wahl M. Marine epibiosis. I. Fouling and antifouling: Some basic aspects Marine Ecology Progress Series. 1989;**58**:175-189. https://doi.org/10.3354/meps058175

[2] Shah RK, Sekulić DP. Fundamentals of Heat Exchanger Design. John Wiley & Sons. 2003. 941 p. DOI: 10.1007/BF00740254

[3] Trueba A, García S, Otero FM. Mitigation of biofouling using electromagnetic fields in tubular heat exchangers-condensers cooled by seawater. Biofouling. 2014;**30**:95-103. DOI: 10.1080/08927014.2013.847926

[4] Trueba A, Otero FM, González JA, Vega LM, García S. Study of the activity of quaternary ammonium compounds in the mitigation of biofouling in heat exchangers–condensers cooled by seawater. Biofouling. 2013;**29**:1139-1151. DOI: 10.1080/08927014.2013.830108

[5] Bott TR. Industrial Biofouling. Elsevier. 2011. 220 p. ISBN: 9780444532244

[6] Paul VJ, Ritson-Williams R, Sharp K. ChemInform abstract: Marine chemical ecology in benthic environments. ChemInform. 2011;**28**(2):345-387. DOI: 10.1002/chin.201121256

[7] Trueba A, García S, Otero FM, Vega LM, Madariaga E. The effect of electromagnetic fields on biofouling in a heat exchange system using seawater. Biofouling. 2015;**31**:19-26. DOI: 10.1080/08927014.2014.994096

[8] García S, Trueba A. Influence of the reynolds number on the thermal effectiveness of tubular heat exchanger subjected to electromagnetic field-based antifouling treatment in an open once-through seawater cooling system. Applied Thermal Engineering. 2018;**140**:531-541. DOI: 10.1016/j.applthermaleng.2018.05.069

[9] García S, Trueba A, Vega LM, Madariaga E. Ceramic coating solution for offshore structures. Progress in Maritime Technology and Engineering. 2018;**1**:453-460. ISBN: 9781138585393

[10] Eguía E, Trueba A. Application of marine biotechnology in the production of natural biocides for testing on environmentally innocuous antifouling coatings. Journal of Coatings Technology and Research. 2007;**4**:191-202. DOI: 10.1007/s11998-007-9022-3

[11] Eguía López E. El problema del biofouling en intercambiadores de Calor-condensadores refrigerados por agua de mar. Universidad de Cantabria. 1998. 157 p

[12] Somerscales EFC, Kassemi M. Fouling due to corrosion products formed on a heat transfer surface. Journal of Heat Transfer. 1987;**109**:267-271. DOI: 10.1115/1.3248061

[13] Huang JY, John Chew YM, Ian Wilson D. A spinning disc study of fouling of cold heat transfer surfaces by gel formation from model food fat solutions. Journal of Food Engineering. 2012;**109**:49-61. DOI: 10.1016/j.jfoodeng.2011.09.034

[14] Bott TR. Freezing fouling or liquid solidification. Fouling of Heat Exchangers. Elsevier. 1995:137-147. 546 p. ISBN: 0080531903

[15] Garcia S, Trueba A, Vega L, Madariaga E. Improvement of electromagnetic fields treatment for biofouling growth control in tubular heat exchanger-condenser cooled by seawater. In: Ocean 2017 - Aberdeen. IEEE. 2017. DOI: 10.1109/OCEANSE.2017.8084576

[16] Boerlage SFE, Kennedy MD, Bonne PAC, Galjaard G, Schippers JC. Prediction of flux decline in membrane systems due to particulate fouling. Desalination. 1997;**113**:231-233. DOI: 10.1016/S0011-9164(97)00134-3

[17] Eguía E, Trueba A, Río-Calonge B, Girón A, Bielva C. Biofilm control in tubular heat exchangers refrigerated by seawater using flow inversion physical treatment. International Biodeterioration and Biodegradation. 2008;**62**:79-87. DOI: 10.1016/j.ibiod.2007.12.004

[18] Berkey K, BenDor TK. A comprehensive solution to the biofouling problem for the endangered Florida manatee and other species. Journal of Environmental Law. 2012;**42**:415-468

[19] Callow JA, Callow ME. Advanced nanostructured surfaces for the control of marine biofouling: The AMBIO project. In: Advances in Marine Antifouling Coatings and Technologies. Woodhead Publishing Series in Metals and Surface Engineering. 2009. p. 647-663. DOI: 10.1533/9781845696313.4.647

[20] Satpathy KK, Mohanty AK, Sahu G, Biswas S, Selvanayagam M. Biofouling and its control in seawater cooled power plant cooling water system—A review. In: Nuclear Power. IntechOpen; 2010. pp. 191-243. DOI: 10.5772/9912. ISBN: 978-953-307-110-7

[21] Holm ER. Barnacles and biofouling. Integrative and Comparative Biology. 2012;**52**:348-355. DOI: 10.1093/icb/ics042

[22] García S, Trueba A, Vega LM, Madariaga E. Impact of the surface roughness of AISI 316L stainless steel on biofilm adhesion in a seawater-cooled tubular heat exchanger-condenser. Biofouling. 2016;**32**:1-9. DOI: 10.1080/08927014.2016.1241875

[23] Zobell CE. The effect of solid surfaces upon bacterial activity. Journal of Bacteriology. 1943;**8**:1-8. DOI: 10.1146/annurev.arplant.58.032806.103848

[24] Jones HC, Roth IL, Sanders WM. Electron microscopic study of a slime layer. Journal of Bacteriology. 1969;**99**:316-325

[25] Costerton JW, Geesey GG, Cheng KJ. How bacteria stick. Scientific American. 1978;**238**:86-95. DOI: 10.1038/scientificamerican0178-86

[26] Flemming HC, Wingender J. Biofilms as a particular form of microbial life. Biofouling and Biocorrosion in Industrial Water Systems. Springer. 1991. DOI: 10.1007/978-3-642-76543-8

[27] Cloete TE, Brözel VS, Von Holy A. Practical aspects of biofouling control in industrial water systems. International Biodeterioration and Biodegradation. Elsevier. 1992;**29**:299-341. DOI: 10.1016/0964-8305(92)90050-X

[28] Mah TFC, O'Toole GA. Mechanisms of biofilm resistance to antimicrobial agents. Trends in Microbiology. 2001;**9**:34-39. DOI: 10.1016/S0966-842X(00)01913-2

[29] Flemming HC, Sriyutha Murthy P, Venkatesan R, Cooksey K. Marine and Industrial Biofouling. Spinger. 2008. 334 p. DOI: 10.1007/978-3-540-69796-1

[30] Characklis WG, Marshall KC, editors. Biofilms: Basis for an Interdisciplinary Approach. New York: Wiley; 1990. pp. 3-15. DOI: 10.1081/E-EBBE-120007264

[31] Costerton JW. Overview of microbial biofilms. Journal of Industrial Microbiology. 1995;**15**:137-140. DOI: 10.1007/BF01569816

[32] O'Toole G, Kaplan HB, Kolter R. Biofilm formation as microbial development. Annual Review of

Microbiology. 2000;**54**:49-79. DOI: 10.1146/annurev.micro.54.1.49

[33] Kraigsley AM, Finkel SE. Adaptive evolution in single species bacterial biofilms. FEMS Microbiology Letters. 2009;**293**:135-140. DOI: 10.1111/j.1574-6968.2009.01526.x

[34] Danese PN, Pratt LA, Kolter R. Biofilm formation as a developmental process. Methods in Enzymology. 2001;**336**:19-26. DOI: 10.1016/s0076-6879(01)36574-6

[35] McFeters GA, Bazin MJ, Mirelman D, Bryers JD, Mitchell R, Caldwell DE, et al. Biofilm development and its consequences. In: Report of the Dahlem Workshop on Microbial Adhesion and Aggregation; Berlin: Springer; 15-20 January 1984. p. 108-124. DOI: 10.1007/978-3-642-70137-5_9

[36] Pedersen K. Biofilm development on stainless steel and PVC surfaces in drinking water. Water Research. 1990;**24**:239-243. DOI: 10.1016/0043-1354(90)90109-J

[37] Railkin AI, Ganf TA, Manylov OG. Marine biofouling colonization processes and defenses. Biofouling. 2004;**20**:129-131. DOI: 10.1038/467531c

[38] Boonaert CJP, Dufrene YF, Rouxhet PG. Adhesion (primary) of microorganisms onto surfaces. In: Encyclopedia Environmental Microbiology. John Wiley & Sons. 2003. DOI: 10.1002/0471263397.env154. ISBN: 9780471354505

[39] Mayer C, Moritz R, Kirschner C, Borchard W, Maibaum R, Wingender J, et al. The role of intermolecular interactions: Studies on model systems for bacterial biofilms. International Journal of Biological Macromolecules. 1999;**26**:3-16. DOI: 10.1016 S0141-8130(99)00057-4

[40] Flemming H. Why microorganisms live in biofilms and the problem of biofouling. Marine and Industrial Biofouling. 2009;**4**:3-12. DOI: 10.1007/7142

[41] Characklis WG, Marshall K. Biofilms: A basis for an interdisciplinary approach. Biofilms. John Wiley & Sons. 1990

[42] Whiteley M, Ott JR, Weaver EA, McLean RJC. Effects of community composition and growth rate on aquifer biofilm bacteria and their susceptibility to betadine disinfection. Environmental Microbiology. 2001;**3**:43-52. DOI: 10.1046/j.1462-2920.2001.00158.x

[43] Singh PK, Parsek MR, Greenberg EP, Welsh MJ. A component of innate immunity prevents bacterial biofilm development. Nature. 2002;**417**:552-555. DOI: 10.1038/417552a

[44] Parikh A, Madamwar D. Partial characterization of extracellular polysaccharides from cyanobacteria. Bioresource Technology. 2006;**97**:1822-1827. DOI: 10.1016/j.biortech.2005.09.008

[45] Chmielewski RAN, Frank JF. A predictive model for heat inactivation of listeria monocytogenes biofilm on stainless steel. Journal of Food Protection. 2016. DOI: 10.4315/0362-028x-67.12.2712

[46] Lazarova V, Manem J. Biofilm characterization and activity analysis in water and wastewater treatment. Water Research. 1995;**29**:2227-2245. DOI: 10.1016/0043-1354(95)00054-O

[47] Chandy JP, Angles ML. Determination of nutrients limiting biofilm formation and the subsequent impact on disinfectant decay. Water Research. 2001;**35**:2677-2682. DOI: 10.1016/S0043-1354(00)00572-8

[48] Griebe T, Flemming HC. Biocide-free antifouling strategy to protect RO membranes from biofouling. Desalination. 1998;**118**:153-156. DOI: 10.1016/S0011-9164(98)00113-1

[49] Flemming HC, Wingender J. Extracellular polymeric substances: Structure, ecological functions, technical relevance. Encyclopedia of Environmental Microbiology. John Wiley & Sons. 2002

[50] Melo LF, Bott TR. Biofouling in water systems. Experimental Thermal and Fluid Science. 1997;**14**:375-381. DOI: 10.1016/S0894-1777(96)00139-2

[51] Schneider RP, Marshall KC. Retention of the Gramnegative marine bacterium SW8 on surfaces—Effects of microbial physiology, substratum nature and conditioning films. Colloids and Surfaces. B, Biointerfaces. 1994;**2**:387-396. DOI: 10.1016/0927-7765(94)80002-2

[52] Huang YL, Dobretsov S, Xiong H, Qian PY. Effect of biofilm formation by Pseudoalteromonas spongiae on induction of larval settlement of the polychaete Hydroides elegans. Applied and Environmental Microbiology. 2007;**73**:6284-6288. DOI: 10.1128/AEM.00578-07

[53] Jin T, Qian PY. Effect of amino acids on larval metamorphosis of the polychaete Hydroides elegans. Marine Ecology Progress Series. 2004;**267**:209-218. DOI: 10.3354/meps267209

[54] Jin T, Qian PY. Amino acid exposure modulates the bioactivity of biofilms for larval settlement of Hydroides elegans by altering bacterial community components. Marine Ecology Progress Series. 2005;**297**:169-179. DOI: 10.3354/meps297169

[55] Huang YL, Dobretsov S, Ki JS, Yang LH, Qian PY. Presence of acyl-homoserine lactone in subtidal biofilm and the implication in larval behavioral response in the polychaete Hydroides elegans. Microbial Ecology. 2007;**54**:384-392. DOI: 10.1007/s00248-007-9210-9

[56] Bott TR. Fouling of Heat Exchangers. Elsevier. 1995. DOI: 10.1016/B978-044482186-7/50005-1

[57] Wimpenny J, Manz W, Szewzyk U. Heterogeneity in biofilms. FEMS Microbiology Reviews. 2000;**24**:661-671. DOI: 10.1016/S0168-6445(00)00052-8

[58] Clare AS, Rittschof D, Gerhart DJ, Maki JS. Molecular approaches to nontoxic antifouling. Invertebrate Reproduction and Development. 1992;**22**:67-76. DOI: 10.1080/07924259.1992.9672258

[59] Dahms HU, Qian PY. Exposure of biofilms to meiofaunal copepods affects the larval settlement of Hydroides elegans (Polychaeta). Marine Ecology Progress Series. 2005;**297**:203-214. DOI: 10.3354/meps297203

[60] Fenchel T. Formation of laminated cyanobacterial mats in the absence of benthic fauna. Aquatic Microbial Ecology. 1998;**14**:235-240. DOI: 10.3354/ame014235

[61] Beech IB, Gubner R, Zinkevich V, Hanjangsit L, Avci R. Characterisation of conditioning layers formed by exopolymeric substances of pseudomonas NCIMB 2021 on surfaces of AISI 316 stainless steel. Biofouling. 2000;**16**:93-104. DOI: 10.1080/08927010009378435

[62] Stewart PS. Mechanisms of antibiotic resistance in bacterial biofilms. International Journal of Medical Microbiology. 2002;**292**:107-113. DOI: 10.1078/1438-4221-00196

[63] Piola RF, Dafforn KA, Johnston EL. The influence of antifouling practices on marine invasions. Biofouling. 2009;**25**:633-644. DOI: 10.1080/08927010903063065

[64] Beech IB, Sunner JA, Hiraoka K. Microbe-surface interactions in

biofouling and biocorrosion processes. International Microbiology. 2005;**8**:157- 168. PMID: 16200494

[65] Caputo AC, Pelagagge PM, Salini P. Joint economic optimization of heat exchanger design and maintenance policy. Applied Thermal Engineering. 2011;**31**:1381-1392. DOI: 10.1016/j.applthermaleng.2010.12.033

[66] Coetser SE, Cloete TE. Biofouling and biocorrosion in industrial water systems. Critical Reviews in Microbiology. 2005;**31**:213-232. DOI: 10.1080/10408410500304074

[67] Barry DM, Kanematsu H. Cooling water. In: Biofilm and Materials Science. Springer. 2015. DOI: 10.1007/978-3-319-14565-5_10

[68] Nwaoha C, Zakaria ZY. Cooling towers. In: Process Plant Equipment: Operation, Control, and Reliability. John Wiley & Sons. 2012. DOI: 10.1002/9781118162569.ch5

[69] Fisher DJ, Burton DT, Yonkos LT, Turley SD, Ziegler GP. The relative acute toxicity of continuous and intermittent exposures of chlorine and bromine to aquatic organisms in the presence and absence of ammonia. Water Research. 1999;**33**:760-768. DOI: 10.1016/S0043-1354(98)00278-4

[70] Murphy HM, Payne SJ, Gagnon GA. Sequential UV- and chlorine-based disinfection to mitigate Escherichia coli in drinking water biofilms. Water Research. 2008;**42**:2083-2092. DOI: 10.1016/j.watres.2007.12.020

[71] Dixon D, Veil J, Wisniewski J. Defining and assessing adverse environmental impact from power plant impingement and entrainment of aquatic organisms. Symposium in Conjunction with the Annual Meeting of the American Fisheries Society, 2001. In Phoenix, Arizona, USA: CRC Press; 2003. 292 p. ISBN: 9789058095176

[72] Qian PY, Xu Y, Fusetani N. Natural products as antifouling compounds: Recent progress and future perspectives. Biofouling. 2009;**26**:223-234. DOI: 10.1080/08927010903470815

[73] Fusetani N. Antifouling marine natural products. Natural Product Reports. 2011;**28**:400-410. DOI: 10.1039/c0np00034e

[74] Hadfield MG. Biofilms and marine invertebrate larvae: What bacteria produce that larvae use to choose settlement sites. Annual Review of Marine Science. 2010;**3**:453-470. DOI: 10.1146/annurev-marine-120709-142753

[75] Thiyagarajan V. A review on the role of chemical cues in habitat selection by barnacles: New insights from larval proteomics. Journal of Experimental Marine Biology and Ecology. 2010;**392**:22-36. DOI: 10.1016/j.jembe.2010.04.030

[76] Wahl M, Goecke F, Labes A, Dobretsov S, Weinberger F. The second skin: Ecological role of epibiotic biofilms on marine organisms. Frontiers in Microbiology. 2012;**23**:292-313. DOI: 10.3389/fmicb.2012.00292

[77] Puglisi MP, Sneed JM, Sharp KH, Ritson-Williams R, Paul VJ. Marine chemical ecology in benthic environments. Natural Product Reports. 2014;**31**:1510-1553. DOI: 10.1039/c4np00017j

[78] Dobretsov S, Abed RMM, Teplitski M. Mini-review: Inhibition of biofouling by marine microorganisms. Biofouling. 2013;**29**:423-441. DOI: 10.1080/08927014.2013.776042

[79] Mussalli YG, Tsou J. Advances in biofouling control technologies. U.S. and Japanese perspectives. Proceedings of the American Power Conference. Illinois-Institute of Technology. 1989;**1**:1094-1099

[80] Wu TY, Guo N, Teh CY, Hay JXW. Advances in Ultrasound Technology for Environmental Remediation. Springer. 2013. ISBN 978-94-007-5533-8

[81] Park JS, Lee JH. Sea-trial verification of ultrasonic antifouling control. Biofouling. 2018;**34**:98-110. DOI: 10.1080/08927014.2017.1409347

[82] Cho YI, Choi BG. Validation of an electronic anti-fouling technology in a single-tube heat exchanger. International Journal of Heat and Mass Transfer. 1999;**42**:1491-1499. DOI: 10.1016/S0017-9310(98)00196-3

[83] Grandison C, Piola R, Fletcher L. A Review of Marine Growth Protection System (MGPS) Options for the Royal Australian Navy. DSTO-TR-2631. 2011;**38**:2011. Available from: http://www.dtic.mil/dtic/tr/fulltext/u2/a557490.pdf

[84] Pouponnot F. Heat exchanger tube inserts an update in new applications with trouble shooting aspects in crude units, residue service, reboilers, u-tubes. In: Proceedings of the 6th International Conference on Heat Exchanger Fouling and Cleaning-Challenges and Opportunities. Engineering Conferences International. 2005. ISBN: 978-1-5108-1946-7

[85] Xiaokai X. Research on the electromagnetic anti-fouling technology for heat transfer enhancement. Applied Thermal Engineering. 2008;**28**:889-894. DOI: 10.1016/j.applthermaleng.2007.07.006

[86] Andritsos N, Karabelas AJ. Calcium carbonate scaling in a plate heat exchanger in the presence of particles. International Journal of Heat and Mass Transfer. 2003;**46**:4613-4627. DOI: 10.1016/S0017-9310(03)00308-9

[87] Lipus LC, Krope J, Crepinsek L. Dispersion destabilization in magnetic water treatment. Journal of Colloid and Interface Science. 2001;**236**:60-66. DOI: 10.1006/jcis.2000.7392

[88] Gabrielli C, Jaouhari R, Maurin G, Keddam M. Magnetic water treatment for scale prevention. Water Research. 2001;**35**:3249-3259. DOI: 10.1016/S0043-1354(01)00010-0

[89] Wilson DI. Heat exchanger fouling and cleaning: Fundamentals and applications. Challenges in Cleaning: Recent Developments and Future Prospects, Engineering Conferences International. ECI Digital Archives. 2003

# 2

# Introduction to Numerical Approaches for Forward and Inverse Heat Transfer Problems

*Sergey Voronin*

**Abstract**

We describe the formulation of forward and inverse problems in heat transfer and illustrate several techniques applicable to their solution.

**Keywords:** heat transfer, finite differences, regularization, Fourier transforms, interpolation

## 1. Introduction

Heat flow is described by the heat (or diffusion) partial differential equation, which takes the form $\frac{\partial u}{\partial t} = k\frac{\partial^2 u}{\partial x^2}$ with $u = u(x,t)$ representing the temperature at time $t$ and at spatial location $x$ (which may have one or more components). The constant $k$ is the ratio of the thermal conductivity in the material over the specific heat times the density of the material. Applying the conservation law principle to the heat flow across the rod yields the parabolic partial differential equation. If a heat source is present or the material is exothermic, a non-homogeneous version of the problem: $\frac{\partial u}{\partial t} - k\frac{\partial^2 u}{\partial x^2} = f(x,t)$ with $f(x,t) \neq 0$ applies. The initial boundary value problem comes from assigning relevant initial and boundary conditions. For example, for a perfectly insulated rod as in **Figure 1**, the conditions could be $u(0,t) = u(L,t) = 0$. Given some initial profile of the temperature at time $t = 0, u(x,0) = h(x)$, the analytical solution in 1D or 2D with a regular geometrical domain is given by separation of variables. Assuming $u(x,t) = F(x)G(t)$, one obtains a system of two ODEs and from that, the series solution $u(x,t) = \sum_{n=1}^{\infty} u_n(x,t) = \sum_{n=1}^{\infty} A_n \exp\left(-\lambda_n^2 t\right) \sin\left(\frac{n\pi}{L}x\right)$ with the coefficients $A_n$ satisfying the initial condition $u(x,0) = \sum_{n=1}^{\infty} A_n \sin\left(\frac{n\pi}{L}x\right) = f(x)$ and the eigenvalues $\lambda_n = \frac{nk\pi}{L}$. The coefficients are evaluated via the sin series integration: $A_n = \frac{2}{L}\int_0^L f(x) \sin\left(\frac{k\pi x}{L}\right)dx$. For non-intregrable cases or more complicated geometries, numerical techniques for approximation are used [1].

### 1.1 Derivation of homogeneous and non-homogeneous equations

We now discuss two simple scenarios which can be used to derive homogeneous and non-homogeneous forms of the heat Equation [2]. Heat is the flow of energy from a warmer to a cooler location. There are several types of heat transfer including conduction (flow of heat through stationary material), convection (flow of heat through fluids), and radiation (flow of heat through electromagnetic waves) [2].

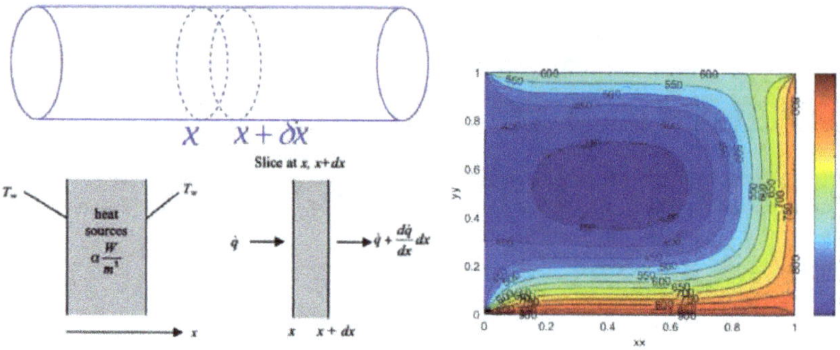


**Figure 1.**
*Perfectly insulated rod used a model of heat transfer, an exothermic slab, and sample modeled heat profile in 2D.*

In this chapter we are concerned mainly with heat conduction for which the differential heat model applies. We take $Q$ to represent the heat transfer rate in units of Watts. It is the rate at which heat is flowing through one of the slabs in **Figure 2**. When heat flows into and out of a mass, the tem perature of the mass changes in accordance with the model: $Q_{\text{in}} - Q_{\text{out}} = mc\frac{dT}{dt}$ where $T$ represents the temperature, $m$ the mass, and $c$ is the specific heat value of the material. It should be noted that in heat problems, $T$ and $u$ are ofter used interchangeably. We define $q = Q/A$ (heat over area) as the heat flux, which by Fourier's law is proportional to the temperature gradient: $q = -\bar{k}\frac{dT}{dx}$, where $\bar{k}$ is the thermal conductivity of the material [2]. Using $Q_{\text{in}} = Q(x)$ and $Q_{\text{out}} = Q(x + \delta x)$ along with heat flux $q = Q/A$, we have $Aq(x) - Aq(x + \delta x) = mc\partial T/\partial t$. Applying Fourier's law yields:

$$\bar{k}A\left[-\frac{\partial T}{\partial x}\bigg|_x + \frac{\partial T}{\partial x}\bigg|_{x+\delta x}\right] = mc\partial T/\partial t$$

Next, using Taylor's approximation $\frac{\partial T}{\partial x}\big|_{x+\delta x} \approx \frac{\partial T}{\partial x}\big|_x + \frac{\partial^2 T}{\partial x^2}\big|_x \delta x$ and expressing $m = \rho A\delta x$ in terms of slab density, area, and width, we obtain:

$$\bar{k}A\frac{\partial^2 T}{\partial x^2} = \rho c\frac{\partial T}{\partial t} \quad \Rightarrow \quad k\frac{\partial^2 T}{\partial x^2} = \frac{\partial T}{\partial t}$$

For the exothermic slab case exhibited in **Figure 2** on the right, we obtain an inhomogeneous equation. We assume the slab is filled with a material which emits heat at a rate $\dot{q}$ with units $W/m^3$ (energy per unit volume). In this case, the in and out energy balance gives $Q(x) + \dot{q}V - Q(x + \delta x) = mc\partial T/\partial t$, which with $q = Q/A, V = A\delta x$, Fourier's law and Taylor's expansion, gives $\bar{k}A\frac{\partial^2 T}{\partial x^2}\delta x + \dot{q}A\delta x = mc\frac{\partial T}{\partial t} \Rightarrow \bar{k}\frac{\partial^2 T}{\partial x^2} + \dot{q} = \rho c\frac{\partial T}{\partial t} \Rightarrow k\frac{\partial^2 T}{\partial x^2} + \frac{\dot{q}}{\rho c} = \frac{\partial T}{\partial t}$. In the case of steady state, $\partial T/\partial t = 0$ and a second order differential equation model results.

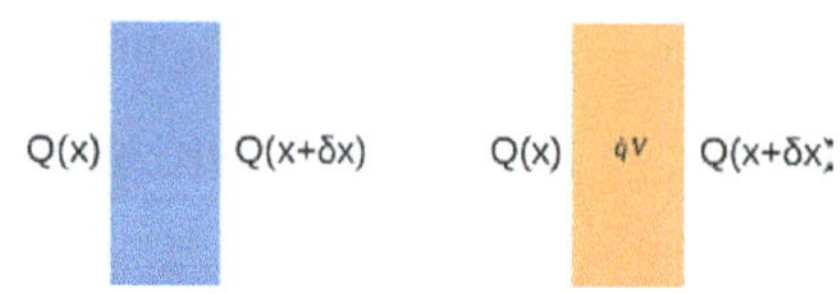


**Figure 2.**
*Heat conducting slabs without and with internal heat generation.*

### 1.2 Numerical schemes

Discretization of the derivative operators yields the finite difference formulation. A number of different schemes are possible based on the difference schemes used to replace the derivatives in the equation resulting in either an explicit method (where the update for $u$ at the next time step is given by a direct calculation from the previous one) or an implicit method (where the update for $u$ is accomplished via a linear solve ). For 2D heat transfer, the model becomes $\alpha^2\left(\frac{\partial^2 u}{\partial x^2}+\frac{\partial^2 u}{\partial y^2}\right)=\frac{\partial u}{\partial t}$ and again, the finite difference method can be used for discretization. The initial value problems are either *forward* or *inverse*, in the sense that they either predict the temperature profile at later time from the knowledge of the initial temperature distribution at time $t=0$ or in the inverse case, predict the temperature distribution at time $t=0$ from knowledge of the temperature at a future time. Sometimes, the initial boundary value problem may be prescribed extra conditions, such as Neumann or other derivate condition type and the inverse problem could instead seek the initial values for such expressions from future data [3]. In general, the inverse problem is more challening than the forward problem due to being ill-posed. The discretized operator matrix which can be assembled from the difference formulation is not well conditioned and as a result the problem is overly sensitive to slight changes in input values. Regularization techniques need to be used for the solution [4].

## 2. Forward problem

The forward problem is concerned by the prediction of heat profile in future time from the initial condition. We first consider the heat equation initial boundary value problem in 1D:

$$u_t = k u_{xx}, \quad 0<x<1, \tag{1}$$

$$u(x,0) = f(x), \quad 0<x<1, \tag{2}$$

$$u(0,t) = u(1,t) = 0, \quad t>0 \tag{3}$$

with constant $k \in \mathbb{R}$. Via a transformation, this problem can be extended to nonzero boundary conditions. For (3), separation of variables gives the solution as the superposition of solutions of the form:

$$\tilde{u}(x,t) = e^{-\pi^2 kt}\sin(\pi x)$$

A quick calculation shows that $\left[\frac{\partial}{\partial t} - k\frac{\partial^2}{\partial x^2}\right]\tilde{u}(x,t) = 0$. In order to form the well-known explicit scheme, we use the forward in time and central in space difference operators:

$$\frac{\partial u}{\partial t}_{\text{fd}}(x_i,t_j) = \frac{u(x_i,t_{j+1}) - u(x_i,t_j)}{l} \tag{4}$$

$$\frac{\partial^2 u}{\partial x^2}_{\text{fd}}(x_i,t_j) = \frac{u(x_{i+1},t_j) - 2u(x_i,t_j) + u(x_{i-1},t_j)}{h^2} \tag{5}$$

which results in the explicit finite difference scheme:

$$u_{i,j+1} = u_{i,j} + \frac{kl}{h^2}\left(u_{i+1,j} - 2u_{i,j} + u_{i-1,j}\right) \tag{6}$$

This can be re-written as $u_{i,j+1} = (1-2\alpha)u_{i,j} + \alpha(u_{i+1,j} + u_{i-1,j})$ with $\alpha = \frac{kl}{h^2}$ , with $l$ the time step ($\Delta t$) and $h$ the spatial step size ($\Delta x$). Von Neumann stability analysis yields the condition $l \leq \frac{h^2}{2k}$. For non- homogeneous cases the general model is $\frac{\partial u}{\partial t} = k\frac{\partial^2 u}{\partial x^2} + f(x,t)$ and the finite difference analysis proceeds in a similar way, with an explicit scheme of the form: $u_{i,j+1} = u_{i,j} + \frac{kl}{l}h^2\ (u_{i+1,j} - 2u_{i,j} + u_{i-1,j})\ + lf(x_i, t_j)$.

If we use the backward time finite difference approximation for $\frac{\partial u}{\partial t}$, we get: $\frac{u_{i,j+1}-u_{i,j}}{k} = \frac{u_{i+1,j+1}-2u_{i,j+1}+u_{i-1,j+1}}{h^2}$, yielding the implicit scheme: $-\alpha u_{i-1,j+1} + (1+2\alpha)u_{i,j+1} - \alpha u_{i+1,j+1} = u_{i,j}$ which can be solved for time $j+\ 1$ from the informa - tion at time $j$ using a linear system with a tridiagonal matrix with elements $(-\alpha, (1+2\alpha), -\alpha)$. The implicit scheme is numerically stable for any $\alpha$ but is costlier than the explicit scheme. Another example, is the so called Crank–Nicolson scheme based on the time derivative approximation at the midpoint of the time stamp. It takes the form:

$$\frac{u_{i,j+1} - u_{i,j}}{l} = k\frac{(u_{i+1,j+1} - 2u_{i,j+1} + u_{i-1,j+1}) + (u_{i+1,j} - 2u_{i,j} + u_{i-1,j})}{2h^2}$$

Again using $\alpha = \frac{kl}{h^2}$, we arrive at the scheme:

$$-\alpha u_{i+1,j+1} + 2(1+\alpha)u_{i,j+1} - \alpha u_{i-1,j+1} = \alpha u_{i+1,j} + 2(1-\alpha)u_{i,j} + \alpha u_{i-1,j}$$

Note that this takes the form of the linear system $Au = b$, where $A$ is a tridiagonal matrix of terms $(-\alpha, 2(1+\alpha), -\alpha)$, as before, $u$ containing the heat distribution at time $j+1$ and $b$ containing the heat distribution information at time $j$. A linear system solves then yields the new time information from previous data.

In 2D, the standard explicit scheme arises from the discretization:

$$\frac{T^{n+1}_{i,j} - T^n_{i,j}}{\Delta t} = k\left(\frac{T^n_{i,j+1} - 2T^n_{i,j} + T^n_{i,j-1}}{(\Delta x)^2} + \frac{T^n_{i+1,j} - 2T^n_{i,j} + T^n_{i-1,j}}{(\Delta y)^2}\right)$$

Like in the 1D scheme, suitable finite difference approximations yield the implicit method.

A practically useful direction to consider is the use of non-standard finite difference (nsfd) schemes [5]. Based on (5), we define the discretized operators:

$$\overline{d}_t f(x,t) = f(x,t+l) - f(x,t) \tag{7}$$

$$\overline{d}^2_x f(x,t) = f(x+h,t) - 2f(x,t) + f(x-h,t) \tag{8}$$

and proceed to evaluate $\left(\overline{d}_t - kl\frac{\overline{d}^2_x}{h^2}\right)\tilde{u}(x,t)$ for $\tilde{u}(x,t) = e^{-\pi^2 kt}\sin(\pi x)$. The standard fd scheme does not set this expression to zero, and hence this analytically valid solution does not satisfy the original partial differential equation. However, based on the calculations below, we can make a replacement for $\frac{kl}{h^2}$, which would force the equation to zero.

$$\begin{aligned} \overline{d}_t[\tilde{u}(x,t)] &= \sin(\pi x)\left[e^{-\pi^2 k(t+l)} - e^{-\pi^2 kt}\right] = \sin(\pi x)e^{-\pi^2 kt}\left(e^{-\pi^2 kl} - 1\right) \\ &= \tilde{u}(x,t)\left(e^{-\pi^2 kl} - 1\right) \end{aligned} \tag{9}$$

$$\overline{d}_x^2[\tilde{u}(x,t)] = e^{-\pi^2 kt}[\sin(\pi(x+h)) - 2\sin(\pi x) + \sin(\pi(x-h))] \tag{10}$$

$$= e^{-\pi^2 kt}\left[\frac{1}{2i}\left(e^{i\pi(x+h)} - e^{-i\pi(x+h)}\right) - 2\sin(\pi x) + \frac{1}{2i}\left(e^{i\pi(x-h)} - e^{-i\pi(x-h)}\right)\right] \tag{11}$$

$$= e^{-\pi^2 kt}\frac{1}{2i}\left[e^{i\pi x}e^{i\pi h} - e^{-i\pi x}e^{-i\pi h} - 4i\sin(\pi x) + e^{i\pi x}e^{-i\pi h} - e^{-i\pi x}e^{i\pi h}\right] \tag{12}$$

$$= e^{-\pi^2 kt}\frac{1}{2i}\left[e^{i\pi x}\left(e^{i\pi h} + e^{-i\pi h}\right) - e^{-i\pi x}\left(e^{-i\pi h} + e^{i\pi h}\right) - 4i\sin(\pi x)\right] \tag{13}$$

$$= e^{-\pi^2 kt}\frac{1}{2i}\left[2\cos(\pi h)\left(e^{i\pi x} - e^{-i\pi x}\right) - 4i\sin(\pi x)\right] \tag{14}$$

$$= e^{-\pi^2 kt}\frac{1}{2i}\left[4i\cos(\pi h)\sin(\pi x) - 4i\sin(\pi x)\right] \tag{15}$$

$$= 2e^{-\pi^2 kt}\sin(\pi x)[\cos(\pi h) - 1] = 2\tilde{u}(x,t)[\cos(\pi h) - 1] \tag{16}$$

Setting $\left(\overline{d}_t - \frac{kl}{h^2}\overline{d}^2{}_x\right)\tilde{u}(x,t) = 0$, we obtain:

$$\tilde{u}(x,t)\left[e^{-\pi^2 kl} - 1\right] - \frac{2kl}{h^2}\tilde{u}(x,t)[\cos(\pi h) - 1] = 0 \quad \Rightarrow \quad \left[e^{-\pi^2 kl} - 1\right] = \frac{2kl}{h^2}[\cos(\pi h) - 1]$$

Hence, the nsfd scheme follows if we make the following replacement in (6):

$$\frac{kl}{h^2} = \frac{1}{2}\left[\frac{e^{-\pi^2 kl} - 1}{\cos(\pi h) - 1}\right]$$

When $h, l \ll 1$, the nsfd substitution coincides with the standard finite difference scheme. We use the small $|q|$ Taylor series approximations $e^q \approx 1 + q$ and $\sin(q) \approx q$ to obtain:

$$\frac{1}{2}\left[\frac{e^{-\pi^2 kl} - 1}{\cos(\pi h) - 1}\right] = \left[\frac{e^{-\pi^2 kl} - 1}{-2\sin^2\left(\frac{\pi h}{2}\right)}\right] \approx \frac{1}{2}\left[\frac{1 - \pi^2 kl - 1}{-\frac{\pi h^2}{2}}\right] = \frac{kl}{h^2}$$

By this it follows that for small $h, l$, the stability constraint $\frac{kl}{h^2} \leq 1$ is satisfied in the nsfd scheme when $l \leq \frac{h^2}{2}$ just like in the original scheme. This also holds for large $h$, as long as $h$ is chosen so that $|\cos(\pi h) \quad 1|$ is not very small.

## 2.1 Numerical experiments

Below, we present some sample results for the case of constant $k \in \mathbb{R}$ in (3), where we found that the described nsfd scheme often outperforms the fd scheme at the same time step. In **Figure 3**, we illustrate the solutions with fd and nsfd methods with the initial condition, $(ux,0) = \sin(\pi x) + x$ for problem 1, with boundary conditions $u(0, t) = 0$ and $u(1, t) = 1$ and with the initial condition $(ux,0) =$ 12 sin $(9\pi x) - 7\sin 4\pi x()$ and zero homogeneous boundary conditions for problem 2. For the second example, errors vs. step size are also compared. In general, we observe more accurate results with the use of the nsfd scheme.

An interesting case is that of heat transfer with an exothermic material, which can be modeled with a heat source distributed throughout [3]. Considering an

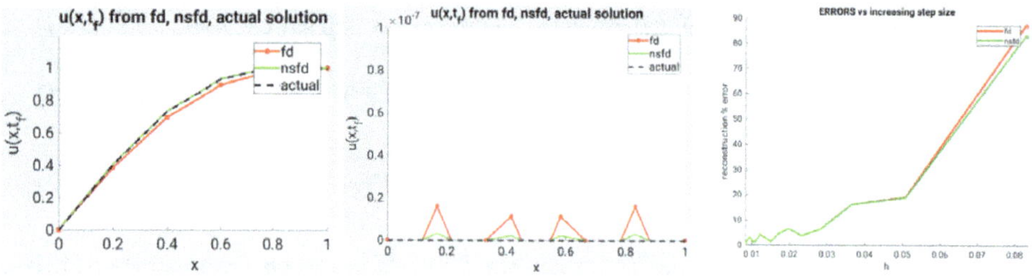


**Figure 3.**
*Comparison of fd and nsfd scheme solutions for the 1-D heat equation for problems 1 and 2 above.*

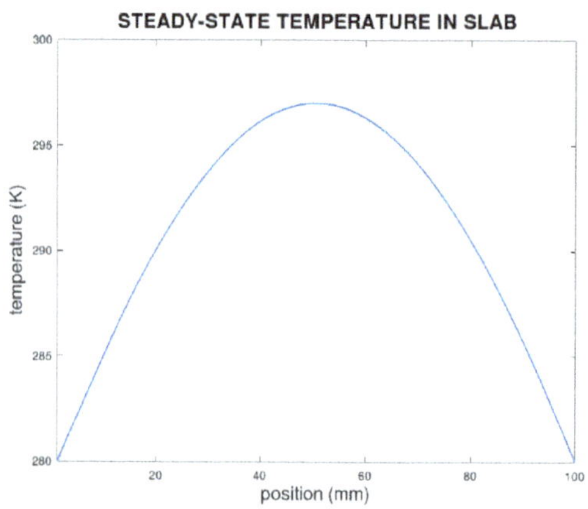


**Figure 4.**
*Temperature distribution inside a thin slab with an exothermic material with homogeneous boundary conditions as simulated by an fd scheme.*

infinitesimal segment as in **Figure 1** yields the energy balance equation: $\dot{q} + \alpha dx - \left(\bar{q} + \frac{d\dot{q}}{dx}dx\right) = 0 \Rightarrow \frac{d\dot{q}}{dx} = \alpha$. It follows that the heat model takes the form $\frac{1}{\alpha}\frac{\partial u}{\partial t} = \frac{\partial^2 u}{\partial x^2} + \frac{\dot{q}}{k}$. The steady state behavior does not depend on time so that $\frac{\partial u}{\partial t} = 0$ and an ordinary differential equation model $\frac{d^2u}{dx^2} + \frac{\alpha}{k} = 0$ results. Upon a double integration, the general solution is $u(x) = -\frac{\alpha}{2k}x^2 + Ax + B$, which with the boundary conditions $T(0) = T(L) = T_w$ becomes $u(x) = -\frac{\alpha}{2k}x^2 + Lx + T_w$, a parabola. We can simulate this steady state example with a finite difference scheme for the time derivative, yielding the temperature profile shown in **Figure 4**.

## 3. Inverse problem

The inverse problem consists of predicting the initial condition from the observation of the heat profile at some later time $t$. In general, the problem is ill-posed and regularization is necessary for its solution [4]. Let us look at a simple example in 1-D. Starting from the initial boundary value problem on $x \in [0, L]$:

$$\frac{\partial u}{\partial t} = k\frac{\partial^2 u}{\partial x^2} \tag{17}$$

$$u(x,0) = f(x), \quad u(0,t) = 0 = u(L,t) \tag{18}$$

In this case, separation of variables can be used and after a well-known sequence of steps, the following series solution can be derived [3]:

$$u(x,t) = \sum_{n=1}^{\infty} A_n \sin\left(\frac{n\pi x}{L}\right) \exp\left(-k\left(\frac{n\pi}{L}\right)^2 t\right)$$

It follows that $u(x,0) = \sum_{n=1}^{\infty} A_n \sin\left(\frac{n\pi x}{L}\right)$ which is a sin series; that is, $A_n = F[u(x_0)] = \frac{2}{L}\int_0^L u(x,0)\sin\left(\frac{n\pi x}{L}\right)$. It follows that the sin coefficients of the solution at $t = T$ are given by the sin coefficients at $t = 0$ multiplied by $\exp\left(-k\left(\frac{n\pi}{L}\right)^2 T\right)$. The values of $u$ at $t = T$ can then be computed via the inverse sin transform applied to the product. In a similar way, it follows that from the transform of the solution at $t = T, F[u(x,T)]$, we can in principle go backwards, to obtain $u(x,0) = F^{-1}\left[F[u(x,T)].*\exp\left(k\left(\frac{n\pi}{L}\right)^2 T\right)\right]$ with the $.*$ representing component- wise operation [3]. However, with the negative sign inside the exponential gone, there is potential for blow up of any large magnitude terms. The issue can be addressed by performing filtering on the large coefficients, setting the largest magnitude portion to zero or some maximum value. The magnitude cut off value can be selected based on quartile statistics of the nonzero absolute values.

A more general approach for inversion is to utilize the inverse problem formulations of the matrix heat propagation formulation. For example, for the standard implicit or Crank-Nicholson scheme, we have the system $Au = b$, where $A$ is a tridiagonal matrix of terms $(-\alpha, 2(1+\alpha), -\alpha)$. This means that $u^1 = A^{-1}b(u^0)$, $u^2 = A^{-1}b(u^1) = A^{-2}b(u^0) \quad \Rightarrow \quad b(u^0) = A^2u^2$. This then implies that the initial condition at time $t = 0$, $u^0$ can be recovered from the application of the inverse matrix raised to a power $n_t$ corresponding to the time point. As another example, if we discretize $u_t = ku_{xx}$ as:

$$\frac{\partial u(t)}{\partial t} - \frac{k}{h^2}\left[u_{i+1}^{j+1}(t) - 2u_i^{j+1}(t) + u_{i-1}^{j+1}(t)\right] = 0 = \frac{u_i^{j+1} - u_i^j}{\Delta t} - Mu^{j+1}(t),$$

with $M = \frac{k}{h^2}\text{tridiag}(-1,2,-1)$, for $j = 0, 1, \ldots, n_t - 1$. Using $u(x,0) = U_0$, we obtain the relations:

$$u^0 = U_0; u^{j+1} = (I + \Delta tM)^{-1}u^j = (I + \Delta tM)^{-(j+1)}U_0 \Rightarrow \overline{L}^{-(j+1)}U_0 = u^{j+1}$$

This means that $U_0$ can in theory be obtained right from $(I + \Delta tM)^{n_t}u^{n_t}$. The matrix needs not be formed explicitly, as the finite differencing scheme can be repeatedly applied to $u^{n_t} = u(T)$. However, as shown in **Figure 5**, raising the typical differencing matrix to a power results in non-linear singular value decay. Hence, the application of the inverse of matrix $(I + \Delta tM)^{-n_t}$ to the approximant for $b = u(T)$, the heat profile at time $T$, would result in the blow-up of any numerical errors due to the ill-conditioning. We must instead consider the inverse formulation $LU_0 = b$ with $L = \overline{L}^{-n_t}$ and utilize a regularization scheme.

This can be accomplished by replacing the naive solution $L^{-1}b$ by the regularized formulation $\min_u\left(\|Lu - b\|_2^2 + \lambda\|u\|_2^2\right)$, which for large enough $\lambda > 0$, alleviates the problem due to the fast decay of singular values. This well-known Tikhonov regularization formulation essentially acts to remove the influence of small singular values on the solution, which would otherwise blow up in value upon inversion.

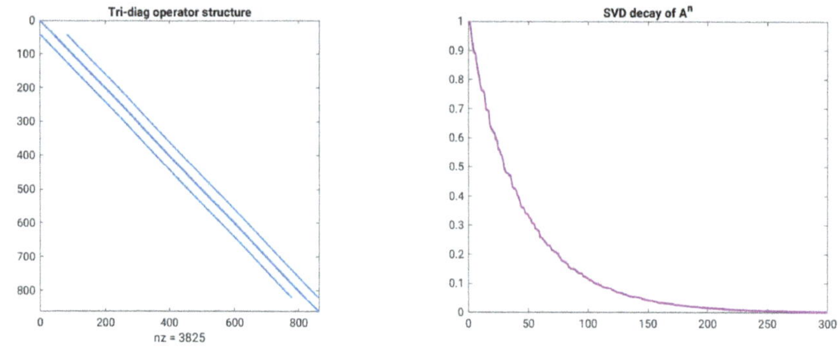


**Figure 5.**
*Tridiagonal structure of differencing operator matrix and singular value decay of operator raised to a power.*

The regularization parameter $\lambda$ should be chosen with care to provide regularization but not to move too far away from the naive inverse operator. For this reason, $\lambda$ is usually selected based on an L-curve criterion [6], whereby the parameter is initially set at some fraction of $||L^T b||_\infty$ and then iteratively lowered to a lower fraction (with the values either linearly or logarithmically spaced). At each $\lambda$ value, the solution to $(L^T L + \lambda I)u_\lambda = L^t b$ is obtained (e.g. with a Conjugate Gradient scheme) using the solution at the previously used $\lambda$ as the initial guess. Typically, the progression to lower $\lambda$ stops once the value of $||Lu_\lambda - b||_2/N$ reaches a plateau. It is possible to employ a *predictor-corrector* scheme to iteratively improve the solution. In this approach, the approximation to $U_0$ can be obtained as above, then the forward operator can be applied to propagate this approximated initial heat profile at $t = 0$ to time $t = T$ and compare with observations [4]. We can compare the residual at time $T$ and attempt some corrections to $U_0$ based on this residual such that the propagated solution at $t = T$ best matches observations. In the case of only a finite subset of observations, interpolation techniques can be utilized to complete the profile at $t = T$ across all values of $x$ [7] prior to regularization to obtain an approximate smooth and continuous $u(T)$ heat profile.

### 3.1 Numerical experiments

As an example, consider the 1D cosine based initial heat profile with $k = 1$ and zero boundary conditions. The profile and subsequent integration results are shown in **Figure 6**. It is clear that the temperature in the rod will approach zero at all locations with increasing time. The inverse problem consists of approximately recovering the initial profile $u(x, 0) = 0.5\cos(2\pi/65x) + 1)$, with support on

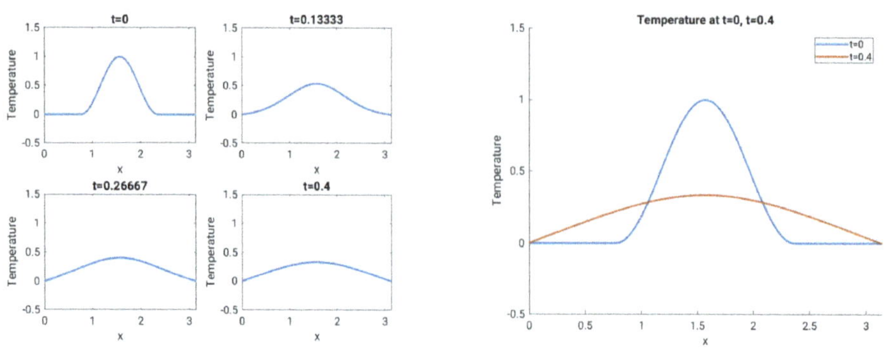


**Figure 6.**
*Homogeneous 1D profile and change of temperature with time.*

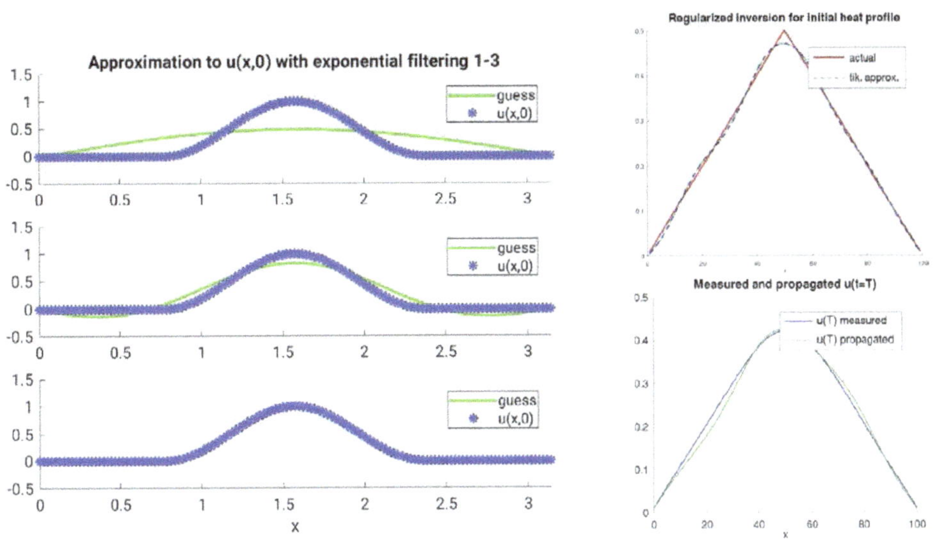


**Figure 7.**
*Initial condition reconstruction with different filtering levels and illustration of inversion for initial condition using Tikhonov regularization with parameter tuned via the L-curve approach.*

$[N/4, 3N/4]$. For this, we take the sin transform of the solution at $t = T$ and multiply by $\exp\left(k\left(\frac{n\pi}{L}\right)^2 T\right)$. In Octave, this can be accomplished with the code:

```
fuT=fft (u (:, n)); % temperature profile at time T
R=exp (k.*k '*T);
filter (R);
init_cond_approx=real (ifft (fuT.*R));
```

The term $R$ may contain large entries, which will blow up when the inverse fft is performed. For this reason, the filtering of small entries is necessary. This can be accomplished simply by replacing all large entries of $R$ with either zero or some max value; e.g. taking the cutoff to be all values above $1e6$ and replacing them either with 0 or $1e2$. The cutoff can be estimated with standard statistical outlier techniques, by taking the quartiles $Q_1, Q_2, Q_3$ of the $|R|$ values, computing $IQR = Q_3 - Q_1$ and filtering all entries of $R$ above $Q_3 + \alpha IQR$ with $\alpha \geq 2$ by setting them to the $Q_2$ value. The process is illustrated in **Figure 7**, for the same example as above with filtering for absolute value of entries above $1e1, 1e3, 1e5$. It can be observed that filtering only the entries of $R$ above $1e5$ in absolute value leads to a good approximation of the initial condition. Tikhonov regularization can be utilized when it is possible to build up the finite difference propagation operator $(L)$, either as an explicit matrix, or as a function. An example is shown to the right of **Figure 7**, with an initial triangular temperature profile that is approximately recovered by a simple continuation scheme. In the same Figure, we also show the profiles of $u(T)$ both as observed and as propagated from the approximated initial values. The smoothing properties of the heat kernel [3] give rise to the observed transformation between $t = 0$ and $t = T$.

## 4. Conclusions

This article discusses numerical approaches for the heat equation, both for the forward and inverse problems with examples for homogeneous and non-homogeneous cases. For the forward heat propagation modeling problem, we have presented a simple non-standard finite difference formulation, which satisfies the differential model and appears to result in improved performance over the standard finite difference scheme with similar magnitude time and spatial step sizes. For the inverse problem, we have described approaches based on Fourier filtering and Tikhonov regularization.

## Author details

Sergey Voronin
Research Scientist, Amathsolutions, USA

*Address all correspondence to: svoronin@amathsolutions.org

## References

[1] Özişik, M. Necati, Helcio RB Orlande, Marcelo J. Colaço, and Renato M. Cotta. Finite difference methods in heat transfer. CRC press, 2017.

[2] Glicksman, Leon R., and John H. Lienhard. Modeling and Approximation in Heat Transfer. Cambridge University Press, 2016.

[3] Taler, Jan, and Piotr Duda. Solving direct and inverse heat conduction problems. Springer Science & Business Media, 2010.

[4] Muniz, Wagner Barbosa, Haroldo F. de Campos Velho, and Fernando Manuel Ramos. "A comparison of some inverse methods for estimating the initial condition of the heat equation." Journal of Computational and Applied Mathematics 103, no. 1 (1999): 145–163.

[5] Mickens, Ronald E. "Nonstandard finite difference schemes for differential equations." Journal of Difference Equations and Applications 8, no. 9 (2002): 823–847.

[6] Hansen, Per Christian, and Dianne Prost O'Leary. "The use of the L-curve in the regularization of discrete ill-posed problems." SIAM journal on scientific computing 14, no. 6 (1993): 1487–1503.

[7] Castro, L. P., Q. Chen, and S. Saitoh. "Source inversion of heat conduction from a finite number of observation data." Applicable Analysis 89, no. 6 (2010): 801–813.

3

# A Numerical Approach to Solving an Inverse Heat Conduction Problem using the Levenberg-Marquardt Algorithm

*Tao Min, Xing Chen, Yao Sun and Qiang Huang*

**Abstract**

This chapter is intended to provide a numerical algorithm involving the combined use of the Levenberg-Marquardt algorithm and the Galerkin finite element method for estimating the diffusion coefficient in an inverse heat conduction problem (IHCP). In the present study, the functional form of the diffusion coefficient is an unknown priori. The unknown diffusion coefficient is approximated by the polynomial form and the present numerical algorithm is employed to find the solution. Numerical experiments are presented to show the efficiency of the proposed method.

**Keywords:** parabolic equation, inverse problem, Levenberg-Marquardt

## 1. Introduction

The numerical solution of the inverse heat conduction problem (IHCP) requires to determine diffusion coefficient from an additional information. Inverse heat conduction problems have many applications in various branches of science and engineering, mechanical and chemical engineers, mathematicians and specialists in many other sciences branches are interested in inverse problems, each with different application in mind [1–15].

In this work, we propose an algorithm for numerical solving an inverse heat conduction problem. The algorithm is based on the Galerkin finite element method and Levenberg-Marquardt algorithm [16–17] in conjunction with the least-squares scheme. It is assumed that no prior information is available on the functional form of the unknown diffusion coefficient in the present study, thus, it is classified as the function estimation in inverse calculation. Run the numerical algorithm to solve the unknown diffusion coefficient which is approximated by the polynomial form. The Levenberg-Marquardt optimization is adopted to modify the estimated values.

The plan of this paper is as follows: in Section 2, we formulate a one-dimensional IHCP. In Section 3, the numerical algorithm is derived. Calculation of sensitivity coefficients will be discussed in Section 4. In order to discuss on some numerical aspects, two examples are given in Section 5. Section 6 ends this paper with a brief discussion on some numerical aspects.

## 2. Description of the problem

The mathematical formulation of a one-dimensional heat conduction problem is given as follows:

$$\frac{\partial u}{\partial t} = \frac{\partial}{\partial x}\left[q(x)\frac{\partial u}{\partial x}\right] + f(x,t), \quad (x,t) \in (0,\mathrm{L}) \times (0,\mathrm{T}], \tag{1}$$

with the initial condition

$$u(\mathrm{x},0) = \mathrm{u}_0(x), \quad 0 \le x \le L, \tag{2}$$

and Dirichlet boundary conditions

$$u(0,\mathrm{t}) = \mathrm{g}_1(t), \quad 0 \le t \le T, \tag{3}$$

$$u(1,\mathrm{t}) = \mathrm{g}_2(t), \quad 0 \le t \le T, \tag{4}$$

where $\mathrm{f}(\mathrm{x},\mathrm{t})$, $u_0(\mathrm{x})$, $g_1(\mathrm{t})$, $g_2(\mathrm{t})$ and $q(\mathrm{x})$ are continuous known functions. We consider the problem (1)–(4) as a direct problem. As we all know, if $u_0(\mathrm{x}), g_1(\mathrm{t})$, $g_2(\mathrm{t})$ are continuous functions and $q(\mathrm{x})$ is known, the problem (1)–(4) has a unique solution.

For the inverse problem, the diffusion coefficient $q(\mathrm{x})$ is regarded as being unknown. In addition, an overspecified condition is also considered available. To estimate the unknown coefficient $q(\mathrm{x})$, the additional information on the boundary $x = x_0$, $0 < x_0 < L$ is required. Let the $u(\mathrm{x},\mathrm{t})$ taken at $x = x_0$ over the time period $[0,\mathrm{T}]$ be denoted by

$$\mathrm{u}(x_0,t) = \mathrm{g}(t) \quad 0 \le t \le T. \tag{5}$$

It is evident that for an unknown function $q(\mathrm{x})$, the problem (1)–(4) is underdetermined and we are forced to impose additional information (5) to provide a unique solution pair $(\mathrm{u}(\mathrm{x},\mathrm{t}), \mathrm{q}(\mathrm{x}))$ to the inverse problem (1)–(5).

We note that the measured overspecified condition $\mathrm{u}(\mathrm{x}_0,\mathrm{t}) = \mathrm{g}(t)$ should contain measurement errors. Therefore the inverse problem can be stated as follows: by utilizing the above-mentioned measured data, estimate the unknown function $q(\mathrm{x})$.

In this work the polynomial form is proposed for the unknown function $q(\mathrm{x})$ before performing the inverse calculation. Therefore $q(\mathrm{x})$ approximated as

$$q(x) \approx \hat{q}(\mathrm{x}) = \mathrm{p}_1 + p_2 x + p_3 x^2 + \cdots + p_{m+1} x^m, \tag{6}$$

where $p_1, p_2, \cdots, p_{m+1}$ are constants which remain to be determined simultaneously. The unknown coefficients $p_1, p_2, \cdots, p_{m+1}$ can be determined by using least squares method. The error in the estimate

$$F(p_1, p_2, \ldots, p_{m+1}) = \sum_{i=1}^{n} \left[u(x_0, t_i, p_1, p_2, \ldots, p_{m+1}) - g(\mathrm{t}_i)\right]^2, \tag{7}$$

is to be minimized. Here, $u(x_0, t_i, p_1, p_2, \ldots, p_{m+1})$ are the calculated results. These quantities are determined from the solution of the direct problem which is given previously by using an approximated $\hat{q}(\mathrm{x})$ for the exact $q(\mathrm{x})$. The estimated values of $p_j$, $j = 1, 2, \cdots, m+1$ are determined until the value of $F(p_1, p_2, \ldots, p_{m+1})$ is minimum. Such a norm can be written as

$$F(\mathbf{P}) = [\mathbf{U}(\mathbf{P})\text{-}\mathbf{G}]^T[\mathbf{U}(\mathbf{P})\text{-}\mathbf{G}], \tag{8}$$

where $\mathbf{P}^T = [p_1, p_2, \cdots, p_{m+1}]$ denotes the vector of unknown parameters and the superscript $T$ above denotes transpose. The vector $[\mathbf{U}(\mathbf{P})\text{-}\mathbf{G}]^T$ is given by

$$[\mathbf{U}(\mathbf{P})\text{-}\mathbf{G}]^T = [u(x_0, t_1, \mathbf{p}) - g(t_1), u(x_0, t_2, \mathbf{p}) - g(t_2), \cdots, u(x_0, t_n, \mathbf{p}) - g(t_n)]. \tag{9}$$

$F(\mathbf{P})$ is real-valued bounded function defined on a closed bounded domain $D \subset R^{m+1}$. The function $F(\mathbf{P})$ may have many local minimum in $D$, but it has only one global minimum. When $F(\mathbf{P})$ and $D$ have some attractive properties, for instance, $F(\mathbf{P})$ is a differentiable concave function and $D$ is a convex region, then a local maximum and problem can be solved explicitly by mathematical programming methods.

## 3. Overview of the Levenberg-Marquardt method

The Levenberg-Marquardt method, originally devised for application to nonlinear parameter estimation problems, has also been successfully applied to the solution of linear ill-conditioned problems. Such a method was first derived by Levenberg (1944) by modifying the ordinary least-squares norm. Later Marquardt (1963) derived basically the same technique by using a different approach. Marquardt's intention was to obtain a method that would tend to the Gauss method in the neighborhood of the minimum of the ordinary least-squares norm, and would tend to the steepest descent method in the neighborhood of the initial guess used for the iterative procedure.

To minimize the least squares norm (8), we need to equate to zero the derivatives of $F(\mathbf{P})$ with respect to each of the unknown parameters $[p_1, p_2, \cdots, p_{m+1}]$, that is

$$\frac{\partial F(\mathbf{P})}{\partial p_1} = \frac{\partial F(\mathbf{P})}{\partial p_2} = \cdots = \frac{\partial F(\mathbf{p})}{\partial p_{m+1}} = 0. \tag{10}$$

Let us introduce the sensitivity or Jacobian matrix, as follows:

$$\mathbf{J}(\mathbf{P}) = \left[\frac{\partial \mathbf{U}^T(\mathbf{P})}{\partial \mathbf{P}}\right]^T = \begin{bmatrix} u_{p_1}(x_0, t_1, \mathbf{p}) & u_{p_2}(x_0, t_1, \mathbf{p}) & \cdots & u_{p_{m+1}}(x_0, t_1, \mathbf{p}) \\ u_{p_1}(x_0, t_2, \mathbf{p}) & u_{p_2}(x_0, t_2, \mathbf{p}) & \cdots & u_{p_{m+1}}(x_0, t_2, \mathbf{p}) \\ \cdots & & & \\ u_{p_1}(x_0, t_n, \mathbf{p}) & u_{p_2}(x_0, t_n, \mathbf{p}) & \cdots & u_{p_{m+1}}(x_0, t_n, \mathbf{p}) \end{bmatrix}, \tag{11}$$

$$\text{or} \quad J_{ij} = u_{p_j}(x_0, t_i, \mathbf{p}) = \frac{\partial u(x_0, t_i, \mathbf{p})}{\partial p_j}, i = 1, 2, \ldots n, j = 1, 2, \ldots m + 1. \tag{12}$$

The elements of the sensitivity matrix are called the sensitivity coefficients, the results of differentiation (10) can be written down as follows:

$$-2\mathbf{J}^T(\mathbf{P})[\mathbf{U}(\mathbf{P}) - \mathbf{G}] = 0. \tag{13}$$

For linear inverse problem the sensitivity matrix is not a function of the unknown parameters. The Eq. (13) can be solved then in explicit form:

$$\mathbf{P} = (\mathbf{J}^T\mathbf{J})^{-1}\mathbf{J}^T\mathbf{G}. \tag{14}$$

In the case of a nonlinear inverse problem, the matrix $\mathbf{J}$ has some functional dependence on the vector $\mathbf{p}$. The solution of Eq. (13) requires an iterative procedure, which is obtained by linearizing the vector $\mathbf{UP}()$with a Taylor series expansion around the current solution at iteration $k$ Such a linearization is given by

$$\mathbf{U}(\mathbf{P}) = \mathbf{U}(\mathbf{P}^k) + \mathbf{J}^k(\mathbf{P}\text{-}\mathbf{P}^k), \tag{15}$$

where $\mathbf{U}(\mathbf{P}^k)$ and $\mathbf{J}^k$ are the estimated temperatures and the sensitivity matrix evaluated at iteration $k$, respectively. Eq. (15) is substituted into (14) and the resulting expression is rearranged to yield the following iterative procedure to obtain the vector of unknown parameters $\mathbf{P}$:

$$\mathbf{P}^{k+1} = \mathbf{P}^k + \left[(\mathbf{J}^k)^T\mathbf{J}^k\right]^{-1}(\mathbf{J}^k)^T\left[\mathbf{G}\text{-}\mathbf{U}(\mathbf{P}^k)\right]. \tag{16}$$

The iterative procedure given by Eq. (16) is called the Gauss method. Such method is actually an approximation for the Newton (or Newton-Raphson) method. We note that Eq. (14), as well as the implementation of the iterative procedure given by Eq. (16), require the matrix $\mathbf{J}^T\mathbf{J}$ to be nonsingular, or

$$\left|\mathbf{J}^T\mathbf{J}\right| \neq 0, \tag{17}$$

where $|\cdot|$ is the determinant.

Formula (17) gives the so called identifiability condition, that is, if the determinant of $\mathbf{J}^T\mathbf{J}$ is zero, or even very small, the parameters $p_j$, for $j = 1, 2, \cdots, \mathrm{m} + 1$, cannot be determined by using the iterative procedure of Eq. (16).

Problems satisfying $\left|\mathbf{J}^T\mathbf{J}\right| \approx 0$ are denoted ill-conditioned. Inverse heat transfer problems are generally very ill-conditioned, especially near the initial guess used for the unknown parameters, creating difficulties in the application of Eqs. (14) or (16). The Levenberg-Marquardt method alleviates such difficulties by utilizing an iterative procedure in the form:

$$\mathbf{P}^{k+1} = \mathbf{P}^k + \left[(\mathbf{J}^k)^T\mathbf{J}^k + \mu^k\Omega^k\right]^{-1}(\mathbf{J}^k)^T\left[\mathbf{G}\text{-}\mathbf{U}(\mathbf{P}^k)\right], \tag{18}$$

where $\mu^k$ is a positive scalar named damping parameter and $\Omega^k$ is a diagonal matrix.

The purpose of the matrix term $\mu^k\Omega^k$ is to damp oscillations and instabilities due to the ill-conditioned character of the problem, by making its components large as compared to those of $\mathbf{J}^T\mathbf{J}$ if necessary. $\mu^k$ is made large in the beginning of the iterations, since the problem is generally ill-conditioned in the region around the initial guess used for iterative procedure, which can be quite far from the exact parameters. With such an approach, the matrix $\mathbf{J}^T\mathbf{J}$ is not required to be nonsingular in the beginning of iterations and the Levenberg-Marquardt method tends to the steepest descent method, that is, a very small step is taken in the negative gradient direction. The parameter $\mu^k$ is then gradually reduced as the iteration procedure advances to the solution of the parameter estimation problem, and then the Levenberg-Marquardt method tends to the Gauss method given by (16). The following criteria were suggested in literature [13] to stop the iterative procedure of the Levenberg-Marquardt method given by Eq. (18):

$$F(\mathbf{p}^{k+1}) < \varepsilon_1, \tag{19}$$

$$\left\| \left(\mathbf{J}^{k}\right)\left[\mathbf{G}\text{-}\mathbf{U}\left(\mathbf{p}^{k}\right)\right]\right\| < \varepsilon_2, \tag{20}$$

$$\left\| \mathbf{p}^{k+1}\text{-}\mathbf{p}^{k} \right\| < \varepsilon_3, \tag{21}$$

where $\varepsilon_1$, $\varepsilon_2$ and $\varepsilon_3$ are user prescribed tolerances and $\|\cdot\|$ denotes the Euclidean norm. The criterion given by Eq. (19) tests if the least squares norm is sufficiently small, which is expected in the neighborhood of the solution for the problem. Similarly, Eq. (20) checks if the norm of the gradient of $F(\mathbf{p})$ is sufficiently small, since it is expected to vanish at the point where $F(\mathbf{p})$ is minimum. The last criterion given by Eq. (21) results from the fact that changes in the vector of parameters are very small when the method has converged. Generally, these three stopping criteria need to be tested and the iterative procedure of the Levenberg-Marquardt method is stopped if any of them is satisfied.

Different versions of the Levenberg-Marquardt method can be found in the literature, depending on the choice of the diagonal matrix $\Omega^k$ and on the form chosen for the variation of the damping parameter $\mu^k$. In this paper, we choose the $\Omega^k$ as

$$\Omega^k = diag\left[\left(\mathbf{J}^k\right)^T\mathbf{J}^k\right]. \tag{22}$$

Suppose that the vector of temperature measurements $\mathbf{G} = [g(t_1), g(t_2), \cdots, g(t_n)]$ are given at times $t_i$, $i = 1, 2, \cdots$, n and an initial guess $\mathbf{P}^0$ is available for the vector of unknown parameters $\mathbf{P}$. Choose a value for $\mu^0$, say, $\mu^0 = 0.001$ and $k = 0$. Then,

Step 1. Solve the direct problem (1)–(4) with the available estimate $\mathbf{P}^k$ in order to obtain the vector $\mathbf{U}(\mathbf{P}^k) = [u(x_0, t_1, \mathbf{p}^k), u(x_0, t_2, \mathbf{p}^k), \cdots, u(x_0, t_n, \mathbf{p}^k)]$.

Step 2. Compute $F(\mathbf{P}^k)$ from the Eq. (8).

Step 3. Compute the sensitivity matrix $\mathbf{J}^k$ from (12) and then the matrix $\Omega^k$ from (22), by using the current value of $\mathbf{P}^k$.

Step 4. Solve the following linear system of algebraic equations, obtained from (18): $\left[\left(\mathbf{J}^k\right)^T\mathbf{J}^k + \mu^k\Omega^k\right]\Delta\mathbf{P}^k = \left(\mathbf{J}^k\right)^T\left[\mathbf{G}\text{-}\mathbf{U}\left(\mathbf{P}^k\right)\right]$ in order to compute $\Delta\mathbf{P}^k = \mathbf{P}^{k+1}\text{-}\mathbf{P}^k$.

Step 5. Compute the new estimate $\mathbf{P}^{k+1}$ as $\mathbf{P}^{k+1} = \mathbf{P}^k + \Delta\mathbf{P}^k$.

Step 6. Solve the exact problem (1)–(4) with the new estimate $\mathbf{P}^{k+1}$ in order to find $U(\mathbf{P}^{k+1})$. Then compute $F(\mathbf{P}^{k+1})$.

Step 7. If $F(\mathbf{P}^{k+1}) \geq F(\mathbf{P}^k)$ replace $\mu^k$ by $10\mu^k$ and return to step 4.

Step 8. If $F(\mathbf{P}^{k+1}) \leq F(\mathbf{P}^k)$, accept the new estimate $\mathbf{P}^{k+1}$ and replace $\mu^k$ by $0.1\mu^k$.

Step 9. Check the stopping criteria given by (19). Stop the iterative procedure if any of them is satisfied; otherwise, replace $k$ by $k + 1$ and return to step 3.

## 4. Calculation of sensitivity coefficients

Generally, there have two approaches for determining the gradient; the first is a discretize-then-differentiate approach and the second is a differentiate-then-discretize approach.

The first approach is to approximate the gradient of the functional by a finite difference quotient approximation, but in general, we cannot determine the sensitivities exactly, so this method may led to larger error.

Here we intend to use differentiate-then-discretize approach which we refer to as the sensitivity equation method. This method can be determined more efficiently with the help of the sensitivities

$$\mathrm{u}_k = \frac{\partial u}{\partial p_k}, k = 1, 2, \cdots m + 1. \tag{23}$$

We first differentiate the flow system (1)–(4) with respect to each of the design parameters $[p_1, p_2, ..., p_{m+1}]$, to obtain the $m+1$ continuous sensitivity systems: for $k = 1, 2, \cdots, m + 1$

$$\begin{cases} \frac{\partial u_k}{\partial t} = \frac{\partial}{\partial x}\left[\left(p_1 + p_2 x + \cdots + p_{m+1}x^m\right)\frac{\partial u_k}{\partial x} + x^{k-1}\frac{\partial u}{\partial x}\right] \\ u_k(x, 0) = 0 \\ u_k(0, t) = 0 \\ u_k(L, t) = 0 \end{cases}. \tag{24}$$

There have $(m + 2)$ equations, we can make them in one system equation and use the finite element methods to solve the system of equation. Here, we give the vector form of the equation as follow:

$$(\mathbf{P1})\begin{cases} \frac{\partial \vec{\mathbf{U}}}{\partial t} + \nabla \cdot \vec{\mathbf{\Gamma}} = \vec{\mathbf{F}} \\ \vec{\mathbf{U}}(x, 0) = \vec{\mathbf{U}}_0(x), \\ \vec{\mathbf{U}}(0, t) = \vec{\mathbf{G}}_1(t) \\ \vec{\mathbf{U}}(L, t) = \vec{\mathbf{G}}_2(t) \end{cases} \tag{25}$$

where

$$\vec{\mathbf{U}} = \begin{bmatrix} u \\ u_1 \\ u_2 \\ \cdots \\ u_{m+1} \end{bmatrix}, \vec{\mathbf{\Gamma}} = \begin{bmatrix} -\left(p_1 + p_2 x + \cdots + p_{m+1}x^m\right)\frac{\partial u}{\partial x} \\ -\left(p_1 + p_2 x + \cdots + p_{m+1}x^m\right)\frac{\partial u_1}{\partial x} - \left(\frac{\partial u}{\partial x}\right) \\ -\left(p_1 + p_2 x + \cdots + p_{m+1}x^m\right)\frac{\partial u_2}{\partial x} - x\left(\frac{\partial u}{\partial x}\right) \\ \cdots \\ -\left(p_1 + p_2 x + \cdots + p_{m+1}x^m\right)\frac{\partial u_{m+1}}{\partial x} - \left(x^m\frac{\partial u}{\partial x}\right) \end{bmatrix}, \vec{\mathbf{F}} = \begin{bmatrix} f(\mathrm{x}, \mathrm{t}) \\ 0 \\ 0 \\ \cdots \\ 0 \end{bmatrix}. \tag{26}$$

$$\vec{\mathbf{U}}_0(x) = [\mathrm{u}_0(x), 0, 0, \cdots 0]^T, \vec{\mathbf{G}}_1(t) = [g_1(t), 0, 0, \cdots 0]^T, \vec{\mathbf{G}}_2(t) = [g_2(t), 0, 0, \cdots 0]^T. \tag{27}$$

We use the Galerkin finite element method approximation for discretizing problem (25). For this, we multiply the Eq. (25) by a test function $v : [0, L] \to R$, $v \in V_0 := H_0^1(0, L)$ and integrate the obtained equation in space form 0 to $L$. We obtain the following equation:

$$\int_0^{\mathrm{L}} \frac{\partial \mathbf{U}(x, t)}{\partial t} \cdot v(x)dx - \int_0^L \nabla\mathbf{\Gamma} \cdot v(x)dx = \int_0^L \mathbf{F}(x, t) \cdot v(x)dx, \tag{28}$$

integrating by parts gives

$$\int_0^L \nabla\mathbf{\Gamma} \cdot v(x)dx = (\mathbf{\Gamma} \cdot v(x))|_0^L - \int_0^L \mathbf{\Gamma} \cdot \frac{\partial v(x, t)}{\partial x}dx. \tag{29}$$

We can change the first derivative in time and the integral. We have $v(0) = 0 = v(L)$, because $v \in V_0$. This leads to an equivalent problem to (**P1**): ∀*t*>0, find $\mathbf{U}(x,t)$ satisfying

$$\frac{d}{dt}\int_0^L \mathbf{U}(x,t)\cdot v(x)dx + \int_0^L \mathbf{\Gamma}\cdot\frac{\partial v(x,t)}{\partial x}dx = \int_0^L \mathbf{F}(x,t)\cdot v(x)dx, \tag{30}$$

for all $v \in V_0 := H_0^1(0,L)$. To simplify the notation we use the scalar product in $L^2(0,L)$

$$(f,g) = \int_0^L f(x)\cdot g(x)dx. \tag{31}$$

We also can define the following bilinear form:

$$a(\mathbf{U},v) = \int_0^L \mathbf{\Gamma}\cdot\frac{\partial v(x,t)}{\partial x}dx = \begin{cases} \int_0^L -(p_1+p_2x+\cdots+p_{m+1}x^m)\frac{\partial u}{\partial x}\frac{\partial v}{\partial x}dx \\ \int_0^L\left(-(p_1+p_2x+\cdots+p_{m+1}x^m)\frac{\partial u_1}{\partial x}\frac{\partial v}{\partial x}-\frac{\partial u}{\partial x}\frac{\partial v}{\partial x}\right)dx \\ \int_0^L\left(-(p_1+p_2x+\cdots+p_{m+1}x^m)\frac{\partial u_2}{\partial x}\frac{\partial v}{\partial x}-x\frac{\partial u}{\partial x}\frac{\partial v}{\partial x}\right)dx \\ \cdots \\ \int_0^L\left(-(p_1+p_2x+\cdots+p_{m+1}x^m)\frac{\partial u_{m+1}}{\partial x}\frac{\partial v}{\partial x}-x^{m+1}\frac{\partial u}{\partial x}\frac{\partial v}{\partial x}\right)dx \end{cases}. \tag{32}$$

Finally, we obtain with this notations the weak problem of (**P1**):

$$(\mathbf{P2})\begin{cases} \frac{d}{dt}(\mathbf{U},v)_{L^2}+a(\mathbf{U},v)=(\mathbf{F},v)_{L^2} \\ \mathbf{U}(x,0)=\mathbf{U}_0(x) \\ \mathbf{U}(0,t)=\mathbf{G}_1(t) \\ \mathbf{U}(L,t)=\mathbf{G}_2(t) \end{cases}. \tag{33}$$

### 4.1 Space-discretization with the Galerkin method

In this section, we search a semi-discrete approximation of the weak problem (**P2**), using the Galerkin finite element method. This leads to a first order Cauchy-problem in time.

Let $V_h$ be a $N_x+1$ dimensional subspace of $V$ and $V_{0,h} = V_h \cap V_0$. Then the following problem is an approximation of the weak problem, find $u_h, u_{1,h}, u_{2,h}\cdots u_{m+1.h} \in V_h$ that satisfies:

$$\begin{cases} \frac{d}{dt}(\mathbf{U}_h,v_h)+a(\mathbf{U}_h,v_h)=(\mathbf{F},v_h) \\ \mathbf{U}_h(\mathrm{x},0)=\mathbf{U}_{0,h}(x) \\ \mathbf{U}_h(0,t)=\mathbf{G}_1(t) \\ \mathbf{U}_h(L,t)=\mathbf{G}_2(t) \end{cases}, \tag{34}$$

for all $v_h \in V_{0,h}$. where $\mathbf{U}_h = [u_h, u_{1,h}, u_{2,h}\cdots u_{m+1.h}]^T$.

The choice of $V_h$ is completely arbitrary. So we can choose it the way that for later treatment, it will be as easy as possible. For example, we subdivide the interval $[0, L]$ into partitions of equal distances $h$:

$$0 = a_1 < a_2 < \cdots < a_{N_x} < a_{N_x+1} = L, , a_i = (i-1) \cdot h \tag{35}$$

$$V_h = \{v_h \in C^0[0, L] : v_h|_{[a_i, a_{i+1}]} \in P_1, \forall i = 1 \cdots N_x\}, \tag{36}$$

$$V_{0,h} = \{v_h \in V_h : v_h(0) = v_h(L) = 0\}. \tag{37}$$

Note, that the finite dimension allows us to build a finite base for the corresponding space. In the case of $V_{0,h}$, we have: $\{\varphi_i\}_{i=2}^{N_x}$ where $\forall i = 2 \cdots N_x$.

$$\varphi_i(x) = \begin{cases} 0 & x \in [a_0, a_{i-1}] \\ \dfrac{x}{h} - (i-2) & x \in [a_{i-1}, a_i] \\ i - \dfrac{x}{h} & x \in [a_i, a_{i+1}] \\ 0 & x \in [a_{i+1}, a_{N_x+1}] \end{cases}, \tag{38}$$

while we add for $V_h$ the two functions $\varphi_1$ and $\varphi_{N_x+1}$ defined as:

$$\varphi_1(x) = \begin{cases} 1 - \dfrac{x}{h}, & if x \in [a_1, a_2] \\ 0, & if x \in [a_2, a_{N_x+1}] \end{cases}, \tag{39}$$

$$\varphi_{N_x+1}(x) = \begin{cases} 0, & if x \in [a_1, a_{N_x}] \\ \dfrac{x}{h} - N_x + 1, & if x \in [a_{N_x}, a_{N_x+1}] \end{cases}, \tag{40}$$

so that we can write $\mathbf{U}_h$ as a linear combination of the basic elements:

$$\mathbf{U}_h(x, t) = \sum_{j=1}^{N_x+1} \tilde{\mathbf{U}}_j(t) \cdot \varphi_j(x), \tag{41}$$

$$\mathbf{U}_{0,h}(x, t) = \sum_{j=1}^{N_x+1} \mathbf{U}_0(x_j) \cdot \varphi_j(x), \tag{42}$$

where $\tilde{\mathbf{U}}_1(t) = \mathbf{G}_0(t)$ and $\tilde{\mathbf{U}}_{N_x+1}(t) = \mathbf{G}_1(t)$. Using that $a(\cdot, \cdot)$ is bilinear form and that Eq. (34) is valid for each element of the base $\{\varphi_i\}_{i=2}^{N_x}$, we obtain

$$\sum_{j=1}^{N_x+1} \frac{d}{dt} \tilde{\mathbf{U}}_j(t) \cdot \left(\varphi_j, \varphi_i\right) + \sum_{j=1}^{N_x+1} \tilde{\mathbf{U}}_j(t) \cdot a\left(\varphi_j, \varphi_i\right) = (\mathbf{F}, \varphi_i), \tag{43}$$

$\forall i = 2 \cdots N_x$. This equation can be written in a vector form. For this we define the vectors $\vec{\mathbf{u}}$ , $\vec{\mathbf{u}}_0$ and $\vec{\mathbf{F}}$ with components

$$\mathbf{F}_i(t) = (\mathbf{F}, \varphi_i)_{L^2}, \mathbf{u}_j(t) := \tilde{\mathbf{u}}_\mathrm{j}(t), \mathbf{u}_{0\,,\,\mathrm{j}} = \mathbf{u}_0(\mathbf{x}_\mathrm{j}), \tag{44}$$

and matrices $\mathbf{M}$ and $\mathbf{A}$ as

$$m_{ij} := \left(\varphi_i, \varphi_j\right)_{L^2}, a_{ij} := a\left(\varphi_i, \varphi_j\right), \tag{45}$$

Note that $\mathbf{M}, \mathbf{A} \in \mathbf{R}^{N_x-1 \times N_x+1}, \vec{\mathbf{u}} \in R^{N_x+1}, and\ \vec{\mathbf{F}} \in \mathbf{R}^{N_x-1}$. So that (43) is equal to the Cauchy problem

$$\begin{cases} \mathbf{M}\dfrac{d}{dx}\vec{\mathbf{u}}(t) + \mathbf{A} \cdot \vec{\mathbf{u}}(t) = \vec{\mathbf{F}}(t) \\ \vec{\mathbf{u}}(t_0) = \vec{\mathbf{u}}_\mathbf{0} \end{cases}, \tag{46}$$

the Crank-Nicolson method can be applied to (46) at time $t_k$, resulting in

$$\mathbf{M} \cdot \left( \frac{\vec{\mathbf{u}}_{k+1} - \vec{\mathbf{u}}_k}{\Delta t} \right) + \frac{1}{2}\mathbf{A} \cdot \vec{\mathbf{u}}_{\mathbf{k}+1} + \frac{1}{2}\mathbf{A} \cdot \vec{\mathbf{u}}_\mathbf{k} = \frac{1}{2}\left( \vec{\mathbf{F}}_\mathbf{k} + \vec{\mathbf{F}}_{\mathbf{k}+1} \right), \tag{47}$$

where $\vec{\mathbf{u}}_\mathbf{k} = \vec{\mathbf{u}}(t_k), \vec{\mathbf{F}}_\mathbf{k} = \vec{\mathbf{F}}(t_k), k = 0, 1, \cdots$.
The Eq. (47) can be written in simple form as

$$\left( \mathbf{M} + \frac{\Delta t}{2}\mathbf{A} \right) \cdot \vec{\mathbf{u}}_{\mathbf{k}+1} = \left( \mathbf{M} - \frac{\Delta t}{2}\mathbf{A} \right) \cdot \vec{\mathbf{u}}_\mathbf{k} + \frac{\Delta t}{2}\left( \vec{\mathbf{F}}_\mathbf{k} + \vec{\mathbf{F}}_{\mathbf{k}+1} \right), \tag{48}$$

the algebraic system (48) is solved by Gauss elimination method.

## 5. Numerical experiment

In this section, we are going to demonstrate some numerical results for $(\mathrm{u}(\mathrm{x}, \mathrm{t}), \mathrm{q}(\mathrm{x}))$ in the inverse problem (1)–(5). Therefore the following examples are considered and the solution is obtained.

**Example 1.** Consider (1)–(4) with

$$u(\mathrm{x}, 0) = \sin\mathrm{x}, \quad 0 \le x \le 1, \tag{49}$$

$$u(0, \mathrm{t}) = 0, \quad 0 \le t \le 1 \tag{50}$$

$$u(1, \mathrm{t}) = \sin(1)\mathrm{e}^{-t}, \quad 0 \le t \le 1, \tag{51}$$

$$f(x, t) = \left( \sin x \left( \frac{x^2}{4} + \frac{x}{2} + 1 \right) \right) e^{-t} - \frac{(x+1)}{2} \cos(x) e^{-t} - \sin x e^{-t}, \quad 0 \le x \le 1 0 \le t \le 1 \tag{52}$$

We obtain the unique exact solution

$$q(\mathrm{x}) = 1 + 0.5\mathrm{x} + 0.25\mathrm{x}^2 \tag{53}$$

And

$$u(\mathrm{x}, \mathrm{t}) = \sin(\mathrm{x})\mathrm{e}^{-t}. \tag{54}$$

We take the observed data $g$ as

$$g(t) = u(0.5, \mathrm{t}) = \sin(0.5)\mathrm{e}^{-t} 0 \le t \le 1 \tag{55}$$

The unknown function $q(\mathrm{x})$ defined as the following form

$$q(\hat{\mathrm{x}}) = p_1 + p_2 x + p_3 x^2, \tag{56}$$

where $p_1, p_2, p_3$ are unknown coefficients.

**Table 1** shows how the Levenberg-Marquardt algorithm can find the best parameters after 12 iterations when it is initialized in four different points.

**Figures 1–4** show the fitness of the estimated parameters and the rate of convergence.

**Figures 5–8** show the comparison between the inversion results $\hat{q}(x)$ and the exact value $q(x)$:

**Table 2** shows the values of $q(\mathrm{j}\Delta \mathrm{x})$ and $u(\mathrm{j}\Delta \mathrm{x}, 0.5)$ in $x = j\Delta x$ with the all the initial values are set 1.

**Example 2.** Consider (1)–(4) with

$$u(\mathrm{x}, 0) = \mathrm{xe}^{-x}, \quad 0 \leq x \leq 1, \tag{57}$$

$$u(0, \mathrm{t}) = \mathrm{te}^{-t}, \quad 0 \leq t \leq 1 \tag{58}$$

$$u(1, \mathrm{t}) = (1 + \mathrm{t})\mathrm{e}^{-1-t}, \quad 0 \leq t \leq 1 \tag{59}$$

$$\begin{aligned} f(\mathrm{x}, \mathrm{t}) = {} & \mathrm{e}^{-(\mathrm{t}+\mathrm{x})} - (\mathrm{t} + \mathrm{x})\mathrm{e}^{-(\mathrm{t}+\mathrm{x})} - e^{x}\left(\mathrm{e}^{-(\mathrm{t}+\mathrm{x})} - (\mathrm{t} + \mathrm{x})\mathrm{e}^{-(\mathrm{t}+\mathrm{x})}\right) \\ & + \mathrm{e}^{x}\left(2\mathrm{e}^{-(\mathrm{t}+\mathrm{x})} - (\mathrm{t} + \mathrm{x})\mathrm{e}^{-(\mathrm{t}+\mathrm{x})}\right), \quad 0 \leq x \leq 1 0 \leq t \leq 1 \end{aligned} \tag{60}$$

| Starting point | 0.5 0.5 0.5 | 1 1 1 | 10 10 10 | 50 50 50 |
|---|---|---|---|---|
| Iteration 12 | 0.999729028233135<br>0.499885876453067<br>0.252009862457275 | 0.999729028233183<br>0.499885876453056<br>0.252009862457315 | 0.999729028233194<br>0.499885876453057<br>0.252009862457325 | 0.999729028307261<br>0.499885876454169<br>0.25200986249336 |
| Error F | $8.7564944405 \times 10^{-14}$ | $8.7564944427 \times 10^{-14}$ | $8.7564944420 \times 10^{-14}$ | $8.7564944420 \times 10^{-14}$ |

**Table 1.**
*Performance of the algorithm when it is run to solve the model using three different parameters guesses.*

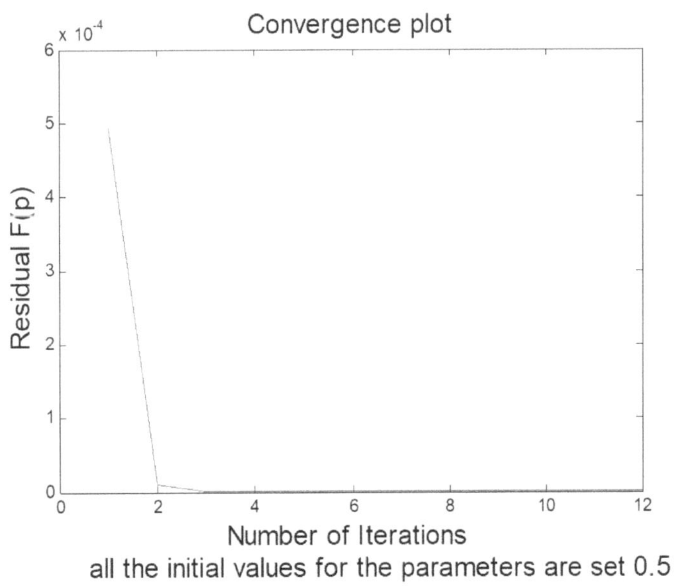


**Figure 1.**
*All the initial values for the parameters are set 0.5.*

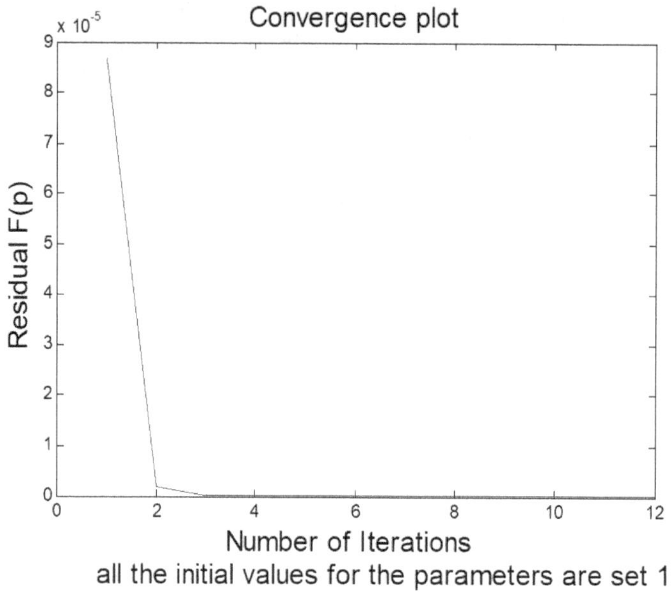


**Figure 2.**
*All the initial values for the parameters are set 1.*

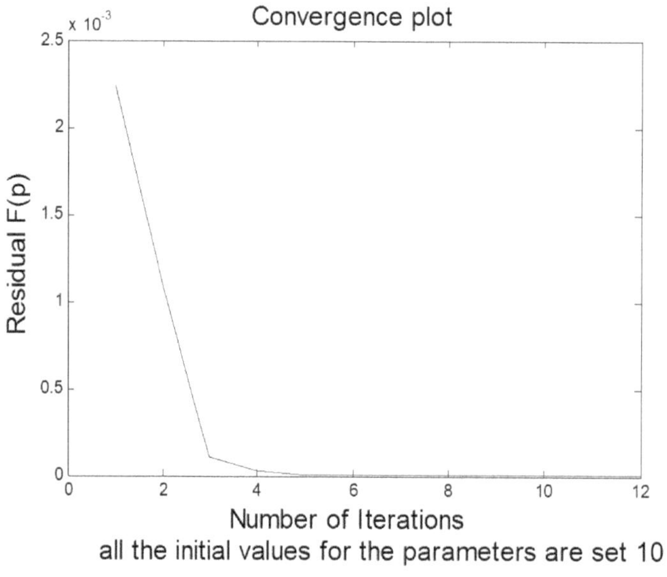


**Figure 3.**
*All the initial values for the parameters are set 10.*

We obtain the unique exact solution

$$q(\mathrm{x}) = e^{x}, \tag{61}$$

And

$$u(\mathrm{x}, \mathrm{t}) = (\mathrm{x} + \mathrm{t})\mathrm{e}^{-x-t}. \tag{62}$$

We take the observed data $g$ as

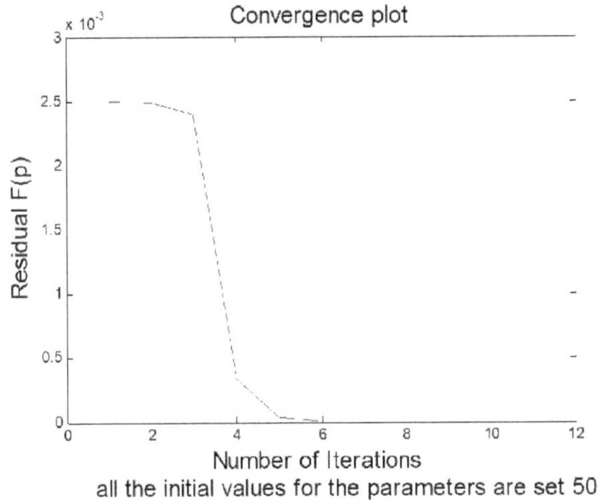


**Figure 4.**
*All the initial values for the parameters are set 50.*

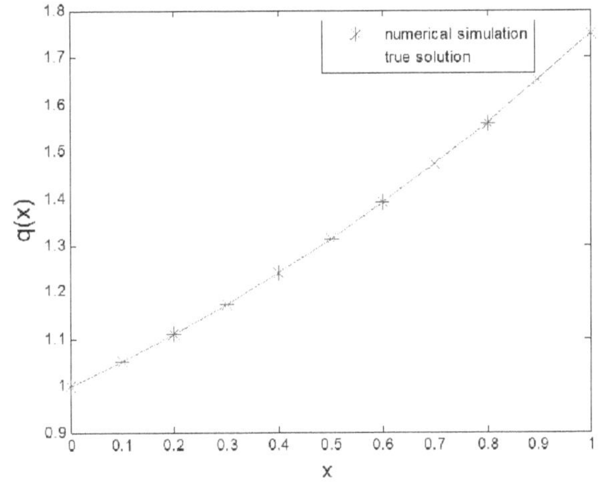


**Figure 5.**
*The comparison chart with all the initial values for the parameters is set 0.5.*

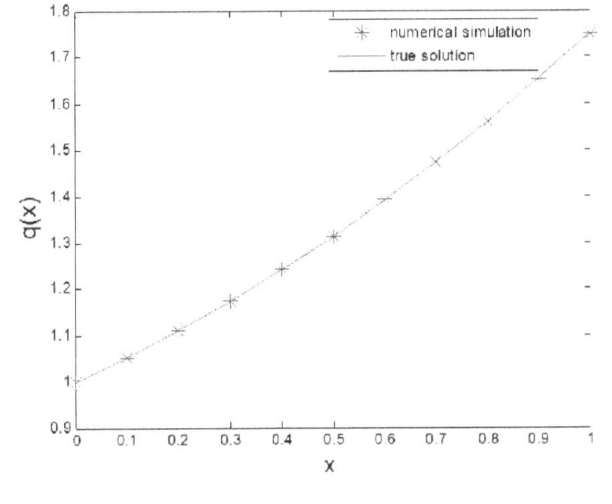


**Figure 6.**
*The comparison chart with all the initial values for the parameters is set 1.*

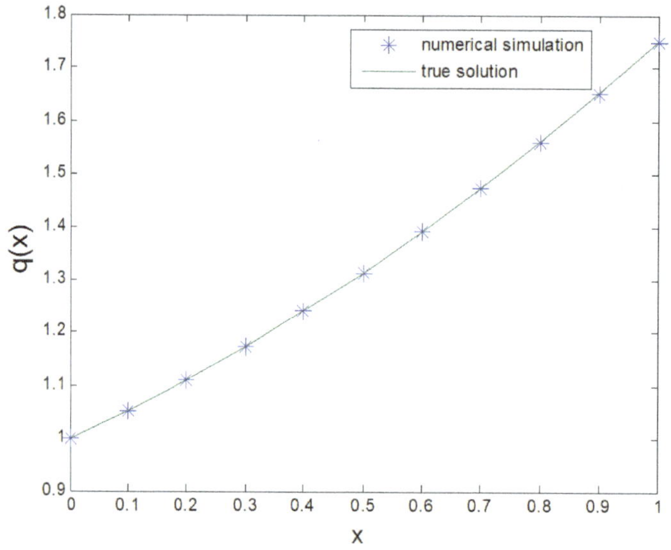


**Figure 7.**
*The comparison chart with all the initial values for the parameters is set 10.*

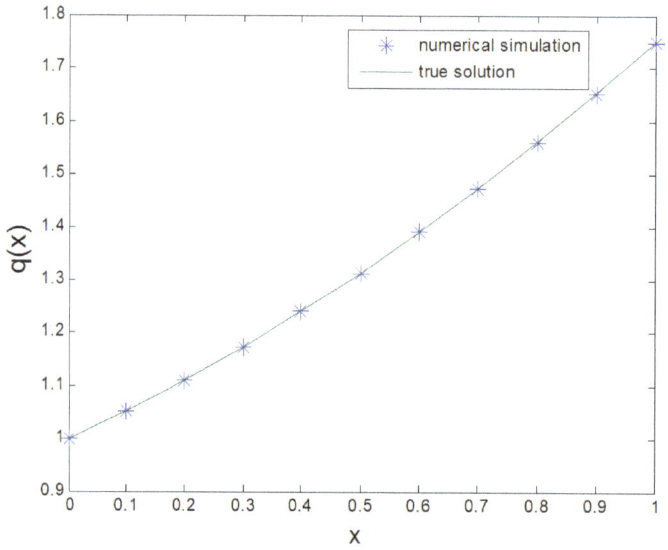


**Figure 8.**
*The comparison chart with all the initial values for the parameters is set 50.*

$$g(t) = u(0.5, \mathrm{t}) = (0.5 + \mathrm{t})\mathrm{e}^{(-0.5-\mathrm{t})} 0 \le t \le 1. \tag{63}$$

The unknown function $q(\mathrm{x})$ defined as the following form $\hat{q(\mathrm{x})} = p_1 + p_2x + p_3x^2 + p_4x^3 + p_5x^4 + p_6x^5 + p_7x^6 + p_8x^7$, where $p_1, p_2, \cdots, p_7, p_8$ are unknown coefficients.

**Table 3** shows how the Levenberg-Marquardt algorithm can find the best parameters after 20 iterations when it is initialized in four different points.

**Figures 9–12** show the fitness of the estimated parameters and the rate of convergence.

**Figures 13–16** show the comparison between the inversion results $\hat{q}(x)$ and the exact value $q(x)$:

| | Numerical | Exact | Numerical | Exact |
|---|---|---|---|---|
| $j$ | $q(j\Delta x)$ | $q(j\Delta x)$ | $u(j\Delta x, 0.5)$ | $u(j\Delta x, 0.5)$ |
| 0 | 0.999729028233183 | 1 | 0 | 0 |
| 1 | 1.05223771450306 | 1.0525 | 0.0605593190239173 | 0.0605520280601669 |
| 2 | 1.10978659802209 | 1.11 | 0.120511797611786 | 0.120499040271796 |
| 3 | 1.17237567879026 | 1.1725 | 0.179257059078521 | 0.179242065904716 |
| 4 | 1.24000495680758 | 1.24 | 0.236207449080344 | 0.236194164064666 |
| 5 | 1.31267443207404 | 1.3125 | 0.290793943250869 | 0.290786288212692 |
| 6 | 1.39038410458965 | 1.39 | 0.342471828361625 | 0.342472971890064 |
| 7 | 1.47313397435441 | 1.4725 | 0.390726114897089 | 0.390737778838824 |
| 8 | 1.56092404136831 | 1.56 | 0.435076630410587 | 0.435098463062163 |
| 9 | 1.65375430563136 | 1.6525 | 0.475082717530532 | 0.475111787267016 |
| 10 | 1.75162476714355 | 1.75 | 0.510347406713368 | 0.510377951544573 |

**Table 2.**
*The values of $q(j\Delta x)$ and $u(j\Delta x, 0.5)$ in $x = j\Delta x$ with the all the initial values being set to 1.*

| **Starting point** | **0.1 0.1 0.1 0.1<br>0.1 0.1 0.1 0.1** | **0.5 0.5 0.5 0.5<br>0.5 0.5 0.5 0.5** | **1 1 1 1<br>1 1 1 1** | **2 2 2 2<br>2 2 2 2** |
|---|---|---|---|---|
| Iteration 20 | 1.01536263526644<br>0.896348846894057<br>0.954285303464511<br>−0.890298938193057<br>1.40131927153131<br>−0.871276408882294<br>0.183785623507722<br>0.0359103726343979 | 1.01536263500695<br>0.896348850692318<br>0.954285278486704<br>−0.890298849338373<br>1.40131909032315<br>−0.871276197318301<br>0.183785492186491<br>0.0359104061952515 | 1.01536263525763<br>0.896348847022403<br>0.954285302637587<br>−0.890298935334171<br>1.40131926588117<br>−0.871276402487896<br>0.18378561965322<br>0.0359103735933848 | 1.01536263525905<br>0.896348846999736<br>0.954285302790922<br>−0.890298935876618<br>1.40131926696099<br>−0.87127640370648<br>0.183785620380814<br>0.0359103734148875 |
| Error $F$ | $7.89749200363512 \times 10^{-11}$ | $7.89749200363504 \times 10^{-11}$ | $7.89749200353888 \times 10^{-11}$ | $7.8974920035389 \times 10^{-11}$ |

**Table 3.**
*Performance of the algorithm when it is run to solve the model using four different parameters guesses.*

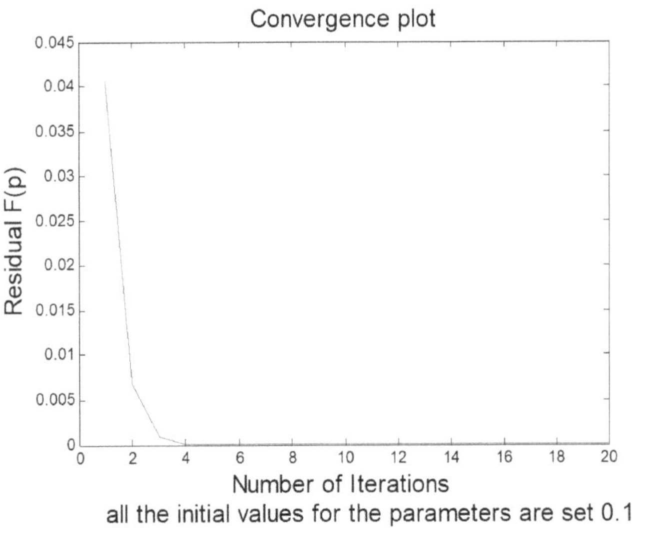


**Figure 9.**
*All the initial values for the parameters are set 0.1.*

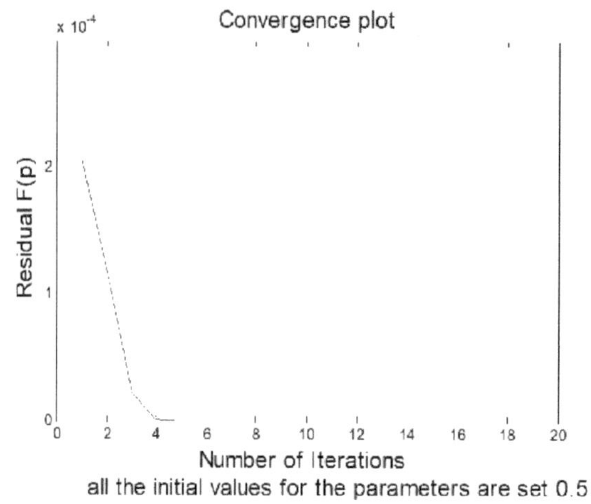


**Figure 10.**
*All the initial values for the parameters are set 0.5.*

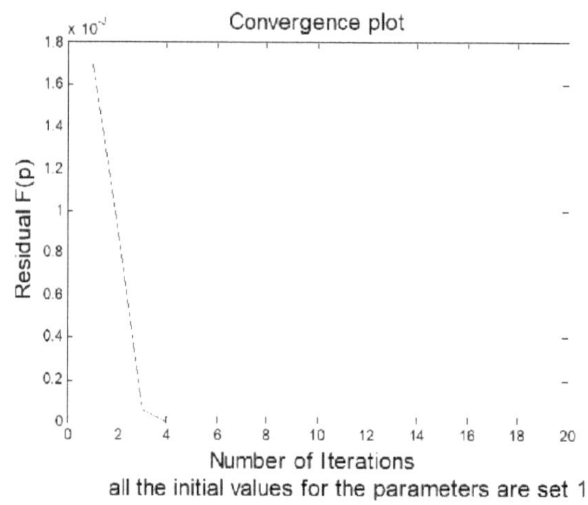


**Figure 11.**
*All the initial values for the parameters are set 1.*

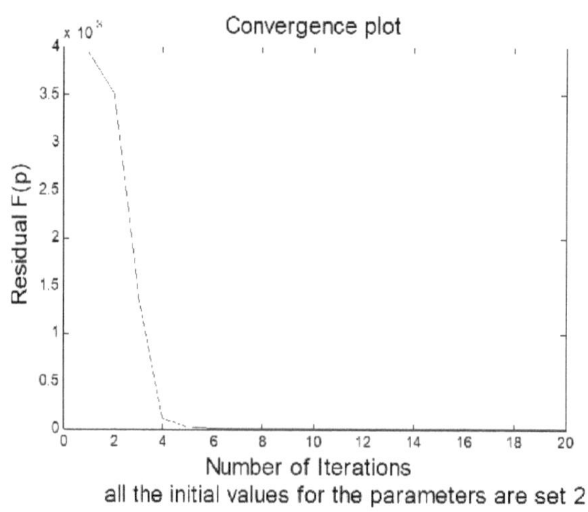


**Figure 12.**
*All the initial values for the parameters are set 2.*

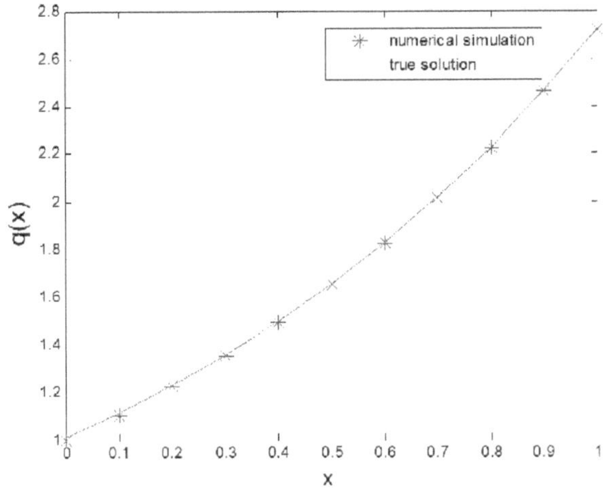


**Figure 13.**
*The comparison chart with all the initial values for the parameters is set 0.1.*

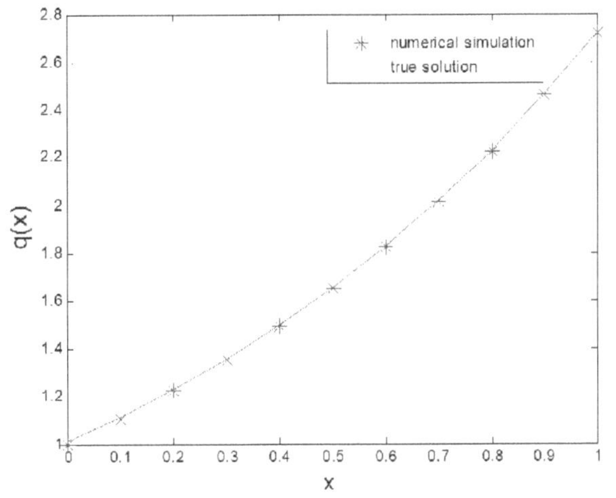


**Figure 14.**
*The comparison chart with all the initial values for the parameters is set 0.5.*

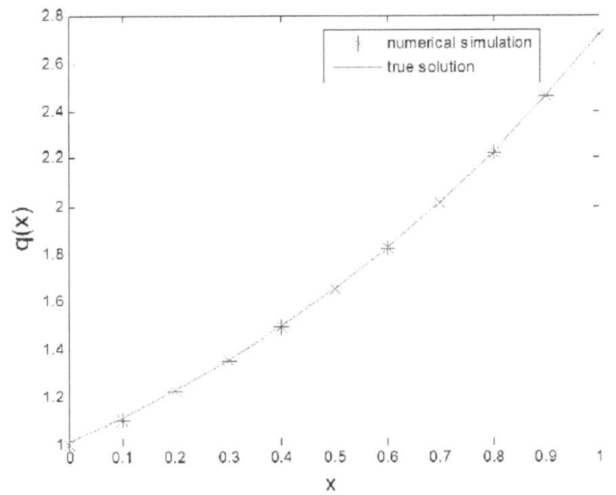


**Figure 15.**
*The comparison chart with all the initial values for the parameters is set 1.*

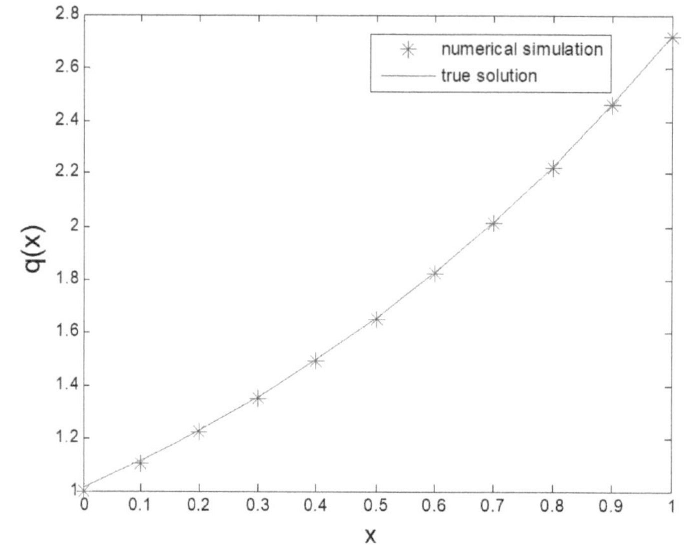


**Figure 16.**
*The comparison chart with all the initial values for the parameters is set 2.*

| | Numerical | Exact | Numerical | Exact |
|---|---|---|---|---|
| $j$ | $q(j\Delta x)$ | $q(j\Delta x)$ | $u(j\Delta x, 0.5)$ | $u(j\Delta x, 0.5)$ |
| 0 | 1.01536263525763 | 1 | 0.303342962644088 | 0.303265329856317 |
| 1 | 1.11378168059013 | 1.10517091807565 | 0.329201964677126 | 0.329286981656416 |
| 2 | 1.22765694959399 | 1.22140275816017 | 0.347492882224886 | 0.347609712653987 |
| 3 | 1.35549021305874 | 1.349858807576 | 0.359347568702678 | 0.359463171293777 |
| 4 | 1.4973722149265 | 1.49182469764127 | 0.365808711321159 | 0.365912693766539 |
| 5 | 1.65432828415206 | 1.64872127070013 | 0.367792378208857 | 0.367879441171442 |
| 6 | 1.82785056869393 | 1.82211880039051 | 0.366091218454182 | 0.366158192067887 |
| 7 | 2.01963499046464 | 2.01375270747048 | 0.361387761473796 | 0.361433054294643 |
| 8 | 2.23154102006879 | 2.22554092849247 | 0.354267869273063 | 0.354291330944216 |
| 9 | 2.46579237015687 | 2.45960311115695 | 0.345233023618059 | 0.345235749518249 |
| 10 | 2.72543670622333 | 2.71828182845905 | 0.334712604803175 | 0.334695240222645 |

**Table 4.**
*The values of $q(j\Delta x)$ and $u(j\Delta x, 0.5)$ in $x = j\Delta x$ with the all the initial values are set 1.*

**Table 4** shows the values of $q(j\Delta x)$ and $u(j\Delta x, 0.5)$ in $x = j\Delta x$ with the all the initial values are set 1.

## 6. Conclusions

A numerical method to estimate the temperature $u(x, t)$ and the coefficient $q(x)$ is proposed for an IHCP and the following results are obtained.

1. The present study, successfully applies the numerical method involving the Levenberg-Marquardt algorithm in conjunction with the Galerkin finite element method to an IHCP.

2. From the illustrated example it can be seen that the proposed numerical method is efficient and accurate to estimate the temperature $u(x, t)$ and the coefficient $q(x)$.

## Acknowledgements

The work of the author is supported by the Special Funds of the National Natural Science Foundation of China (Nos. 51190093 and 51179151). The author would like to thank the referees for constructive suggestions and comments.

## Author details

Tao Min[1,2*], Xing Chen[1], Yao Sun[1] and Qiang Huang[2]

1 School of Science, Xi'an University of Technology, Xi'an Shaanxi, China

2 State Key Laboratory of Eco-Hydraulic Engineering in Shaanxi, Xi'an University of Technology, Xi'an Shaanxi, China

*Address all correspondence to: mintao@xaut.edu.cn

## References


[1] Shidfar A, Karamali GR. Numerical solution of inverse heat conduction problem with nonstationary measurements. Applied Mathematics and Computation. 2005;**168**(1):540-548

[2] Shidfar A, Karamali GR, Damirchi J. An inverse heat conduction problem with a nonlinear source term. Nonlinear Analysis: Theory Methods & Applications. 2006;**65**(3):615-621

[3] Shidfar A, Pourgholi R. Numerical approximation of solution of an inverse heat conduction problem based on Legendre polynomials. Applied Mathematics and Computation. 2006; **175**(2):1366-1374

[4] Shidfar A, Azary H. An inverse problem for a nonlinear diffusion equation. Nonlinear Analysis: Theory Methods & Applications. 1997;**28**(4): 589-593

[5] Kurpisza K, Nowaka AJ. BEM approach to inverse heat conduction problems. Engineering Analysis with Boundary Elements. 1992;**10**(4):291-297

[6] Han H, Ingham DB, Yuan Y. The boundary-element method for the solution of the back ward heat conduction equation. Journal of Computational Physics. 1995;**116**(2): 292-299

[7] Skorek J. Applying the least squares adjustment technique for solving inverse heat conduction problems. In: Taylor C, editor. Proceedings of the 8th Conference on Numerical Methods in Laminar and Turbulent Flow. Swansea: Pineridge Press; 1993. pp. 189-198

[8] Pasquetti R, Le Niliot C. Boundary element approach for inverse heat conduction problems: Application to a bidimensional transient numerical experiment. Numerical Heat Transfer, Part B. 1991;**20**(2):169-189

[9] Ingham DB, Yuan Y. The solution of a nonlinear inverse problem in heat transfer. IMA Journal of Applied Mathematics. 1993;**50**(2):113-132

[10] Hensel E. Inverse Theory and Applications for Engineers. N.J. ISBN: 0135034590: Prentice Hall; 1991

[11] Özisik MN. Inverse Heat Transfer: Fundamentals and Applications. New York, USA: Taylor and Francis. ISBN: 1-56032-838-X; 2000

[12] Pourgholi R, Azizi N, Gasimov YS, Aliev F, Khala HK. Removal of numerical instability in the solution of an inverse heat conduction problem. Communications in Nonlinear Science and Numerical Simulation. 2009;**14**(6): 2664-2669

[13] Shidfar A, Pourgholi R, Ebrahimi M. A numerical method for solving of a nonlinear inverse diffusion problem. Computers and Mathematics with Applications. 2006;**52**(6–7):1021-1030

[14] Wang J, Zabaras N. A Bayesian inference approach to the inverse heat conduction problem. International Journal of Heat and Mass Transfer. 2004;**47**(17–18):3927-3941

[15] Dehghan M. Determination of an unknown parameter in a semi-linear parabolic equation. Mathematical Problems in Engineering. 2002;**8**(2): 111-122

[16] Lihua J, Changfeng M. On L-M method for nonlinear equations. Journal of Mathematics. Wuhan University. 2009;**29**(3):253-259

[17] Chen P. Why not use the Levenberg–Marquardt method for fundamental matrix estimation? The Institution of Engineering and Technology. 2010;**4**(4):286-288

4

# Applications of Heat Transfer Enhancement Techniques: A State-of-the-Art Review

*Suvanjan Bhattacharyya, Devendra K. Vishwakarma, Sanghati Roy, Ranjib Biswas and Mohammad Moghimi Ardekani*

**Abstract**

The fundamentals of heat transfer and its applications, the classification of heat transfer technology and different heat transfer techniques, and the needs for augmentation and its benefits and the different combinations of two or more inserts and integral roughness elements for heat transfer augmentation purpose have been introduced and discussed in this chapter. It is shown that most of the compound techniques performed better than the individual inserts for heat transfer enhancement. This chapter has also been dedicated to understanding the basic concepts of vortex generators for heat transfer enhancement in plate-fin heat exchangers. The performance of transverse, longitudinal, and wing-type vortex generators has been discussed as well.

**Keywords:** heat transfer, review, enhancement, heat exchanger, vortex generators, twisted tape, ribs, combine techniques

## 1. Introduction

The phenomenon of heat transfer has always been a topic of interest to research-ers and manufacturers alike. The previous researchers have addressed heat transfer characteristics of wide varieties of fields like bio-heat transfer, semiconductors, various cooling techniques, and natural phenomenon like oceanic currents and other important and relevant areas.

This chapter aims to cover all the relevant research papers about heat transfer published till 2018; few are there containing numerical and analytical aspects of heat transfer, while others are highlighted for its applications in engineering.

## 2. Free stream and flows over a surface

The chapters have been classified into categories like compressible and high-speed flows, externally influenced flows, flow related to films and interfaces, instable flow effects, flows with special fluid types, and flow related to reactions.

### 2.1 Effect of external surface

The effect of turbulence on free stream during heat transfer enhancement caused by the destruction of the viscous sublayer in the gaseous cavitation of $CO_2$-saturated water was recognized. The influence of roughness and wall temperature on the turbulent boundary layers was investigated [1, 2]. A model was developed to evaluate fluxes in urban boundary layers using the naphthalene sublimation technique [3].

### 2.2 Effect of geometry

Heat transfer enhancement is the process of improving the rate of heat deposition or removal on a surface. It is a subject of interest to the researchers as it results in savings in energy as well as cost. Heat transfer can be enhanced by using different types of swirl generators. Geometry plays a vital role in heat transfer enhancement. Transverse ribs with twisted tape and helical tape; axial rib with screw tape; and inclined limb in cylindrical dust have been studied for friction factor and Nusselt number [4–8]. Heat transfer augmentation techniques have been used to the study the effect of heat transfer and pressure drop due to insertion of twisted tape, inclined turbulator, corrugated tube with spring tape, diamond shape cylinder, wavy turbulator for short length and full length, center-trimmed twisted tape, flow around hexagonal cylinder, wavy channel, rhombus duct, square duct, and double pipe [9–20] as shown in **Figure 1**.

### 2.3 High-speed flow

A computational fluid dynamic (CFD) model has been developed to understand the hypersonic flow fields for reentry vehicles; facility was created for modeling

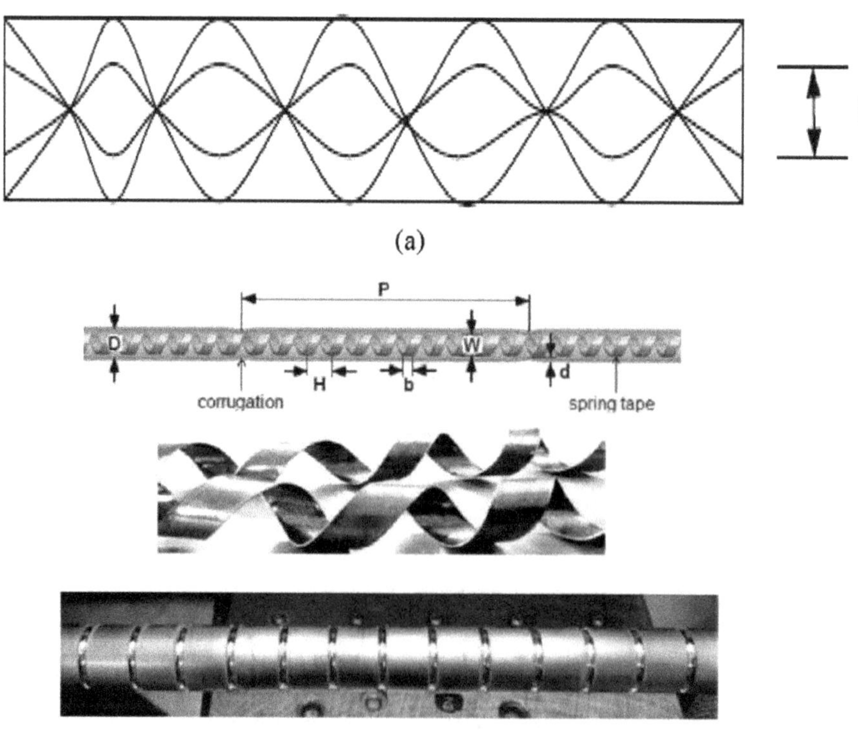

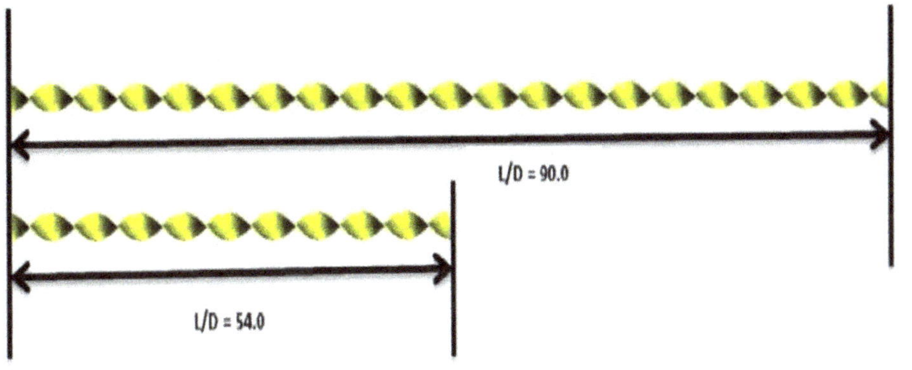
L/D = 90.0
L/D = 54.0

(c)

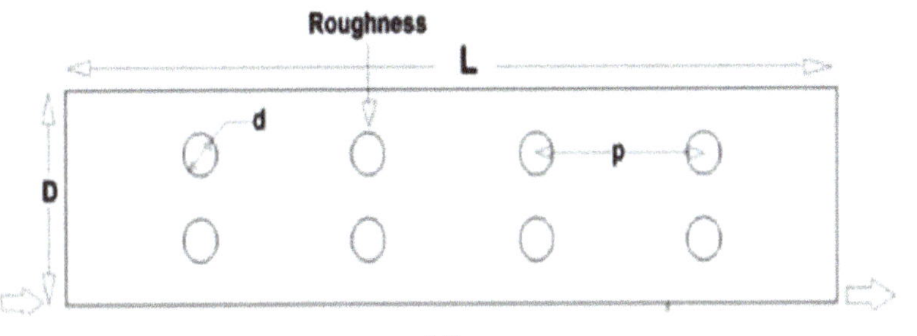
Roughness
L
d
p
D

(d)

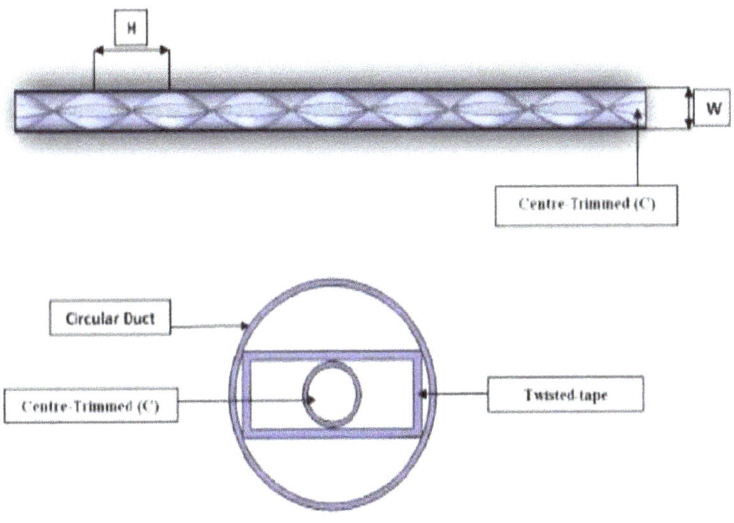
H
W
Centre-Trimmed (C)
Circular Duct
Centre-Trimmed (C)
Twisted tape

(e)

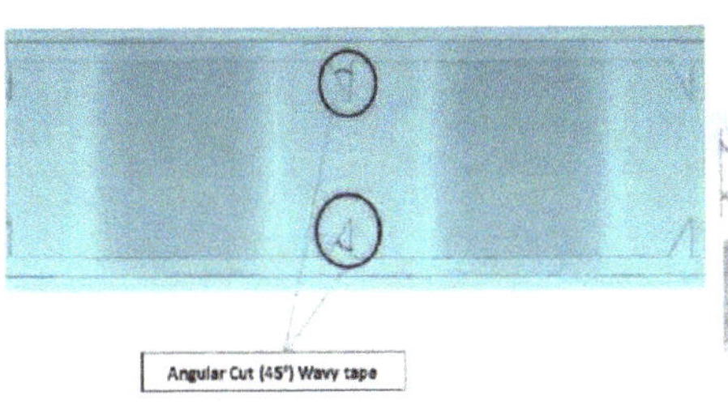
Angular Cut (45°) Wavy tape

(a)

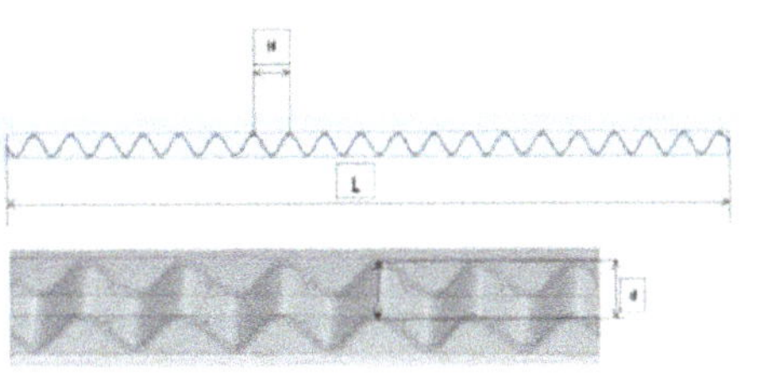
L

(b)

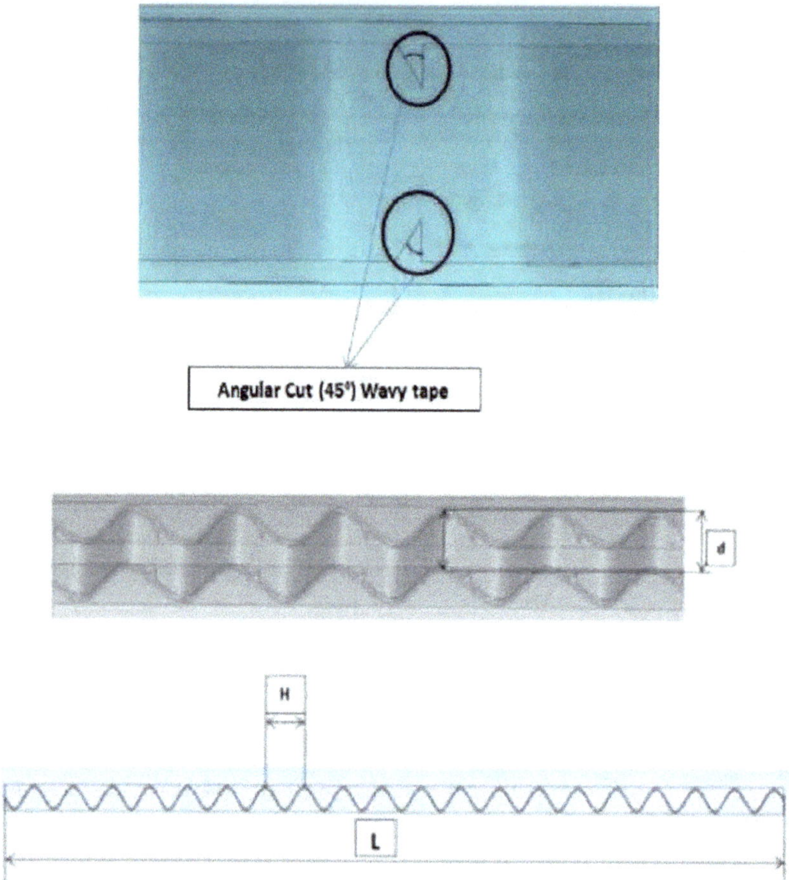


(f)

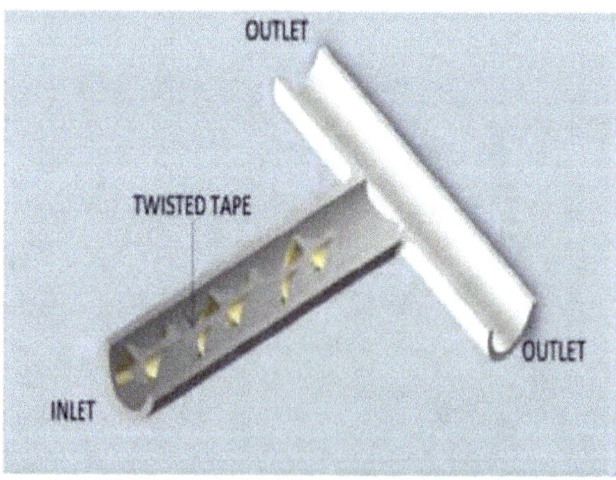


(g)

**Figure 1.**
*Different types of vortex generators used for enhancement of heat transfer. (a) Center-cleared twisted tape [4]. (b) Spring tape insert [9]. (c) Twisted tape [10]. (d) Swirl generator [12]. (e) Twisted tape with clearance at the center [13]. (f) Wavy tape with angular cuts [14]. (g) Full-length twisted tape [16].*

the projectile flight heating upon reentry. Simulation model for heat transfer due to convection and heat penetration was proposed [21], and comparative study has been conducted using of the European Atmospheric Reentry Demonstrator.

## 3. Channel flows

### 3.1 Straight wall passage

Initially, thermal characteristics in straight wall passages have been considered to analyze the heat transfer phenomenon in channel flows. Using the finite elements method, Nusselt number and friction factor were calculated for laminar regime. An investigation on laminar-turbulent transition inside a heated horizontal tube was conducted [22]. An analytical study for joule heating in a parallel plate channel with thermally developed flow has been conducted [23]. A circular tube was examined using various different conditions for viscous flow [24]. A novel method was devel-oped for evaluating the Nusselt number for hydrodynamic flow conditions [25]. Horizontal, inclined channels and vertical plane passages were examined for mixed convective heat transfer [26, 27]. A prediction was presented for Nusselt number for the in-tube cooling of supercritical carbon dioxide [28].

### 3.2 Microscale heat transfer

The study of fine scale heat transfer was done with various channel configurations. 3D flow and heat transfer were examined in microchannels [29]. Theoretical analysis for heat transfer in laminar flow between two parallel plates separated by a very small space in micron range was conducted. The momentum and energy equations are solved for the hydraulic and fully developed thermal flow in the microchannel [30]. This method was also used to simulate rarefied gas flow and heat transfer in micro-channels in a particular Knudsen number range [31]. Water was used as the working fluid in microchannel of rectangular shaped heat sinks, and computational studies were carried out [32]; also, their thermal performance was optimized minus water [33]. Convective heat transfer of fully developed flow both thermally and hydrody-namically in a rectangular microchannel is investigated [34]. A simulation model of low-power microchannel thermal reactor was presented [35]. Fractal branching used for the cooling of electronic chips was investigated [36]. Slotted microchannels were studied analytically on the basis of conduction and convection [37]. The performance of thermal fluid in a small capillary was studied experimentally [38].

### 3.3 Irregular geometries

A variety of papers covering numerous geometries have been taken into consid-eration in this section. Narrow-spaced fuel element configuration in multichannel was modeled numerically [39]. Rhombus and ellipse shape ducts were studied using Galerkin integral method [40, 41]. The heat transfer in a pin fin at the end of the wall was investigated [42]. The heat transfer in a milliscale thrust nozzle was studied numerically [43]. Viscous flow convections and heat transfer were studied in corru-gated ducts [44]. For square ducts a combined study was undertaken to understand the thermal characteristics in different shapes [45]. Experimental study was done with regard to two-pass internal coolant passages in gas turbines [46]. An increase in heat transfer due to rolling and pitching action in swirling ducts was found experi-mentally [47]. Flow and heat transfer for metal honeycomb geometry was inspected [48]. The effect of viscous forced convection in branching ducts was studied [49].

## 4. Flow separation

A study showing the separation of energy in free shear zone was carried out, and the role of pressure in the flow separation was studied [50]. In vertical ducts and other cross sections, mixed convection was investigated numerically [51]. Three-dimensional flow studies for understanding the effect of step height were also available. Large eddy simulation (LES) was undertaken for the turbulent flow over a backward-facing step [52]. A laminar airfoil was taken, and adiabatic and heating conditions were investigated at modest subsonic Mach numbers [53]. Correlations among the heat transfer coefficients of dull-edged flat plates and square channels were studied [54]. A finite volume method was used to investigate the 2D natural convection in a heated cylinder, and the significance of aspect ratio, Prandtl number, and boundary conditions on thermal characteristics were studied [55].

## 5. Experimental methods

Precise measurement results in good outcome in every research work. In heat transfer, measurement plays a pivotal role in the analysis of thermal system. Even after the numerical modeling of a heat flow system, it is not possible to define all the parameters with full accuracy leading to the failure of many thermal systems. This includes most of the engineering devices such as spacecraft, cryogenic engines, satellites, etc. Modeling of a turbulent flow and transition zone is very complicated, and hence it is difficult to predict very accurately. For the accuracy and relevancy of data, precise measurement is required which gives rise to the development of precise system with better accuracy.

### 5.1 Heat transfer

Heat flux measurement is an important aspect for understanding the physics related to transport of heat; distinguish among conduction, convection, and radiation mechanism; analyze energy balance; derive material properties; and understand the flow regimes, etc. The physical and mathematical models are presented to investigate the evolution of surface waves for free-falling turbulent wavy films with varying Reynolds number [56]. Thermochromic liquid crystal (TLC) is one of the best techniques to visualize the transient heat flow over the surface. The change in color of TLC from red to green to blue helps understand the flow of heat flux over a surface in supersonic wind tunnel [57]. Luminescent coating is another option for measurement of heat flow over a surface. The method has been used to determine the heat flow in shorter duration of less than 10 ms hypersonic flow [58]. Heater foils can be used to measure the bulk temperature development in time-based heat flow measurement by using a simplified model in which temperature development has been characterized [59].

### 5.2 Temperature measurement

For the temperature measurement during rapid contact solidification at the surface of substrate, an interfacial temperature sensor of 1 μm diameter has been fabricated [60]. A telecentric objective has been used for the first time to phase out the dependence on the angles in color determination in fluid-based TLCs for precision measurement [61]. An acoustic thermal scan has been evaluated using numerical methodology to evaluate the spatial resolution [62].

### 5.3 Velocity measurement

A multiple hot-film sensor (MHFS) arrays were used to evaluate the skin friction along the surfaces of two-dimensional streamlined objects (circular cylinder) [63]. A high time resolution ultrasonic velocity profiler (UVP) system has been devel-oped to determine 1D velocity profile on an ultrasonic beamline [64]. A numerical investigation [65] was conducted to determine the thermal response of hot wire for the measurement of sudden shift in the velocity in turbulent flow.

### 5.4 Miscellaneous

Bubble cluster pattern has been reported in the turbulent bubbly flow using rake of resistive flow and signal processing associated with it [66]. Air and liquid flows have been measured individually in two-phase air liquid flow [67]. Ultrasound Doppler velocimetry has been used to measure the thickness and velocity of the liquid film [68]. A novel pressure-sensitive paint (PSP) has been formulated and used for pressure measurement in cryogenic wind tunnel [69]. A high-sensitivity thermal conductivity detector has been developed from different materials which can be used in the diagnosis of fault in transformer, oil exploration, etc. [70].

## 6. Phase change

This part of the chapter deals with melting and freezing of materials. The section is divided into several subsections such as phase change materials (PCMs); formation of ice and its melting; melting and freezing of radial objects; melting and solidification of metals, nonmetals, and composites; crystallization; and globule, spray, and plunge cooling for better understanding.

### 6.1 Phase change materials

The inability to recover latent heat after super cooling of PCMs has been pointed out, and the method to recover latent heat has been discussed [71]. The melting mechanism of PCMs in magnetic field in low-gravity atmosphere has been dis-cussed [72]. Other works include fabrication of carbon brushes which can be used to enhance the thermal conductivity in phase change materials, the role of ultra-sonic vibration on melting characterization of PCMs, and detailed examination of solid liquid phase change heat flow enhancement.

### 6.2 Formation of ice and its melting

Researches on the formation of ice and its melting include the thermal behavior of ice under constant heat flow per unit area and melt removal, melting of ice using natural convection, ice making by cooling water-oil emulsion with stirring, and numerical simulation of melting of ice in water under the influence of natural convection and cooling effect produced by melting of ice [73–75].

### 6.3 Melting and freezing of radial objects

A lot of work has been presented on phase change in radial objects such as sphere, cylinders, and slabs. A mathematical model using numerical analysis has been developed to study the melting process of PCMs in sphere [76]. A novel packed bed of spheres has been developed using graphite/PCM composite for increasing the thermal conductivity which resulted in reduction in melting and freezing time significantly [77].

### 6.4 Melting and solidification of metals, non-metals, and composites

It has been observed that the supercooling properties of sodium acetate trihydrate can be improved by addition of nano-Cu [78]. A simulation model of melting and solidification of PCM in metallic porous foam has been investigated in the heat exchanger. The Hunt-Trivedi model has been used to simulate the solidification process of AISI 304 stainless steel [79].

### 6.5 Crystal growth

The crystal growth involves controlled growth of microstructure using opti-cal heating, modeling of mass crystallization in magma chamber, effect of crystal growth on solute distribution, and simulation of crystal growth for binary melting process [80, 81].

### 6.6 Globules, spray, and plunge cooling

This subsection deals with the deposition of metal droplets on the premolten pool and wavy surfaces, numerical analysis using FEM to study fluid mechanics and heat transfer of solder droplet on flat surface, numerical analysis of microdroplet deposition over a novel micromanufacturing process, and utilization of impulse atomization technique to produce controlled size droplets [82–84].

## 7. Numerical methods

Numerical methods are used to develop the mathematical models to solve complex numerical problems. The technique is used widely in research for modeling and optimization of the physical work which otherwise required rigorous work. The research work done in the field of heat transfer using numerical methods has been depicted in this section.

### 7.1 Heat conduction

A hybrid 3D model has been developed for the analysis of transient heat conduction in a functionally graded material (FGM) using generalized finite difference method [85], Cattaneo-Vernotte model (CV model) was used to develop numerical simulation of non-Fourier heat conduction for a fin attached to a microelectronic surface [86], Galerkin-vector theory and numerical method are used to develop a mathematical model to study heat conduction in nonhomogenous materials [87], and heat conduction model was developed using numerical methods to understand the flow of heat in the granular materials [88].

### 7.2 Inverse analysis

Systematic and local error has been identified using WKB method through numerical analysis [89], numerical inverse Laplace transform was used to solve nonlinear differential Equation [90], and numerical inverse method has been developed to extract heat flux in heat-sensitive coating region [91].

### 7.3 Fluid flow

The lattice Boltzmann method is used for simulation model of non-Newtonian fluid flow, two fluid method, and discrete particle method used for simulating the gas-solid flow of rough particles. A CFD model can be used effectively to study the hydrofluidization freezing, and a numerical simulation of fluid flow with thermal hydraulic mechanical coupling method on an uneven surface was developed [89].

### 7.4 Turbulent flow

Numerical methods can also be utilized to predict the turbulent flow. k-ϵ and LES model were used to study turbulent flow field around rows of tree and building, turbulence in flow field and temperature can be predicted, renormalization is used to determine the eddy diffusion in turbulence flow, intermittency model was developed for studying the laminar boundary transition at supersonic and hypersonic condition, and LES is used to forecast the heat transfer coefficient and blade metal temperature [92].

## 8. Heat exchanger and thermosyphons

### 8.1 Applications

The sheer variety of heat transfer operations has been demonstrated by a number of researchers in their works dealing with thermoacoustic and thermoelectric devices, rotating heat exchangers, commercial blood oxygenators, soil and deep bore heat exchangers, space craft radiators, and pressurized bubble columns.

### 8.2 Enhancement of heat transfer

The procedure to ease heat transfer has been stated by many researchers. The fin technology of extension is quite prevalent in the recent times. An investigation was carried out with fin tubes using liquid crystal display technology and plate finned tube exchanger by infrared thermal imaging, and performance measurement has been reported for a finned tube surface and annular fins. Fins having curly surfaces are examined for humid airflow. In addition to this film-wise condensation on plane low finned tubes, transient conduction in a fin, performance of extruded-serrated and extruded-finned tube bundles, and the features of a multi-pass heat exchanger have also been reported.

### 8.3 Microscale heat transfer

A number of applications now employ miniaturization of heat transfer devices: micro-heat pipe arrays, electronic cooling, microturbine, evaporation and boiling in microfin, microheat pipes, microscale temperature measurements, and modeling of microchannel flows [93].

### 8.4 Effect of fouling

An investigation has been done to study the effect of gas-side fouling in cross flow. Calcium carbonate fouling effect was studied with a microscale image; mineral fouling in extended tube heat exchangers was studied; the use of polyacrylic acid as anti-scaling and antifouling agent was studied [94–96].

### 8.5 Systems based on thermosyphon

Thermosyphons found applications in a variety of heat transfer complications such as space radiators and cooling of structures, solar water heaters, nuclear reactors and system based on geothermal energy, evaporators, preheaters, tiny heat pipes used to cool PC, laptops, and other electronic components [97–100].

## 9. Heat transfer: general applications

The relationship between the parameters of a fluidized bed and the heat transfer to a body engulf in it [101]. The SIMPLE algorithm [102] was used to simulate a blast furnace, and a relationship has been developed for the heat transfer coef-ficients on extension walls and hydrowalls of the boilers [103]. A porous radiant recirculated burner (PRRB) concept is developed to reduce losses due to open-flame combustion [104]. Leong studied the effect of latent heat of fusion on thin plates and numerical analysis of temperature change in biscuits using Monte Carlo (MC) method [105].

Multiple papers investigated the thermohydraulics of the cooling flow in nuclear reactors. A model was developed to study heat flux for low flow rates [106]. Inter-wrapper flow was studied, and its effects were analyzed numerically [107]. In the case of ceramic-coated turbine blades, the heat transfer coefficient does not significantly affect metal temperatures when thermal radiation is in the picture.

## 10. Insolation

### 10.1 Solar radiation

Various perspectives to evaluate solar data using modified modeling have been conducted by researchers. A new correlation between sunshine duration and radiation on the surface of the earth has been derived by Suehrcke [108]. It was found that the correlation is very well established for average value. A correction factor was proposed by Muneer [109] for calibrating the shadow band pyranometer. A model was using upper air humidity to estimate global solar radiation [110].

### 10.2 Solar air heater

Numerical solutions were also developed for absorbers in a porous medium, Nusselt number and Reynolds-Rayleigh number correlations for natural convection in an open-ended rectangular channel and models for solar air heater with fins [111]. Collector efficiency was predicted in a simplified manner.

### 10.3 Solar water heaters

Novelties in the design of solar water heating application are presented in this subsection. Fourier transform technique has been used to estimate the heat transfer and efficiency of a flat solar plate collector [112]. Double-sided flat plate collector was used to experimentally investigate the reduction in heat losses in comparison to conventional solar collector [113]. An experimental investigation on ICS solar water heart with compound parabolic concentrating integral collector storage system was designed and tested [114]. A 2D concentrator was developed aiming to store solar energy [115].

## 11. Plasma heat transfer and MHD

### 11.1 Investigation and application

In this section, heat transfer in thermal plasma reactor for nanoparticle synthesis has been investigated through different models [116]. A 3D model of heat transfer in thermal plasma system has been developed to show 3D effect of carrier gas [117]; effects of nucleation temperature were investigated by radio frequency; 2D numeri-cal simulation was developed to show flow and heat transfer in argon gas plasma, temperature gradient, velocity, and concentration to study the nitridation of $MoSi_2$ which was carried in thermal plasma reactor; and numerical simulation model was developed to show the effect of radial injection of gas (with and without swirl) on flow and temperature field [118]. Plasma induced between two electrodes with and without swirl has been investigated for heat transfer with fluid flow [119].

### 11.2 Magnetohydrodynamics (MHD)

A simple monoenergetic operator and the Bhatnagar-Gross-Krook model were presented to estimate heat transfer in a rare gas between parallel plates [120]. A mathematical model of 2D magnetohydrodynamic Prandtl fluid flow over a sheet is examined [121], 3D magnetohydrodynamic Cauchy problem has been investigated [122], 2D pseudo-steady compressible magnetohydrodynamic system is studied for expansion of gas in vacuum [123], bio-convection flow of nanofluid is studied in magnetic field [124], and numerical analysis of MHD flow over vertical rotating cone is investigated. Also, the volume of fluid method is used to investigate the MHD of incompressible flow, and MHD fluid behavior is studied for flow and heat transfer [125].

## 12. Conclusion

In this review article, an effort has been made to study the recent development in the field of heat transfer enhancement. A lot of experimental and numerical research have been done to study the aspect of heat transfer in different fields such as channel flow, crystal growth, heat exchangers, thermosyphons, phase change materials, temperature and velocity measurement, solar energy, etc. The effect of geometry such as channel modification through inserts, roughness, etc. and external power such as magnetic field, electric field, ultrasound, etc. on the thermal performance and augmentation of heat transfer has been studied. In addition to this, the lattice Boltzmann method, WKB method, numerical inverse method, k-epsilon, Cattaneo-Vernotte model, Hunt-Trivedi model, and LES model have been studied for different heat transfer applications. Overall this review gives a full-scale summary of heat transfer applications.

## Author details

Suvanjan Bhattacharyya[1*], Devendra K. Vishwakarma[2], Sanghati Roy[2], Ranjib Biswas[3] and Mohammad Moghimi Ardekani[4]

1 Center for Renewable Energy and Environment Development (CREED), Department of Mechanical Engineering, Birla Institute of Technology and Science, Pilani, Vidyavihar, Rajasthan, India

2 Department of Mechanical Engineering, Birla Institute of Technology and Science, Pilani, Vidyavihar, Rajasthan, India

3 Department of Mechanical Engineering, MCKV Institute of Engineering, Howrah, West Bengal, India

4 Department of Design and Engineering, Staffordshire University, Stoke-on-Trent, UK

*Address all correspondence to: suvanjan.bhattacharyya@pilani.bits-pilani.ac.in

## References

[1] Lee SY, Choi YD. Turbulence enhancement by ultrasonically induced gaseous cavitation in the $CO_2$ saturated water. KSME Journal. 2002;**16**(2):246-254

[2] Avelino MR, Freire APS. On the displacement in origin for turbulent boundary layers subjected to sudden changes in wall temperature and roughness. International Journal of Heat and Mass Transfer. 2002;**45**(15):3143-3153

[3] Barlow JF, Belcher SE. A wind tunnel model for quantifying fluxes in the urban boundary layer. Boundary-Layer Meteorology. 2002;**104**(1):131-150

[4] Bhattacharyya S, Saha S, Saha SK. Laminar flow heat transfer enhancement in a circular tube having integral transverse rib roughness and fitted with centre-cleared twisted-tap. Experimental Thermal and Fluid Science. 2012;**44**:727-735

[5] Bhattacharyya S, Saha SK. Thermohydraulics of laminar flow through a circular tube having integral helical rib roughness and fitted with centre-cleared twisted-tape. Experimental Thermal and Fluid Science. 2012;**42**:154-162

[6] Saha SK, Bhattacharyya S, Pal PK. Thermohydraulics of laminar flow of viscous oil through a circular tube having integral axial rib roughness and fitted with center-cleared twisted-tape. Experimental Thermal and Fluid Science. 2012;**41**:121-129

[7] Bhattacharyya S, Chattopadhyay H, Benim AC. 3D CFD simulation on heat transfer enhancement in turbulent channel flow with twisted tape inserts. Progress in Computational Fluid Dynamics, An International Journal. 2017;**17**(3):193-197

[8] Bhattacharyya S, Chattopadhyay H, Benim AC. Numerical investigation on heat transfer in a circular tube with inclined ribs. Progress in Computational Fluid Dynamics, An International Journal. 2017;**17**(6):390-396

[9] Bhattacharyya S, Benim AC, Chattopadhyay H, Banerjee A. Experimental investigation of the heat transfer performance of a corrugated tube with spring tape turbulator inserts. Experimental Heat Transfer. 2019;**32**(5):411-425

[10] Bhattacharyya S. Experimental study on effect of heat transfer enhancement of heat exchanger tube inserted with short length spring tapes. Iranian Journal of Science and Technology Transactions of Mechanical Engineering. DOI: 10.1007/s40997-018-0251-0

[11] Bhattacharyya S, Benim AC, Chattopadhyay H, Banerjee A. Experimental and numerical analysis of forced convection in a twisted tube. Journal of Thermal Science. 2019;**23**(4):1043-1052

[12] Bhattacharyya S, Chattopadhyay H. Experimental investigation on heat transfer enhancement by swirl generators in a solar air heater duct. International Journal of Heat and Technology. 2016;**34**(2):191-196

[13] Bhattacharyya S, Chattopadhyay H, Bandhopadhyay S. Numerical study on heat transfer enhancement through a circular duct fitted with centre-trimmed twisted tape. International Journal of Heat and Technology. 2016;**34**(3):401-406

[14] Bhattacharyya S, Chattopadhyay H, Benim AC. Heat transfer enhancement of laminar flow of ethylene glycol through a square channel fitted with

angular cut wavy strip. Procedia Engineering. 2016;**157**:19-28

[15] Bhattacharyya S, Chattopadhyay H, Swami A, Uddin MK. Convective heat transfer enhancement and entropy generation of laminar flow of water through a wavy channel. International Journal of Heat and Technology. 2016;**34**(4):727-733

[16] Bhattacharyya S, Chattopadhyay H, Haldar A. Design of twisted tape turbulator at different entrance angle for heat transfer enhancement in a solar heater. Beni-Suef University Journal of Basic and Applied Sciences. 2018;**7**(1):118-126

[17] Nandi TK, Bhattacharyya S, Gobinda Das S, Chattopadhyay H. Thermohydraulic transport characteristics in wavy microchannel under pulsating inlet flow condition. Chemical Engineering Transactions. 2017;**62**:271-276

[18] Bhattacharyya S, Das S, Sarkar A. Numerical simulation of flow and heat transfer around hexagonal cylinder. International Journal of Heat and Technology. 2017;**35**(2):360-363

[19] Bhattacharyya S, Khan AI, Maity DK, Pradhan S. Hydrodynamics and heat transfer of turbulent flow around a rhombus cylinder. Chemical Engineering Transactions. 2017;**62**:373-378

[20] Bhattacharyya S, Sarkar A, Das S, Mullick A. Computational studies of heat transfer enhancement in a circular wavy micro channel. Chemical Engineering Transactions. 2017;**62**:361-366

[21] Meese EA, Norstrud H. Simulation of convective heat flux and heat penetration for a spacecraft at re-entry. Aerospace Science and Technology. 2002;**6**(3):185-194

[22] Koizumi H. Laminar-turbulent transition behavior of fully developed air flow in a heated horizontal tube. International Journal of Heat and Mass Transfer. 2002;**45**(5):937-949

[23] Lahjomri J, Oubarra A, Alemany A. Heat transfer by laminar Hartmann flow in thermal entrance region with a step change in wall temperatures: The Graetz problem extended. International Journal of Heat and Mass Transfer. 2002;**45**(5):1127-1148

[24] Mahulikar SP, Tso CP. A new classification for thermal development of fluid flow in a circular tube under laminar forced convection. Proceedings of the Royal Society of London Series A: Mathematical, Physical and Engineering Sciences. 2002;**458**(2019):669-682

[25] Nickolay M, Martin H. Improved approximation for the Nusselt number for hydrodynamically developed laminar flow between parallel plates. International Journal of Heat and Mass Transfer. 2002;**45**(15):3263-3266

[26] Ozsunar A, Baskaya S, Sivrioglu M. Experimental investigation of mixed convection heat transfer in a horizontal and inclined rectangular channel. Heat and Mass Transfer. 2002;**38**(3):271-278

[27] Wang J, Li J, Jackson JD. Mixed convection heat transfer to air flowing upwards through a vertical plane passage: Part 3. Chemical Engineering Research and Design. 2002;**80**(A3):252-260

[28] Pitla SS, Groll EA, Ramadhyani S. New correlation to predict the heat transfer coefficient during in-tube cooling of turbulent supercritical $CO_2$. International Journal of Refrigeration - Revue internationale du Froid. 2002;**25**(7):887-895

[29] Toh KC, Chen XY, Chai JC. Numerical computation of fluid flow

and heat transfer in microchannels. International Journal of Heat and Mass Transfer. 2002;**45**(26):5133-5141

[30] Zhu X, Xin MD, Liao Q. Analysis of heat transfer between two unsymmetrically heated parallel plates with microspacing in the slip flow regime. Microscale Thermophysical Engineering. 2002;**6**(4):287-301

[31] Wang X, Wang QW, Tao WQ, Zheng P. Simulation of rarefied gas flow and heat transfer in microchannels. Science in China, Series E Technological Sciences. 2002;**45**(3):321-327

[32] Qu WL, Mudawar I. Analysis of three-dimensional heat transfer in micro-channel heat sinks. International Journal of Heat and Mass Transfer. 2002;**45**(19):3973-3985

[33] Ryu JH, Choi DH, Kim SJ. Numerical optimization of the thermal performance of a microchannel heat sink. International Journal of Heat and Mass Transfer. 2002;**45**(13):2823-2827

[34] Tunc G, Bayazitoglu Y. Heat transfer in rectangular microchannels. International Journal of Heat and Mass Transfer. 2002;**45**(4):765-773

[35] Erickson D, Li DQ. Numerical simulations of a low power microchannel thermal cycling reactor. International Journal of Heat and Mass Transfer. 2002;**45**(18):3759-3770

[36] Chen YP, Cheng P. Heat transfer and pressure drop in fractal tree-like microchannel nets. International Journal of Heat and Mass Transfer. 2002;**45**(13):2643-2648

[37] Leont'ev AI, Polyakov AF. Formulation and solution of the problem on convection-conduction heat transfer in a system of slotted microchannels at a uniform temperature of the frame. High Temperature. 2002;**40**(4):577-585

[38] Celata GP, Cumo M, Guglielmi M, Zummo G. Experimental investigation of hydraulic and single-phase heat transfer in 0.130-mm capillary tube. Microscale Thermophysical Engineering. 2002;**6**(2):85-97

[39] Huang J, Wang QW, Tao WQ. Numerical simulation of turbulent flow and heat transfer in multi-channel, narrow-gap fuel element. Engineering Computations (Swansea, Wales). 2002;**19**(3-4):327-345

[40] Lee YM, Lee PC. Heat transfer coefficients of laminar flow in a rhombic duct with constant wall temperature. Numerical Heat Transfer: Part A Applications. 2002;**42**(3):285-296

[41] Sakalis VD, Hatzikonstantinou PM, Kafousias N. Thermally developing flow in elliptic ducts with axially variable wall temperature distribution. International Journal of Heat and Mass Transfer. 2002;**45**(1):25-35

[42] Hwang JJ, Lui CC. Measurement of endwall heat transfer and pressure drop in a pin-fin wedge duct. International Journal of Heat and Mass Transfer. 2002;**45**(4):877-889

[43] Alexeenko AA, Levin DA, Gimelshein SF, Collins RJ, Markelov GN. Numerical simulation of high-temperature gas flows in a millimeter-scale thruster. Journal of Thermophysics and Heat Transfer. 2002;**16**(1):10-16

[44] Niu JL, Zhang LZ. Heat transfer and friction coefficients in corrugated ducts confined by sinusoidal and arc curves. International Journal of Heat and Mass Transfer. 2002;**45**(3):571-578

[45] Wang LB, Wang QW, He YL, Tao WQ. Experimental and numerical study of developing turbulent flow and heat transfer in convergent/divergent square ducts. Heat and Mass Transfer. 2002;**38**(4-5):399-408

[46] Chanteloup D, Juaneda Y, Bolcs A. Combined 3-D flow and heat transfer measurements in a 2-pass internal coolant passage of gas turbine airfoils. Journal of Turbomachinery-Transactions of the ASME. 2002;**124**(4):710-718

[47] Chang SW, Zheng Y. Enhanced heat transfer with swirl duct under rolling and pitching environment. Journal of Ship Research. 2002;**46**(3):149-166

[48] Brautsch A, Griffin T, Schlegel A. Heat transfer characterization of support structures for catalytic combustion. International Journal of Heat and Mass Transfer. 2002;**45**(15):3223-3231

[49] El-Shaboury AMF, Soliman HM, Ormiston SJ. Laminar forced convection in two-dimensional equalsided and reduced branching ducts. Numerical Heat Transfer: Part A Applications. 2002;**42**(5):487-512

[50] Han B, Goldstein RJ, Choi HG. Energy separation in shear layers. International Journal of Heat and Mass Transfer. 2002;**45**(1):47-55

[51] Cheng CH, Huang SY, Aung W. Numerical predictions of mixed convection and flow separation in a vertical duct with arbitrary cross section. Numerical Heat Transfer: Part A Applications. 2002;**41**(5):491-514

[52] Nie JH, Armaly BF. Three-dimensional convective flow adjacent to backward-facing step-effects of step height. International Journal of Heat and Mass Transfer. 2002;**45**(12):2431-2438

[53] Stock HW. Wind tunnel-flight correlation for laminar wings in adiabatic and heating flow conditions. Aerospace Science and Technology. 2002;**6**(4):245-257

[54] Sun ZF. Correlations for mass transfer coefficients over blunt boards based on modified boundary layer theories. Chemical Engineering Science. 2002;**57**(11):2029-2033

[55] Roychowdhury DG, Das SK, Sundararajan T. Numerical simulation of natural convective heat transfer and fluid flow around a heated cylinder inside an enclosure. Heat and Mass Transfer. 2002;**38**(7-8):565-576

[56] Ye XM, Yan WP, Jiang ZY, Li CX. Hydrodynamics of free-falling turbulent wavy films and implications for enhanced heat transfer. Heat Transfer Engineering. 2002;**23**(1):48-60

[57] Mee DJ, Chiu HS, Ireland PT. Techniques for detailed heat transfer measurements in cold supersonic blowdown tunnels using thermochromic liquid crystals. International Journal of Heat and Mass Transfer. 2002;**45**(16):3287-3297

[58] Hubner JP, Carroll BF, Schanze KS. Heat-transfer measurements in hypersonic flow using luminescent coating techniques. Journal of Thermophysics and Heat Transfer. 2002;**16**(4):516-522

[59] von Wolfersdorf J. Bulk temperature development in transient heat transfer measurements using heater foils. Journal of Heat Transfer-Transactions of the ASME. 2002;**124**(5):982-985

[60] Wang W, Qiu HH. Interfacial thermal conductance in rapid contact solidification process. International Journal of Heat and Mass Transfer. 2002;**45**(10):2043-2053

[61] Gunther A, von Rohr PR. Influence of the optical configuration on temperature measurements with fluid dispersed TLCs. Experiments in Fluids. 2002;**32**(5):533-541

[62] Bograchev KM, Passechnik VI. Evaluation of the spatial resolution of

passive acoustic thermal tomography. Acoustical Physics. 2002;**48**(4):406-411

[63] Desgeorges O, Lee T, Kafyeke F. Multiple hot-film sensor array calibration and skin friction measurement. Experiments in Fluids. 2002;**32**(1):37-43

[64] Ozaki Y, Kawaguchi T, Takeda Y, Hishida K, Maeda M. High time resolution ultrasonic velocity profiler. Experimental Thermal and Fluid Science. 2002;**26**(2-4):253-258

[65] Durst F, Shi JM, Breuer M. Numerical prediction of hot-wire corrections near walls. Journal of Fluids Engineering, Transactions of the ASME. 2002;**124**(1):241-250

[66] Salesse A, de Tournemine AL, Roig V. Development of bubble cluster detection and identification method. Experimental Thermal and Fluid Science. 2002;**26**(2-4):163-171

[67] Dong W, Hu LZ. Gas-liquid two-phase flow measurement using ESM. Experimental Thermal and Fluid Science. 2002;**26**(6-7):827-832

[68] Zhaojun F, Jing C, Shitao S. Investigating the liquid film characteristics of gas-liquid swirling flow using ultrasound Doppler velocimetry. AICHE Journal. 2017;**63**:2348-2357

[69] Asai K, Amao Y, Iijima Y, Okura I, Nishide H. Novel pressure-sensitive paint for cryogenic and unsteady wind-tunnel testing. Journal of Thermophysics and Heat Transfer. 2002;**16**(1):109-115

[70] Wu YE, Chen K, Chen CW, Hsu KH. Fabrication and characterization of thermal conductivity detectors (TCDs) of different flow channel and heater designs. Sensors and Actuators A: Physical. 2002;**100**(1):37-45

[71] Zhao Y, Zhang X, Xu X, Zhang S. Research progress in nucleation and supercooling induced by phase change materials. Journal of Energy Storage. 2020;**27**:101156

[72] Asako Y, Goncalves E, Faghri M, Charmchi M. Numerical solution of melting processes for fixed and unfixed phase change material in the presence of magnetic field-simulation of low-gravity environment. Numerical Heat Transfer: Part A Applications. 2002;**42**(6):565-583

[73] Tsuchida D, Kang C, Okada M, Matsumoto K, Kawagoe T. Ice formation process by cooling water-oil emulsion with stirring in a vessel. International Journal of Refrigeration - Revue internationale du Froid. 2002;**25**(2):250-258

[74] Bansal H, Niktriyuk P. Arbitrary shaped ice particle melting under the influence of natural convection. AICHE Journal. 2017;**63**:3158-3176

[75] Yeh HM, Ho CD, Wang WP. Cool thermal discharge by melting ice and producing chilled air. Heat Transfer Engineering. 2002;**23**(1):61-68

[76] Wei L, Sui-Guang L, Shikai G, Yunhao W, Xueling L. Numerical study on melt fraction during melting of phase change material inside a sphere. International Journal of Hydrogen Energy. 2017;**42**(29):18232-18239

[77] Al-Shannaq R, Young B, Farid M. Cold energy storage in a packed bed of novel graphite/PCM composite spheres. Energy. 2019;**171**(15):296-305

[78] Wenlong C, Yanping Y, Liangliang S, Xiaoling C, Xiaojiao Y. Experimental studies on the supercooling and melting/freezing characteristics of nano-copper/sodium acetate trihydrate composite phase change materials. Renewable Energy. 2016;**99**:1029-1037

[79] Esapour M, Hamzehnezhad A, Ali Rabienaaj A, Darzi MJ. Melting and solidification of PCM embedded in porous metal foam in horizontal multi-tube heat storage system. Energy Conversion and Management. 2018;**171**:398-410

[80] Bazarov LS, Gordeeva VI, Shevchenko VS, Petrushin EI. Experimental modeling of mass crystallization processes in the volume of a flat magma chamber. Petrology. 2002;**10**(5):469-478

[81] Fukui K, Maeda K. Effects of crystal growth rate and heat and mass transfer on solute distribution. Chemical Engineering Science. 2002;**57**(15):3133-3140

[82] Butty V, Poulikakos D, Giannakouros J. Three dimensional presolidification heat transfer and fluid dynamics in molten micro droplet deposition. International Journal of Heat and Fluid Flow. 2002;**23**(3):232-241

[83] Haferl S, Poulikakos D. Transport and solidification phenomena in molten microdroplet pileup. Journal of Applied Physics. 2002;**92**(3):1675-1689

[84] Henein H. Single fluid atomization through the application of impulses to a melt. Materials Science & Engineering, A: Structural Materials - Properties, Microstructure and Processing. 2002;**326**(1):92-100

[85] Wenzhen Q , Chia-Ming F, Yaoming Z. Analysis of three dimensional heat conduction in functionally graded materials by using a hybrid numerical method. International Journal of Heat and Mass Transfer. 2019;**145**:118771

[86] Yi L, Ling L, Yuwen Z. Numerical simulation of non-Fourier heat conduction in fins by lattice Boltzmann method. Applied Thermal Engineering. 2020;**145**:114670

[87] Xiujiang S, Qian W, Liqin W. New Galerkin-vector theory and efficient numerical method for analyzing steady-state heat conduction in inhomogeneous bodies subjected to a surface heat flux. Applied Thermal Engineering. 2019;**161**:113838

[88] Jun H, Quansheng L, Zhijun W, Xiangyu X. Modelling transient heat conduction of granular materials by numerical manifold method. Engineering Analysis with Boundary Elements. 2018;**86**:45-55

[89] Ying X, Zhinxue S, Zhuang L, Jun Y, Dongyan F, Kelvin B, et al. Numerical simulation of fluid flow and heat transfer in EGS with thermal-hydraulic-mechanical coupling method based on a rough fracture model. Energy Procedia. 2019;**158**:6038-6045

[90] Mathey P, Jullien P. Numerical analysis of a WKB inverse method in view of index profile reconstruction in diffused waveguides. Optics Communications. 1996;**122**(4-6):127-134

[91] Rani D, Mishra V. Numerical inverse Laplace transform based on Bernoulli polynomials operational matrix for solving nonlinear differential equations. Results in Physics. 2020;**16**:102836

[92] Merci B, Dick E, Vierendeels J, De Langhe C. Determination of epsilon at inlet boundaries. International Journal of Numerical Methods for Heat and Fluid Flow. 2002;**12**(1):65-80

[93] Medic G, Durbin PA. Toward improved prediction of heat transfer on turbine blades. Journal of Turbomachinery-Transactions of the ASME. 2002;**124**(2):187-192

[94] Wand L, Song F. Development of an intermittency equation for the modeling

of the supersonic/hypersonic boundary layer flow transition. Flow, Turbulence and Combustion. 2011;**87**(1):165-187

[95] Tafti DK, He L, Nagendra K. Large eddy simulation for predicting turbulent heat transfer in gas turbines. Philosophical Transactions. Series A, Mathematical, Physical, and Engineering Sciences. 2014;**372**(2022):20130322

[96] Gui-Qing W, He-Ping L, Cheng- Yu B, Xi C. Modeling of the heat transfer and flow features of the thermal plasma reactor with counter-flow gas injection. International Journal of Heat and Mass Transfer. 2009;**52**(3-4):760-766

[97] Wang CC, Lo J, Lin YT, Wei CS. Flow visualization of annular and delta winlet vortex generators in fin-and-tube heat exchanger application. International Journal of Heat and Mass Transfer. 2002;**45**(18):3803-3815

[98] Wang YX, Peterson GP. Optimization of micro heat pipe radiators in a radiation environment. Journal of Thermophysics and Heat Transfer. 2002;**16**(4):537-546

[99] Schaller M, Hoffmann KH, Rivero R, Andresen B, Salamon P. The influence of heat transfer irreversibilities on the optimal performance of diabatic distillation columns. Journal of Non-Equilibrium Thermodynamics. 2002;**27**(3):257-269

[100] Bojic M, Lukic N. Controlling evaporative three finger thermosyphon. Energy Conversion and Management. 2002;**43**(5):709-720

[101] Al-Busoul MA. Bed-to-surface heat transfer in a circulating fluidized bed. Heat and Mass Transfer. 2002;**38**(4-5):295-299

[102] de Castro JA, Nogami H, Yagi J. Three-dimensional multiphase mathematical modeling of the blast furnace based on the multifluid model. ISIJ International. 2002;**42**(1):44-52

[103] Dutta A, Basu P. Overall heat transfer to water walls and wing walls of commercial circulating fluidized-bed boilers. Journal of the Institute of Energy. 2002;**75**(504):85-90

[104] Jugjai S, Rungsimuntuchart N. High efficiency heat-recirculating domestic gas burners. Experimental Thermal and Fluid Science. 2002;**26**(5):581-592

[105] Leong KC, Lu GQ , Rudolph V. The effect of latent heat of fusion on heat transfer in fluidized-bed coating of thin plates. Chemical Engineering and Processing. 2002;**41**(7):567-576

[106] Caro-Corrales J, Cronin K, Abodayeh K, Gutierrez-Lopez G, Ordorica-Falomir C. Analysis of random variability in biscuit cooling. Journal of Food Engineering. 2002;**54**(2):147-156

[107] Holowach MJ, Hochreiter LE, Cheung FB, Aumiller DL. Critical heat flux during reflood transients in small-hydraulic-diameter geometries. Nuclear Technology. 2002;**140**(1):18-27

[108] Driesse A, Thevenard D. A test of Suehrcke's sunshine-radiation relationship using a global data set. Solar Energy. 2002;**72**(2):167-175

[109] Muneer T, Zhang X. A new method for correcting shadow band diffuse irradiance data. Journal of Solar Energy Engineering-Transactions of the ASME Journal. 2002;**124**(1):34-43

[110] Yang K, Koike T. Estimating surface solar radiation from upper-air humidity. Solar Energy. 2002;**72**(2):177-186

[111] Chamkha AJ, Issa C, Khanafer K. Natural convection from an inclined plate embedded in a variable porosity porous medium due to solar radiation. International Journal of Thermal Sciences. 2002;**41**(1):73-81

[112] El-Adawi MK. New approach to modelling a flat plate collector: The Fourier transform technique. Renewable Energy. 2002;**26**(3):489-506

[113] Groenhout NK, Behnia M, Morrison GL. Experimental measurement of heat loss in an advanced solar collector. Experimental Thermal and Fluid Science. 2002;**26**(2-4):131-137

[114] Tripanagnostopoulos Y, Souliotis M, Nousia T. CPC type integrated collector storage systems. Solar Energy. 2002;**72**(4):327-350

[115] Rincon EA, Osorio FA. A new troughlike nonimaging solar concentrator. Journal of Solar Energy Engineering-Transactions of the ASME. 2002;**124**(1):51-54

[116] Belessiotis V, Mathioulakis E. Analytical approach of thermosyphon solar domestic hot water system performance. Solar Energy. 2002;**72**(4): 307-315

[117] Chen YT, Chong KK, Lim CS, Lim BH, Tan KK, Aliman O, et al. Report of the first prototype of non-imaging focusing heliostat and its application in high temperature solar furnace. Solar Energy. 2002;**72**(6):531-544

[118] Xi C, He-Ping L. Three-dimensional flow and heat transfer in thermal plasma systems. Surface and Coating Technology. 2003;**171**(1-3):124-133

[119] Ramachandran K, Nishiyama H. Three-dimensional effects of carrier gas and particle injections on the thermo-fluid fields of plasma jets. Journal of Physics D: Applied Physics. 2002;**35**(4):307-317

[120] Christlieb AJ, Hitchon WNG. Three-dimensional solutions of the Boltzmann equation: Heat transport at long mean free paths. Physical Review E. 2002:6708

[121] Zaman S, Mahwish G. Magnetohydrodynamic bioconvective flow of Williamson nanofluid containing gyrotactic microorganisms subjected to thermal radiation and Newtonian conditions. Journal of Theoretical Biology. 2019;**479**:22-28

[122] Sulochana C, Samrat SP, Sandeep N. Numerical investigation of magnetohydrodynamic (MHD) radiative flow over a rotating cone in the presence of Soret and chemical reaction. Propulsion and power Research. 2018;**7**(1):91-101

[123] Zhang J, Ming-Jiu N. Direct numerical simulations of incompressible multiphase magnetohydrodynamics with phase change. Journal of Computational Physics. 2018;**375**:717-746

[124] Kumaran G, Sandeep N, Animasaun IL. Computational modeling of magnetohydrodynamic non-Newtonian fluid flow past a paraboloid of revolution. Alexandria Engineering Journal. 2018;**57**(3):1859-1865

[125] Rahimah J, Roslinda N, Ioan P. Magnetohydrodynamic rotating flow and heat transfer of ferrofluid due to an exponentially permeable stretching/ shrinking sheet. Journal of Magnetism and Magnetic Materials. 2018;**465**:365-374

# 5

# Influence of Input Parameters on the Solution of Inverse Heat Conduction Problem

*Rakhab C. Mehta*

**Abstract**

A one-dimensional transient heat conduction equation is solved using analytical and numerical methods. An iterative technique is employed which estimates unknown boundary conditions from the measured temperature time history. The focus of the present chapter is to investigate effects of input parameters such as time delay, thermocouple cavity, error in the location of thermocouple position and time- and temperature-dependent thermophysical properties. Inverse heat conduction problem IHCP is solved with and without material conduction. A two-time level implicit finite difference numerical method is used to solve nonlinear heat conduction problem. Effects of uniform, nonuniform and deforming computational grids on the estimated convective heat transfer are investigated in a nozzle of solid rocket motor. A unified heat transfer analysis is presented to obtain wall heat flux and convective heat transfer coefficient in a rocket nozzle. A two-node exact solution technique is applied to estimate aerodynamic heating in a free flight of a sounding rocket. The stability of the solution of the inverse heat conduction problem is sensitive to the spatial and temporal discretization.

**Keywords:** analytical solution, inverse heat conduction problem, numerical analysis, deforming grid, heat transfer coefficient, heat flux, random search method

## 1. Introduction

The basic theory of heat and structure of solid body is associated with the internal energy of matter which in the first law of thermodynamics is referred to as the internal energy concerned with the physical state of the material. The first law of thermodynamics defines that the flowing heat energy is conserved in the absence of heat sources and sinks. It is, therefore, important to study the influence of thermocouple lead wires and distortion due to the thermocouple cavity in solution of the inverse heat conduction problem. According to the second law of thermodynamics, the heat will be transferred from one body to another body only when the bodies are at two different temperatures level and the heat will flow from the point of higher to the point of lower temperature.

A direct solution of transient heat conduction equation with prescribed initial and boundary conditions yields temperature distribution inside a slab of finite thickness. The direct solution is mathematically considered as well-posed because the solution exists, unique and continuously depends on input data. The estimation of unknown parameters from the measured temperature history is called as inverse problem of heat conduction. It is mathematically known as an ill-posed problem since the solution now does not continuously depend on the input data. Measurement data error in temperature, thermal lagging, thermocouple's cavity, signal noise, etc. makes stability problem in the estimation of unknown parameters.

Numerical inversion of the integral solution [1], exact solution [2], numerical techniques [3], least-squares method [4], transform methods [5], different series approach [6], variable time-step size [7] have been applied to solve inverse heat conduction problems. Solutions of the ill-posed inverse heat conduction problem have been presented in detail by Beck et al. [8] and Özisik et al. [9]. Tikhonov regularization method [10] has been described for cross-validation criterion for selecting the regularization parameter to obtain a stable approximation to the solution. Kurpisz et al. [11] have presented series with derivatives with temperature to solve inverse thermal problem. Hensel [12] has described space marching numerical methods to solve inverse heat transfer problem. Various mathematical methods and numerical algorithms for solving inverse heat conduction problems are described and compared by Alifanov [13]. Taler and Duda [14] have presented solutions of direct and inverse heat conduction problems.

Inverse heat conduction analysis provides an efficient tool for estimating the thermophysical properties of materials, the boundary conditions, or the initial conditions. Estimation of surface heat flux has been carried out without [15] and with [16] heat conduction and comparison between them shows discrepancies as high as about 27% [17]. Moving window optimization method [18] has been applied to predict the aerodynamic heating in a free-flight of sounding rocket by comparing numerically calculated and measured temperature history. Howard [19] developed a numerical procedure for estimating the heat flux with variable thermal properties using a single embedded thermocouple. Simultaneous identification of the temperature-dependent thermal conductivity and the asymmetry parameter of the Henyey-Greenstein scattering phase function have been shown by Zmywaczyk and Koniorczyk [20].

The conjugate gradient method with adjoint problem for function estimation iterative technique is used to solve IHCP to estimate heat flux and internal wall temperature of the throat section of the rocket nozzle [21]. Heisler [22] have reported supplementary "short-time" temperature-time charts for the center, mid-location and surface of large plates, long cylinders and spheres for the dimensionless time sub-domain. Convective heat transfer coefficient and combustion temperature in a rocket nozzle is determined using transient-temperature response chart [23].

The solution of transient IHCP can be obtained using analytical or numerical schemes in conjunction with measured temperature-time history. The estimation of the unknown parameters can be carried out by employing gradient or non-gradient methods to predict the unknown parameters in a prescribed tolerance limit. The focus of the present work is to investigate the influence of various parameters on the solution of inverse heat conduction problem.

## 2. Measurement errors

Experimental difficulties [24] are noticed in implanting thermocouples at the surface for temperature measurements. Temperature response delays have been studied to solve IHCP applied to cooled rocket thrust chamber [25]. The temperature measured inside the slab may delay and damp depending on $x_m$ as illustrated in **Figure 1**. A thermocouple indicates temperature lag behind the actual temperature. The effect of the thermocouple sensor dynamics on prediction of a triangular heat

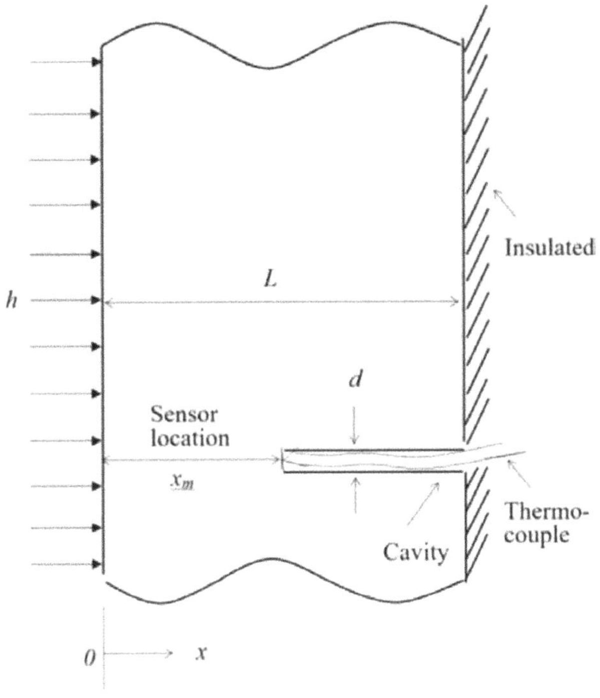


**Figure 1.**
*Geometry of the specimen.*

flux history has been analyzed with simulated data in a one-dimensional domain by Woodbury [26].

Chen and Danh [27] have carried out experimental studies to obtain transient temperature distortion and thermal delay in a slab due to presence of thermocouple cavity. The distortion of temperature profiles inside the slab may be influenced by the dissimilar thermophysical properties of thermocouple and surrounding materials and by the diameter and depth of the cavity. The temperature distortion [28] inside a slab is a function of the thermocouple cavity diameter $d$ and location $x_m$.

Standard statistical analysis consists of error in the measurement as an additive of true plus random, in zero mean, in constant variance, uncorrelated, normal, bell shaped probability density function, constant variance known, errors in the depen-dent variables and no-prior information about the parameters. The error in mea- surement can be obtain using exact analytical solution [29] as

$$E = \cos(x/L)(\cos\varepsilon - 1)e^{-\delta\tau} \quad (1)$$

$$E = \sin(x/L)(\cos\varepsilon - 1)e^{-\delta\tau} \quad (2)$$

where $\varepsilon$ and $\delta\tau$ refer to error in measurement of thermocouple location and in time recording, respectively. One of the important points that must be mentioned, here, is the use of a starting solution. In the case of a solid rocket motor where boundary conditions are suddenly imposed by a wall, there will be high intensity of heat flux on cold wall, and the heat flux during the first few steps in time may not be very accurate. The numerical solution is initiated using an exact analytical solution instead of starting from the initial constant condition. Such solutions can be obtained from to exact analytical solution [30] of transient heat conduction equation. Heat transfer rates to the calorimetric probe are estimated from measurements of temper-ature and rate of temperature change using energy conservation considerations [31].

An optimization method based on a direct and systematic search region reduction optimization method [32] can be employed to estimate the unknown convective heat transfer coefficient in a typical rocket nozzle. The most attractive feature of the direct search scheme is the simplicity of computer programming. The pseudo-random algorithm, an effective tool for optimization, does not require computation of derivatives but depends only on function evaluation. It works even when the differentiability requirements cannot be ensured in the feasible domain. For initiating the search only an estimate of the feasible domain is needed. Therefore, another advantage of the method is that the starting condition is not crucial; any reasonable value will do.

## 3. Heat conduction equation

### 3.1 Analytical solution

The computation of the turbulent convective heat transfer coefficient from combustion gases to the rocket nozzle wall is based on the Bartz's equation [33] incorporating the effects of compressibility, throat curvature and variation of transport properties in the boundary layer. The transient heat conduction in a one-dimensional Cartesian coordinate system having two parallel plane surfaces $S_n$ ($n$ = 1, 2) of a slab may be written in dimensional form [34] as.

$$\frac{\partial \theta(X,\tau)}{\partial \tau} = \frac{\partial}{\partial X}\left[K(\theta)\frac{\partial \theta(X,\tau)}{\partial X}\right], \text{ in region } R, \tau > 0 \tag{3}$$

with following initial and boundary conditions:

$$\theta(X,0) = f_i(X), \text{ in region } R, \tag{4}$$

$$-\frac{\partial \theta(S_n,\tau)}{\partial X} = Bi_n[\theta(S_n,\tau)], \text{ or } \frac{q_w L}{(T_g - T_i)} \text{ on boundary } S_n, \tau > 0 \tag{5}$$

where $f_i$ is initial temperature distribution in the region $R$ of the slab. Eq. (4) represents both convective heat transfer or heat flux condition as applied to the inner surface.

We now consider the constant thermal property solution and can be written in terms of eigen function $\psi(\lambda_m, X)$ as

$$\theta(X,\tau) = \sum_{m=1}^{\infty} \exp\left[-\lambda_m^2 \tau\right]\psi(\lambda_m, X)\int_R \psi(\lambda_m \overline{X}) f(\overline{X}) d\overline{X} \tag{6}$$

In the above Eq. (5), $Bi$ or $q_w$ is the unknown parameter to be determined using measured temperature time history at location $x_m$ as depicted in **Figure 1**. In estimating the unknown condition, one has to minimize the absolute difference between the calculated and measured temperature at specified location and time ($x_m$, $\tau$) in a prescribed tolerance value using an iteration procedure. The iteration scheme is described in the following sections.

### 3.2 Inverse algorithm

The IHCP is solved by comparing calculated and measured temperature using an iterative technique [30]. In estimating $q_w$, one minimizes

$$F(q_w) \approx |\theta_c(X_m, \tau) - \theta_m(X_m, \tau)| \quad (7)$$

where $\theta_c$ and $\theta_m$ are the calculated and measured temperatures at ($X_m$, $\tau$), respectively. The computed temperature is a nonlinear function of unknown parameters such as wall heat flux or convective heat transfer coefficient. Temperature is calculated using Eq. (6) and compared with the measured temperature as expressed in Eq. (7). The inverse problem starts with initial guess value of the unknown parameter. The second step is to correct the previous guessed unknown parameter using the Newton-Raphson method. The sensitivity coefficient can be obtained by differentiating temperature with respect to wall heat flux $q_w$. The iteration procedure will continue until $|F(q_w)| \leq 10^{-4}$. This iterative scheme estimates the component of the $q_w$ at a time and thus may be considered on-line method.

The inverse method for solving a value of $q_w(0, \tau)$ is as follows. Initiate with an initial guess value of $q_w$, satisfy the convergence criterion, and implement the Newton-Raphson to obtain the estimate value.

Now, it is possible to estimate convective heat transfer coefficient and combustion gas temperature in conjunction with measured temperature history [35]. The equation for converting the calculated heat flux to the heat transfer coefficient is

$$Bi = \frac{Lq_w}{K(\theta)(T_g - T_i)} \quad (8)$$

In the foregoing equation,$T_g$ is an unknown quantity and can be estimated using again the above-mentioned minimization and iteration methods. The convergence criterion for the iterative scheme remains same as mentioned above.

### 3.3 Numerical methods

It is not always feasible to obtain analytical solution of temperature-dependent thermal conductivity and radiation boundary condition. The Crank-Nicolson finite difference method with two-time level implicit numerical scheme [36] has been employed to solve the nonlinear conduction problem with the Newton-Raphson method to consider the radiation boundary condition.

Deforming or moving finite elements method [37] is used to solve linear heat conduction equation. The moving finite element [38] is used to consider the time delay in the measurement of back wall temperature.

### 3.4 Two-nodes system of transient heat conduction equation

For only two nodes the system of [39] equations reduce to the following pair of equations:

$$\frac{d\theta_0}{d\tau} = \frac{1}{(\Delta X)^2}[-2(\theta_1 + Bi\Delta X)\theta_0 + 2\theta_1] \quad (9)$$

$$\frac{d\theta_1}{d\tau} = \frac{1}{(\Delta X)^2}[2\theta_0 - 2\theta_1] \quad (10)$$

where 0 and 1 represent node in a slab of finite thickness. These are the exact solutions to the system of two ordinary differential equations which resulted from a two-node finite-difference approximation to the original problem.

$$\theta_0 = \frac{1}{\lambda_2 - \lambda_1}\left[(\lambda_2 + 2Bi)e^{\lambda_1 \tau} - (\lambda_1 + 2Bi)e^{\lambda_2 \tau}\right] \tag{11}$$

$$\theta_1 = \frac{1}{\lambda_2 - \lambda_1}\left[\lambda_2 e^{\lambda_1 \tau} - \lambda_1 e^{\lambda_2 \tau}\right] \tag{12}$$

where $\lambda_1 = -2 - Bi + \sqrt{Bi^2 + 4}$ and $\lambda_2 = -2 - Bi - \sqrt{Bi^2 + 4}$.

Solution of the above simultaneous equation calculates the temperature with a given value of $Bi$. The solution is now solving simultaneously Eqs. (11) and (12) to determine the unknown parameter.

## 4. Inverse problem of heat conduction applied to a rocket nozzle

The influence of constant (average) thermal conductivity, temperature-dependent thermal conductivity, computational grid in numerical solver, nonlinear boundary condition, cylindrical coordinate and the estimation of the wall heat flux and convective heat transfer is carried out by employing measured temperature history of a rocket nozzle of a solid motor. Solution of linear heat conduction equation is used to estimate the convective heat transfer coefficient with the measured temperature data of outer wall of a rocket nozzle. The running time of rocket motor is 16 s. The nozzle wall thickness $L$ = 0.0211 m. The thermo-physical properties of the material are: $\rho$ = 7900 kg m$^{-3}$, $C_p$ = 545 J kg$^{-1}$ K$^{-1}$, $K$ (average) = 35 Wm$^{-1}$ K$^{-1}$. Initial temperature $T_i$ = 300 K and combustion gas temperature $T_g$ = 2946.2 K are used in the solution of the heat conduction equation.

### 4.1 Average thermal conductivity

Prediction of convective heat transfer coefficient is carried out in conjunction with the calculated and measured temperature history at outer surface of nozzle divergent in a solid rocket motor static test. The constant thermal conductivity solution of the linear transient heat conduction problem [30] is

$$\theta(X,\tau) = 1 - 2\sum_{n=1}^{\infty} \frac{Bi}{\left(Bi^2 + \lambda_n^1 + Bi\right)} \frac{\cos\left[\lambda_n(1-X)\right]}{\cos\lambda_n} e^{-\lambda_n^2 \tau} \tag{13}$$

$$\lambda \tan\lambda = Bi \tag{14}$$

For estimating unknown boundary condition, the heat conduction equation is and solved with the following boundary and initial conditions.

$$\frac{\partial\theta(0,\tau)}{\partial X} = Bi[\theta(0,\tau) - 1], \tau>0 \tag{15}$$

$$\frac{\partial\theta(1,\tau)}{\partial X} = 0, \tau>0 \tag{16}$$

and

$$\theta(X,0) = 0, \text{for all } X \tag{17}$$

Exact analytical solution of transient heat conduction as written in Eq. (13) is used to estimate convective heat transfer on the inner surface of the rocket nozzle.

An iterative scheme is used to solve inverse problem [30]. The iteration is carried out till the absolute difference between calculated and measured temperature is less than or equal to $10^{-4}$. **Table 1** exhibits the comparison between the estimated values of the convective heat transfer coefficient based on the exact solution of heat conduction equation with the calculated values of Bartz [33]. Bartz's equation calculates conservative estimates for the convective heat transfer to the wall [40].

### 4.2 Temperature-dependent thermal conductivity

An iteration procedure [41] is employed in conjunction with exact solution to predict convective heat transfer coefficient from the measured temperature-time data at the outer wall of the nozzle as shown in **Table 2** The expression for temperature-dependent conductivity is $K(T) = k_0 - \beta T$. The value of $k_0$ and $\beta$ are 57 $Wm^{-1} K^{-1}$ and 2.718 $Wm^{-1} K^{-2}$, respectively. The advantage of using the exact solution is found directly at specified location and time as compared to the numerical method which needs the computation from the initial state.

### 4.3 Numerical solution with various computational grids

Deforming or moving finite element is used to consider the time delay in temperature at the outer wall of the slab [37]. Estimated values of wall heat flux and heat transfer coefficient are tabulated in **Table 3**. It can be observed from the table that the estimated wall quantities are having significant influence on the predicted unknown boundary conditions. This example is extended to consider spatial grid changed and temporal dependence on the numerical solution using moving finite element method [38].

### 4.4 Nonlinear boundary condition

Numerical analysis of nonlinear heat conduction with a radiation boundary condition [36] is carried out to estimate wall heat flux using temperature history on the back wall of the rocket nozzle. The high temperature variation alters thermophysical properties of the material of mild steel. **Table 4** shows comparison

| $t$, s | $\theta_0$ at inner surface | $\theta_c$ at outer surface | $\theta_m$ at outer surface | $h$, W/m$^2$K | $h_B$, W/m$^2$K |
|---|---|---|---|---|---|
| 6 | 0.2950 | 0.0098 | 0.0096 | 1821.9 | 2254.2 |
| 7 | 0.3109 | 0.0159 | 0.0158 | 1810.0 | 2254.2 |
| 8 | 0.2996 | 0.0212 | 0.0211 | 1610.3 | 2254.2 |
| 9 | 0.3244 | 0.0301 | 0.0302 | 1690.9 | 2254.2 |
| 10 | 0.3340 | 0.0386 | 0.0385 | 1669.7 | 2254.2 |
| 11 | 0.3416 | 0.0473 | 0.0472 | 1641.9 | 2254.2 |
| 12 | 0.3302 | 0.0529 | 0.0529 | 1497.6 | 2254.2 |
| 13 | 0.3312 | 0.0602 | 0.0604 | 1443.1 | 2254.2 |
| 14 | 0.3409 | 0.0677 | 0.0676 | 1387.0 | 2254.2 |
| 15 | 0.3442 | 0.0781 | 0.0782 | 1413.0 | 2254.2 |
| 16 | 0.3475 | 0.0862 | 0.0861 | 1383.7 | 2254.2 |

**Table 1.**
*Solution of inverse heat conduction problem.*

| $t$, s | $\theta(0, \tau)$ | | | | $h$, W/m$^2$K | |
|---|---|---|---|---|---|---|
| | Iterative method | Beck method | $\theta_c(1, \tau)$ | $\theta_m(1, \tau)$ | Iterative method | Beck method |
| 6 | 0.0883 | 0.0838 | 0.0099 | 0.0098 | 536.6 | 581.7 |
| 7 | 0.1067 | 0.1075 | 0.0158 | 0.0159 | 600.6 | 587.0 |
| 8 | 0.1144 | 0.1116 | 0.0220 | 0.0212 | 592.6 | 598.4 |
| 9 | 0.1367 | 0.1367 | 0.0302 | 0.0302 | 674.2 | 685.3 |
| 10 | 0.1522 | 0.1545 | 0.0386 | 0.0385 | 712.9 | 693.2 |
| 11 | 0.1654 | 0.1690 | 0.0472 | 0.0472 | 737.4 | 730.0 |
| 12 | 0.1686 | 0.1639 | 0.0529 | 0.0529 | 718.2 | 721.9 |
| 13 | 0.1773 | 0.1777 | 0.0605 | 0.0605 | 723.6 | 725.8 |
| 14 | 0.1844 | 0.1813 | 0.0677 | 0.0676 | 723.0 | 725.1 |
| 15 | 0.1944 | 0.2040 | 0.0781 | 0.0782 | 753.6 | 765.0 |
| 16 | 0.2083 | 0.2174 | 0.0862 | 0.0862 | 758.3 | 770.0 |

**Table 2.**
*Comparison between iterative and Beck methods.*

| $t$, s | $T_m$ K at $X = 1$ | Uniform grid | | Non-uniform grid | | Moving grid | |
|---|---|---|---|---|---|---|---|
| | | $q_w \times 10^6$, W/m$^2$ | $h_c$, W/m$^2$K | $q_w \times 10^6$, W/m$^2$ | $h_c$, W/m$^2$K | $q_w \times 10^6$, W/m$^2$ | $h_c$, W/m$^2$K |
| 6 | 326 | 3.715 | 1964.5 | 3.846 | 2044.9 | 4.517 | 2412.1 |
| 7 | 342 | 2.700 | 1408.8 | 2.848 | 1449.6 | 2.818 | 1485.9 |
| 8 | 356 | 2.698 | 1436.9 | 2.840 | 1531.6 | 2.820 | 1512.8 |
| 9 | 380 | 2.704 | 1463.0 | 2.589 | 1569.8 | 2.842 | 1552.8 |
| 10 | 402 | 2.705 | 1491.4 | 2.858 | 1603.3 | 2.846 | 1586.7 |
| 11 | 425 | 2.704 | 1518.9 | 2.852 | 1632.6 | 2.845 | 1618.5 |
| 12 | 440 | 2.691 | 1539.7 | 2.805 | 1636.2 | 2.812 | 1630.8 |
| 13 | 460 | 2.683 | 1564.6 | 2.776 | 1649.7 | 2.791 | 1650.6 |
| 14 | 479 | 2.673 | 1588.1 | 2.738 | 1657.1 | 2.764 | 1665.9 |
| 15 | 507 | 2.094 | 1226.4 | 2.015 | 1190.6 | 2.091 | 1235.4 |
| 16 | 528 | 2.086 | 1231.8 | 1.981 | 1178.5 | 2.067 | 1231.6 |

**Table 3.**
*Wall heat flux at various grid arrangements.*

between the estimated convective heat transfer coefficients with the Bartz solution [33]. Effects of nonlinear IHCP with radiation boundary condition are investigated and results are presented in **Table 4**.

### 4.5 Heat conduction in a hollow cylinder

A grid point shift strategy [42] is adapted to solve inverse conduction problem in a radial coordinate of rocket nozzle with inner and outer radius of rocket nozzle. The inner and outer radius of the nozzle is 0.0839 m and 0.0105 m, respectively. The purpose of the present example to investigate the influence of radial coordinate

| $t$, s | $T_o$, K at $X = 0$ | $T_m$ K at $X = 1$ | $q_c \times 10^6$ W/m$^2$ | $h$ W/m$^2$K | $h_B$ W/m$^2$K | $T_{gc}$ K | $T_g$ K |
|---|---|---|---|---|---|---|---|
| 6 | 659.8 | 326 | 2.3547 | 950.0 | 2254.2 | 3137 | 2946.2 |
| 7 | 801.0 | 342 | 2.3899 | 1019.6 | 2254.2 | 3122 | 2946.2 |
| 8 | 900.7 | 356 | 2.2211 | 992.4 | 2254.2 | 3115 | 2946.2 |
| 9 | 996.3 | 380 | 2.6489 | 1237.1 | 2254.2 | 3113 | 2946.2 |
| 10 | 1050.5 | 402 | 2.3670 | 1135.5 | 2254.2 | 3108 | 2946.2 |
| 11 | 1066.4 | 425 | 1.7100 | 827.3 | 2254.2 | 3104 | 2946.2 |
| 12 | 1201.8 | 440 | 2.8144 | 1459.2 | 2254.2 | 3099 | 2946.2 |
| 13 | 1320.0 | 460 | 2.6559 | 1467.0 | 2254.2 | 3098 | 2946.2 |
| 14 | 1354.8 | 479 | 1.7595 | 991.7 | 2254.2 | 3095 | 2946.2 |
| 15 | 1383.4 | 507 | 1.3810 | 791.4 | 2254.2 | 3094 | 2946.2 |
| 16 | 1414.9 | 528 | 1.1684 | 681.8 | 2254.2 | 3094 | 2946.2 |

**Table 4.**
*Solution with nonlinear boundary condition.*

| $t$, s | $T_o$ K at $X = 0$ | $T_m$ K at $X = 1$ | $q_c \times 10^6$ W/m$^2$ | $h$, W/m$^2$K | $h_B$, W/m$^2$K | $\theta_g$, K | $\theta_{gc}$, K |
|---|---|---|---|---|---|---|---|
| 6 | 1260.2 | 326 | 3.6805 | 1789.6 | 2254.2 | 3316 | 2946 |
| 7 | 1175.9 | 342 | 3.3995 | 1628.0 | 2254.2 | 3264 | 2946 |
| 8 | 1160.7 | 356 | 2.4745 | 1181.4 | 2254.2 | 3255 | 2946 |
| 9 | 1165.8 | 380 | 2.5385 | 1194.7 | 2254.2 | 3290 | 2946 |
| 10 | 1196.0 | 402 | 2.5348 | 1261.1 | 2254.2 | 3206 | 2946 |
| 11 | 1192.3 | 425 | 2.3385 | 1166.4 | 2254.2 | 3197 | 2946 |
| 12 | 1205.8 | 440 | 2.2094 | 1114.8 | 2254.2 | 3187 | 2946 |
| 13 | 1211.0 | 460 | 2.1333 | 1229.5 | 2254.2 | 2946 | 2946 |
| 14 | 1222.1 | 479 | 2.0441 | 1187.5 | 2254.2 | 3943 | 2946 |
| 15 | 1237.1 | 507 | 2.0626 | 1206.7 | 2254.2 | 2946 | 2946 |
| 16 | 1249.1 | 528 | 2.0027 | 1180.9 | 2254.2 | 2945 | 2946 |

**Table 5.**
*Inverse problem in a hollow cylinder.*

on the estimated values of heat transfer coefficient. **Table 5** shows the effect of geometrical parameters on the predicted heat transfer coefficient.

## 4.6 Estimation of heat flux and heat transfer coefficient

The calculated convective heat transfer coefficients and inner wall temperature are used to determine the wall heat flux and the combustion temperature using Eq. (8). The iterative scheme is based on relation between wall heat flux and convective heat transfer coefficient [35]. **Table 6** shows the predicted values of wall heat flux and convective heat transfer coefficient. The IHCP is extended to determine wall heat flux in conjunction with convective heat transfer coefficient. A similar IHCP but referring to the 122 mm medium-range missile during correction engine operation has been considered by Zmywaczyk et al. [43].

| $t$, s | $T_0$ K at $X = 0$ | $T_m$ K at $X = 1$ | $q_c \times 10^6$, W/m$^2$ | $h$, W/m$^2$K | $h_B$, W/m$^2$K | $T_g$, K | $T_{gc}$, K |
|---|---|---|---|---|---|---|---|
| 6 | 1355.6 | 326 | 3.2502 | 2631.2 | 2254.2 | 3351 | 2946 |
| 7 | 1287.8 | 342 | 3.2950 | 1805.3 | 2254.2 | 3113 | 2946 |
| 8 | 1315.6 | 356 | 3.2974 | 1861.5 | 2254.2 | 3087 | 2946 |
| 9 | 1368.9 | 380 | 3.2967 | 1885.9 | 2254.2 | 3117 | 2946 |
| 10 | 1414.4 | 402 | 3.2837 | 1962.1 | 2254.2 | 3088 | 2946 |
| 11 | 1463.6 | 425 | 3.2718 | 2049.5 | 2254.2 | 3060 | 2946 |
| 12 | 1370.8 | 440 | 2.3825 | 1476.0 | 2254.2 | 2985 | 2946 |
| 13 | 1360.9 | 460 | 2.4140 | 1502.1 | 2254.2 | 2968 | 2946 |
| 14 | 1370.3 | 479 | 2.3625 | 1520.6 | 2254.2 | 2924 | 2946 |
| 15 | 1382.5 | 507 | 2.3675 | 1517.2 | 2254.2 | 2943 | 2946 |
| 16 | 1399.3 | 528 | 2.3645 | 1540.7 | 2254.2 | 2934 | 2946 |

**Table 6.**
*Wall heat flux and convective heat transfer coefficient.*

## 5. Estimation of heat flux with two-nodes in a sounding rocket

A two-node exact solution is used to calculate the back-wall temperature as described in Section 3.4. The iterative method described above has been used for estimating aerodynamic heating for a sounding rocket in free flight test. Here, the wall heat flux is estimated using the measured temperature history in conjunction with the iterative technique [30]. The aerodynamic heating rate is estimated for a typical sounding rocket as depicted in **Figure 2**. The location of thermocouple is marked in the diagram. The thermophysical properties of Inconel and wall thickness are $k = 18$ Wm$^{-1}$ K$^{-1}$, $\alpha = 4.47 \times 10^{-6}$ m$^2$/s, $L = 0.7874 \times 10^{-3}$ m. **Figure 3** depicts the measured temperature time history at different locations measured from the tip of the cone in the free flight of a sounding rocket as delineated in **Figure 2**. It can be observed from temperature history that the initial time delay in thermal response is 6 s. The unknown $q_w$ are estimated using an iterative technique which starts with an initial value of wall heat flux and is repeated until $|F(q_w)| \leq 10^{-4}$.

A two-node exact solution is used to calculate the wall temperature distribution. The unknown $q_w$ are estimated using an iterative technique which starts with an initial value of wall heat flux and is repeated until $|F(q_w)| \leq 10^{-4}$. **Figure 4** displays the estimated variation of the wall heat flux as a function of flight time of the sounding rocket.

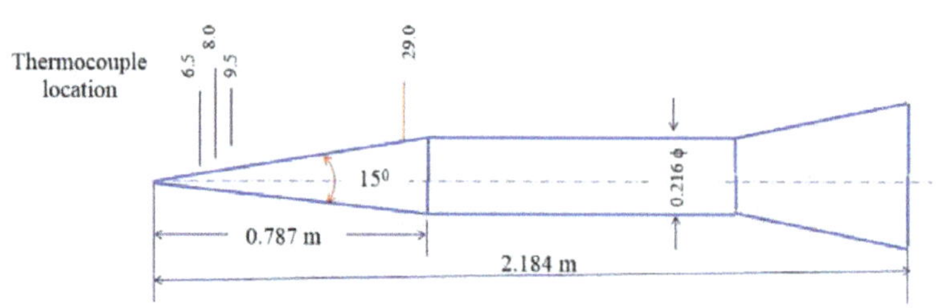


**Figure 2.**
*Schematic sketch of sounding rocket showing location of thermocouple.*

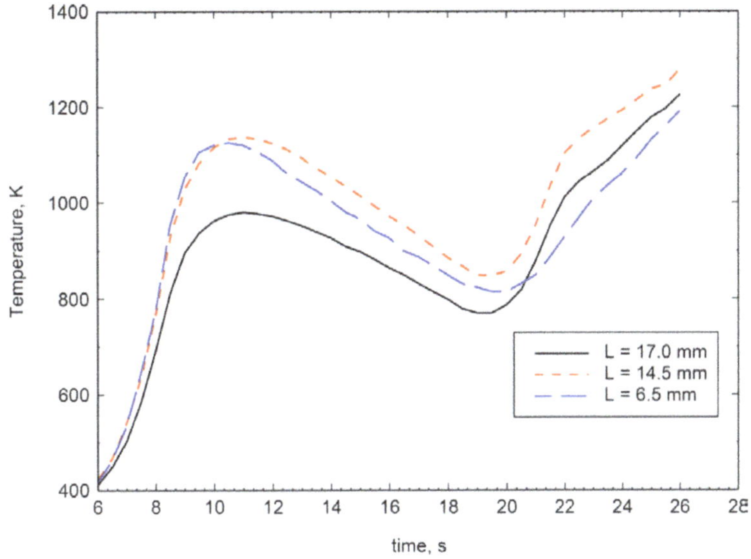


**Figure 3.**
*Measured temperature history in free flight of the sounding rocket.*

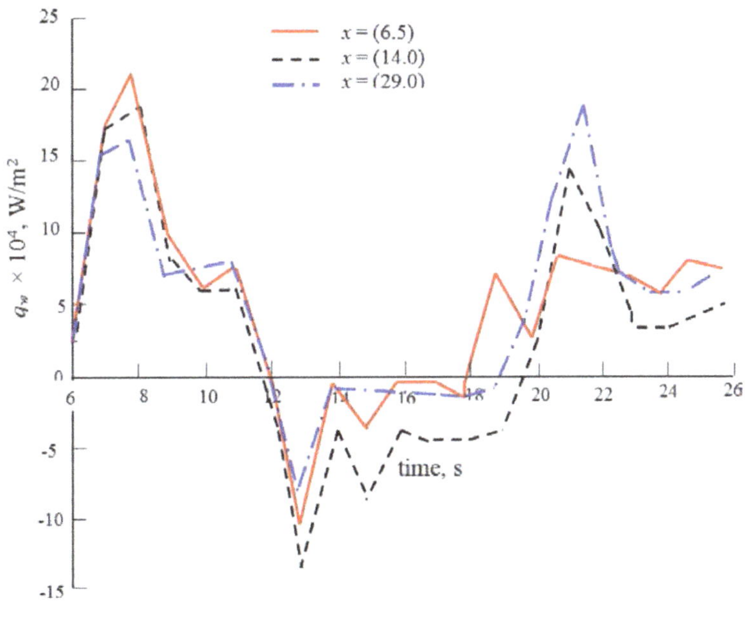


**Figure 4.**
*Variations of wall heat flux vs. flight time.*

The wall heat flux variation depends on the sounding rocket speed. The increase and decrease of the aerodynamic heating are a function of flight Mach number.

The estimated wall heat flux is compared with Van Driest's results [44]. **Table 7** depicts the estimated values of wall heat flux as a function of flight time at thermocouple location 29 as shown in **Figure 2**. It can be observed from the table that highest aerodynamic heating occurs during 7–8 s, another significant peak wall heat flux was found at 22 s.

| $t$, s | $T_m$, K | $T_c$, K | $q_w \times 10^4$, W/m$^2$ | $q_w^{van} \times 10^4$, W/m$^2$ |
|---|---|---|---|---|
| 6 | 313.0 | 320.0 | 2.756 | 8.246 |
| 7 | 341.3 | 349.4 | 16.695 | 14.82 |
| 8 | 408.0 | 412.3 | 19.387 | 22.647 |
| 9 | 469.7 | 469.3 | 8.657 | 18.876 |
| 10 | 495.2 | 495.8 | 8.145 | 10.117 |
| 11 | 504.1 | 513.3 | 9.302 | 4.569 |
| 12 | 502.4 | 521.2 | 2.757 | 1.176 |
| 13 | 495.2 | 506.8 | −7.695 | −0.977 |
| 14 | 487.4 | 487.6 | −0.040 | −2.320 |
| 15 | 477.4 | 477.0 | −0.037 | −3.134 |
| 16 | 467.4 | 467.0 | −0.028 | −3.578 |
| 17 | 457.4 | 457.5 | −0,034 | −3.769 |
| 18 | 445.8 | 447.2 | −0.905 | −3.789 |
| 19 | 438.6 | 437.5 | −0,005 | −1.628 |
| 20 | 444.1 | 445.9 | 5.173 | 4.135 |
| 21 | 469.1 | 467.5 | 13.518 | 11.524 |
| 22 | 533.0 | 521.2 | 20.618 | 19.732 |
| 23 | 556.3 | 558.4 | 9.180 | 13.122 |
| 24 | 567.1 | 573.5 | 6.514 | 8.581 |
| 25 | 584.1 | 587.2 | 8.592 | 5.467 |
| 26 | 596.9 | 602.2 | 7.018 | 3.314 |

**Table 7.**
*Comparison between calculated and Van Driest's heat flux at location 29.*

## 6. Conclusions

Analytical, transient numerical and two-node methods are used to compute temperature distribution in a finite slab. Numerical solution is carried out with temperature-dependent thermal conductivity. Implicit finite difference scheme with two-time level technique is implemented to solve nonlinear problem of heat conduction. Time delay is studied using finite element method with deforming grid strategy. A boundary shifting numerical scheme is used to solve transient heat conduction in radial coordinate. Evidence of temporal accuracy and dependence on time-step is demonstrated in the numerical solving of IHCP. Influence of thermocouple cavity and measurement errors in location and time are discussed. The IHCP is applied to predict the wall heat flux in a rocket nozzle of a solid motor. Wall heat flux is estimated in a free flight of a sounding rocket using the two-node method.

## Nomenclature

| | |
|---|---|
| $Bi$ | Biot number, $hL/k$ |
| $C_p$ | specific heat |
| $h$ | heat transfer coefficient |
| $K(\theta)$ | thermal conductivity, $k(\theta)/k_0$ |

| | |
|---|---|
| $k_0$ | reference thermal conductivity at $T_i$ |
| $L$ | slab thickness |
| $p$ | nondimensional parameter, $\alpha\Delta T/(\Delta x)^2$ |
| $q_w$ | wall heat flux |
| $T$ | temperature |
| $t$ | time |
| $x$ | distance from the inner surface |
| $X$ | dimensionless coordinate |
| $\alpha$ | thermal diffusivity |
| $\rho$ | density |
| $\theta$ | nondimensional temperature, $= (T - T_g)/(T_g - T_i)$ |
| $\tau$ | nondimensional time, $\alpha t/L^2$ |

**Subscripts**

| | |
|---|---|
| $B$ | Bartz |
| $c$ | computed |
| $g$ | combustion gas temperature |
| $i$ | initial value |
| $m$ | measured |
| $o$ | outer wall |
| $w$ | wall |
| $\beta$ | constant thermal conductivity coefficient |

## Author details

Rakhab C. Mehta
Department of Aeronautical Engineering, Noorul Islam Centre for Higher Education, India

*Address all correspondence to: drrakhab.mehta@gmail.com

## References

[1] Stolz G. Numerical solutions to an inverse problem of heat conduction for simple shapes. Journal of Heat Transfer. 1960;**82**(1):20-26

[2] Burggraf OR. An exact solution of the inverse problem in heat conduction theory and applications. Journal of Heat Transfer. 1964;**86**(2):373-382

[3] Beck JV. Nonlinear estimation applied to the nonlinear heat conduction problem. International Journal of Heat and Mass Transfer. 1970;**13**:703-716

[4] Frank I. An application of least squares method to the solution of the inverse problems of heat conduction. Journal of Heat Transfer. 1963;**85**: 378-379

[5] Imber M, Kan J. Prediction of transient temperature distributions with embedded thermocouples. AIAA Journal. 1972;**10**:784-789

[6] Shumakov NV. A method for the experimental study of the process heating a solid body. Journal of Technical Physics of the Academic of Sciences USSR. 1957;**2**:771-777

[7] Vogel J, Sára L, Krejčí L. A simple inverse heat conduction method with optimization. International Journal of Heat and Mass Transfer. 1993;**36**(17): 4215-4220

[8] Beck JV, Blackwell B, St. Clair CR Jr. Inverse Heat Conduction, Ill-Posed Problems. New York: Wiley-Interscience Publication; 1977

[9] Özisik MN, Orlande HRB. Inverse Heat Transfer: Fundamentals and Applications. New York: Taylor and Francis; 2000

[10] Tikhonov AN, Arsenin VY. Solution of Ill-Posed Problems. Washington, DC: Winston & Sons; 1977

[11] Kurpisz K, Nowak AJ. Inverse Thermal Problems. Southampton, UK: Computational Methods Publications; 1995

[12] Hensel E. Inverse Theory and Applications for Engineers. New Jersey: Prentice Hall; 1991

[13] Alifanov OM. Inverse Heat Transfer Problems. Berlin, Germany: Springer-Verlag; 1994

[14] Taler J, Duda P. Solving Direct and Inverse Heat Conduction Problems. Berlin: Springer; 2006

[15] Rumsey CB, Lee DB. Measurements of aerodynamic heat transfer on a 15-degree cone-cylinder, flare configuration in free-flight at Mach numbers up to 4.7. NASA TN D-824; 1961

[16] Mehta RC. Estimation of aerodynamic heat transfer in free-flight at Mach number upto 4.7. Wärm-und Stüffubertrang. 1986;**20**:27-31

[17] Mehta RC, Jayachandran T, Sastri VMK. Finite element analysis of conduction and radiative heating of a thin skin calorimeter. Wärm-und Stüffubertrang. 1988;**22**:227-230

[18] Dąbrowski A, Dąbrowski L. Inverse heat transfer problem solution of sounding rocket using moving window optimization. PLoS One. 2019;**14**(6):1-24

[19] Howard FG. Single thermocouple method for determining heat flux to a thermally thick wall. NASA TN D 4737; 1968

[20] Zmywaczyk J, Koniorczyk P. Numerical solution of inverse radiative-conductive transient heat transfer problem in a grey participating medium. International Journal of Thermophysics. 2009;**30**:1438-1451

[21] Karnal M. High temperature measurements at the internal nozzle wall of the zephyr [MS thesis]. Germany: Department of Computer Science, Electrical and Space Engineering, Lulea University of Technology; 2014

[22] Heisler MP. Temperature charts for induction and constant temperature heating. Journal of Heat Transfer, Transactions of the ASME. 1947;**69**: 227-236

[23] Mehta RC, Jayachandran T. Determination of heat transfer coefficient using transient temperature response chart. Wärm-und Stüffubertrang. 1990;**26**:1-5

[24] Li DI, Wells MA. Effects of subsurface thermocouple installation on the discrepancy of the measured thermal history and predicted surface heat flux during a quench operation. Metallurgical and Materials Transactions B. 2005;**36**(3):343-354

[25] Perakis N, Haidin OJ. Inverse heat transfer method applied to capacitively cooled rocket thrust chambers. International Journal of Heat and Mass Transfer. 2018;**131**:150-166

[26] Woodbury KA. Effect of thermocouple sensor dynamics on surface heat flux prediction obtained via inverse heat conduction analysis. International Journal of Heat and Mass Transfer. 1990;**33**(12):2641-2649

[27] Chen CJ, Danh TM. Transient temperature distortion in a slab to thermocouple cavity. AIAA Journal. 1978;**14**(7):979-981

[28] Anil SL, Mehta RC. Effect of thermocouple cavities on heat transfer measurements. In: Proceedings of XVII National and VI ISHMT/ASME Heat and Mass Transfer Conference, HMT-2004-C68, IGCAR, Kalpakam, India. Jan. 2004

[29] Carslaw HD, Jaeger JC. Conduction of Heat in Solids. London, UK: Oxford University Press; 1959

[30] Mehta RC. Solution of the inverse conduction problem. AIAA Journal. 1977;**15**(9):1355-1358

[31] Mehta RC. Estimation of heating rate using a calorimeter probe. Review of Scientific Instrumentation. 1987; **52**(11):1782-1784

[32] Mehta RC, Tiwari SB. Controlled random search technique for estimation of convective heat transfer coefficient. Heat and Mass Transfer Journal. 2007; **43**:1171-1177

[33] Bartz DR. A simple equation for rapid estimation of rocket nozzle convective heat transfer coefficients. Jet Propulsion. 1957;**27**:49-51

[34] Özisik MN. Boundary Value Problem of Heat Conduction. USA: International Text Book Co; 1968

[35] Mehta RC. Estimation of heat transfer coefficient in a rocket nozzle. AIAA Journal. 1981;**19**(8): 1085-1086

[36] Mehta RC. Numerical solution of nonlinear inverse heat conduction problem with radiation boundary conditions. International Journal of Numerical Methods in Engineering. 1984;**20**:1057-1066

[37] Mehta RC, Jayachandran T. Deforming finite element for the numerical solution of the nonlinear inverse heat conduction problem. Communication in Applied Numerical Methods. 1987;**3**:167-172

[38] Mehta RC, Jayachandran T. Deforming grid method applied to the inverse problem of heat conduction. Journal of Thermophysics and Heat Transfer. 1989;**3**(2):226-229

[39] Myers GF. Analytical Method in Conduction Heat Transfer. New York, USA: McGraw-Hill; 1971

[40] Brinsmade AF, Desmon LG. Hypothesis for correlating rocket nozzle throat convective heat transfer. Heat Transfer-Cleveland, Chemical Engineering Progress Symposium. 1965; **61**(59):88-98

[41] Mehta RC. Extension of the solution of inverse conduction problem. International Journal of Heat and Mass Transfer. 1979;**22**:1149-1150

[42] Mehta RC. An efficient numerical method for solving inverse conduction problem in a hollow cylinder. AIAA Journal. 1984;**22**(6):860-862

[43] Zmywaczyk J, Koniorczyk P, Preiskorn M, Machowski B. An inverse approach to estimate heat transfer coefficients of 122 mm medium-range missile during correction engine operation, problems of mechatronics armament, aviation. Safety Engineering. 2014;**5**(15):25-40

[44] Van Driest ER. In: Lin CC, editor. Turbulent Flows and Heat Transfer, 6. Princeton, NJ, USA: Princeton University Press; 1959

6

# Tubular Heat Exchanger Fouling in Phosphoric Acid Concentration Process

*Rania Jradi, Ali Fguiri, Christophe Marvillet and Mohamed Razak Jeday*

**Abstract**

Fouling in phosphoric acid concentration is a persistent operational problem that compromises energy recovery in this process. Progress is hampered by the lack of quantitative knowledge of fouling dynamic effects on heat exchanger transfer. The object of this work is an experimental determination of the thermal fouling resistance in the tubular heat exchanger of phosphoric acid preheated installed in phosphoric acid concentration process. By measuring the inlet and outlet temperatures of phosphoric acid, steam temperature, suction and discharge pressure of the pump and acid density measurement, the overall heat transfer coefficient has been determined. The determination of the overall heat transfer coefficient with clean and fouled surfaces, allowed calculating the fouling resistance. The results from the heat exchanger studies showed that the fouling resistance increased with time and presented an asymptotic evolution in compliant with the proposed model by Kern and Seaton, with the existence of fluctuation. The poorly cleaned heat exchanger implied the absence of the induction period and caused, consequently, high values of the fouling resistance in a relatively short-time period.

**Keywords:** fouling, heat exchanger, heat transfer coefficient, fouling resistance, phosphoric acid concentration unit

## 1. Introduction

The main issues which various industries are facing are the accumulation of undesired substances or materials dissolved or presented as a suspension in the fluid on the heat transfer surfaces [1]. This phenomenon which is called as fouling affects the equipment operation by reducing their thermal effectiveness. This causes a significant economic loss due to the installations of regular cleaning [2, 3].

Fouling in heat transfer process is often inevitable and reduces energy efficiency and plant operability. Mitigation of fouling, and effective cleaning strategies, both require understanding the mechanisms involved in deposition and cleaning [4]. Many researches on fouling in heat transfer processes are dealt with, by reducing the efficiency of heat transfer and limiting productivity [5]. Phosphoric acid fouling in concentration process preheat exchangers is a persistent operational problem that compromises energy recovery in these process. Progress is hampered by the lack of quantitative knowledge of the dynamic effects of fouling on heat transfer exchanger [6]. Generally, phosphoric acid, which is the cold fluid, flows through the tube side while steam, which is the hot stream, flow through the shell side in heat exchangers [7]. The solution of concentrated phosphoric acid is supersaturated with calcium sulfate, resulting in the deposition on the contact material [8]. Given that the thermal conductivity of these scales is low, even a thin layer of scale can drastically reduce the overall heat transfer coefficient [9]. Furthermore, fluorosilicate and fluoroaluminate deposits on the acid ducts of clarifier tanks and evaporators can be imbedded in gypsum scale, which reduces pipe diameter and flow rate. In spite of considerable research efforts at the phosphoric acid type scale, no viable commercial solution has been found [10–12]. Behbahani et al. [13] have done a high number of fouling experiments in a side-stream of a phosphoric acid plant for various flow velocities, surface temperatures and concentrations in order to determine the mechanisms which control the deposition process. After identifying the effects of operational parameters on the deposition process, a fouling kinetic model by crystallization has been developed in Behbahani et al. [8]. A mathematical model has been elaborated to predict the fouling resistance in concentrating phosphoric acid [14]. The predicted fouling resistances were compared with the experimental data. Majority engineering calculations on heat transfer use the experimental heat transfer coefficients [15].

In this survey, we will examine the fouling phenomenon of the heat exchanger tubes for the preheat circuit of the phosphoric acid. The heat exchanger used for heating phosphoric acid is exposed to the fouling problem at the tube side of heat exchangers. In this context an experimental determination of the thermal fouling resistance by measuring the inlet and outlet temperatures of phosphoric acid, the temperature of steam, suction and discharge pressure of the pump and acid density measurement, the overall heat transfer coefficient has been determined. The determination of the overall heat transfer coefficient for the heat exchanger with clean and fouled surfaces makes it possible to calculate the fouling resistance.

## 2. Fouling mechanisms

Fouling can be divided into a number of distinct mechanisms. In general, many of these fouling mechanisms occur at the same time and each requires a different prevention technique. Among these different mechanisms, some represent different stages of the fouling process. The main mechanisms or stages of fouling include:

1. Period of initiation or delay. This is the clean surface period before dirt accumulation. This accumulation of relatively small deposits can even improve heat transfer over a clean surface and give the appearance of a "negative" fouling rate and a total negative fouling amount.

2. Particle fouling and formation, aggregation and flocculation.

3. Mass transport and migration of fouling agents to fouling sites.

4. Separation and deposition phase involving nucleation or initiation of fouling sites and attachment leading to deposit formation.

5. Growth, aging and hardening and increase of deposit resistance or auto-retardation, erosion and elimination.

## 3. Rate of fouling

Fouling is defined as a phenomenon that occurs with or without a temperature gradient in many natural, domestic and industrial processes. A surface is "dirty" when unwanted material accumulates there.

The fouling rate is normally defined as the average deposit surface loading per unit of surface area in a unit of time. Depending on the fouling mechanism and conditions, the fouling rate may be linear, falling, asymptotic or saw-tooth, as the case may be. **Figures 1** and **2** shows the different types of fouling rate.

1. Linear fouling is the type of fouling where the rate of fouling can be stable over time with the increase of fouling resistance and deposit thickness. It usually occurs when the temperature of the deposition in contact with the flowing fluid remains constant.

Ebert and Panchal [16] presented a fouling model expressing the average (linear) fouling rate under given conditions following two competing terms, namely a deposit term and an attenuated term.

$$\text{Fouling rate} = (\text{deposit term}) - (\text{anti} - \text{deposit term})$$

$$\frac{dRf}{dt} = \alpha\ Re^{\beta}\ Pr^{\delta} \exp\left(\frac{-E}{RT_{film}}\right) - \gamma\tau_w \tag{1}$$

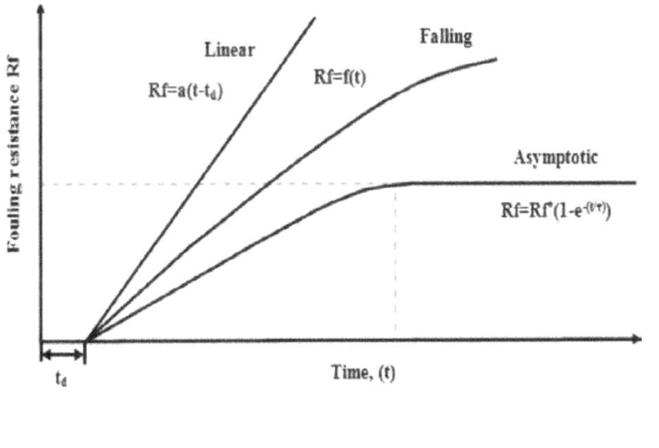


**Figure 1.**
*Fouling curves.*

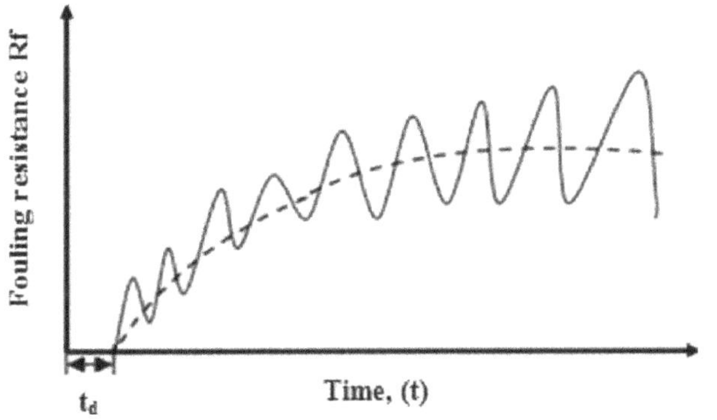


**Figure 2.**
*Practical fouling curve.*

where $\alpha$, $\beta$, $\gamma$ and $\delta$ are parameters determined by regression, $\tau_w$ is the shear stress at the tube wall and $T_{\text{film}}$ is the temperature of the fluid film (average of local bulk and local wall fluid temperatures). The relationship in Eq. (1) indicates the possibility of identifying combinations of temperature and velocity below where fouling rates will be negligible. Ebert and Panchal [16] present this as the "threshold condition." The model in Eq. (1) suggests that the geometry of the heat exchanger which affects the surface and film temperatures, velocities and shear stresses can be effectively applied to maintain the conditions below "threshold conditions" in a given heat exchanger.

1. Falling fouling is the type of fouling where the fouling rate decreases with time, and the deposit thickness does not reach a constant value, although the fouling rate never drops below a certain minimum value. In general falling fouling is due to an increase of removal rate with time. Its progress can often be described by two numbers: the initial fouling rate and the fouling rate after a long period of time.

2. Asymptotic fouling rate is where rate decreases with time until it becomes negligible after a period of time when the deposition rate becomes equal to the deposit removal rate and the deposit thickness remains constant. This type of fouling generally occurs where the tube surface temperature remains constant while the temperature of the flowing fluid drops as a result of increased resistance of fouling material to heat transfer. Asymptotic fouling may also result from soft or poorly adherent suspended solid deposits upon heat transfer surfaces in areas of fast flow where they do not adhere strongly to the surface with the result that the thicker the deposit becomes, the more likely it is to wash off in patches and thus achieve some average asymptotic value over a period of time. The asymptotic fouling resistance increases with increasing particle concentration and decreasing fluid bulk temperature, flow velocity, and particle diameter. The asymptotic fouling model was first described by Kern and Seaton [17]. In this model, the competing fouling mechanisms result in asymptotic fouling resistance beyond which any additional increase in fouling does not happen.

3. Saw-tooth fouling occurs where part of the deposit is detached after a critical residence time or once a critical deposit thickness has been reached. The fouling layer then builds up and breaks off again. This periodic variation could be due to pressure pulses, scaling, trapping of air inside the surface deposits during shutdowns or other reasons. It often corresponds to the moments of system shutdowns, startups or other transients during operation.

## 4. Fouling measurement and monitoring

The fouling resistances can be measured experimentally or analytically. The main measurement methods include:

1. Direct weighing: the simplest method for assessing the extent of deposition on laboratory test surfaces is to weigh directly. The method requires an exact balance in order to be able to detect relatively small changes in the mass of deposits. It may be necessary to use thin walled tube to reduce the tare mass in order to increase the accuracy of the method.

2. Thickness measurement: in many examples of fouling the thickness of the deposit is relatively small, perhaps less than 50 μm, so that a direct measurement is not easy to obtain. A relatively simple technique provided there is reasonable access to the deposit, consists in measuring the thickness. By using a removable coupon or plate, the thickness of a hard deposit such as a scale, can be obtained using a micrometer or traveling microscope. For a deformable deposit containing a large proportion of water, e.g., a biofilm it is possible to use an electrical conductivity technique.

3. Heat transfer measurements: in this method, the fouling resistance can be determined according to the changes in heat transfer during the deposition process. The equation for the following operations will be Eq. (11). The data can be reported in terms of changes in overall heat transfer coefficient. A major hypothesis of this method is that the presence of the deposit does not affect the hydrodynamics of the flowing fluid. However, during the first stages of deposition, the surface of the deposit is generally rougher than the metal surface so that the turbulence in the fluid is greater than when it is flowing on a smooth surface. As a result the fouling resistance calculated from the data will be lower than if the increased turbulence level had been taken into account. It is possible that the increased turbulence offsets the thermal resistance of the deposit and negative values of thermal resistance will be calculated.

4. Pressure drop: as an alternative to direct heat transfer measurements it is possible to use changes in pressure drop caused by the presence of the deposit. The pressure drop is increased for a given flow rate due to the reduced flow area in the fouled condition and the roughness of the deposit. The shape of the curve relating pressure drop with time will generally, follow an asymptotic shape so that the time to achieve asymptotic fouling resistance can be determined. The method is often associated with the direct measurement of thickness of the deposit layer. Friction factor changes can also be used to indicate fouling of a flow channel.

5. Other techniques for assessing fouling: with regard to their effect on heat exchanger performance the measurement of heat transfer reduction or increase in pressure drop provide a direct indication. The simple methods of measuring deposit thickness described above are useful, but in general they require that the experiment be completed in order to allow access to the test sections. Ideally non-intrusive techniques would allow further deposition while maintaining experimental conditions without disturbance. Such techniques include the use of radioactive tracers and optical methods. Laser techniques can be used to study the accumulation and removal of deposits. In addition, infra-red systems are used to study the development and removal of biofilms from tubular test sections. Microscopic examination of deposits may provide further evidence of the mechanisms of fouling, but this is usually a "backup" system rather than providing quantitative data.

## 5. Prediction of fouling factor

As noted above, fouling has the effect of forming on the heat transfer surface a substantially solid deposit of low thermal conductivity, through which heat is to be transferred by conduction. But as the thermal conductivity of the fouling layer and its thickness are not generally known, the only possible solution to the heat transfer problem is to introduce a fouling factor to take into account the additional resistance to heat transfer and possible calculation of the overall coefficient of heat transfer. A fouling coefficient is also sometimes specified, it is the reciprocal value of the fouling factor. When carrying out heat transfer calculations, the selection of fouling factors must be made with caution, especially when the fouling resistances completely dominate the thermal design.

The influence of inherent uncertainties in fouling factors is generally greater than that of uncertainties in other design parameters such as fluid properties, flow rates and temperatures [18]. An important fouling factor is sometimes adopted as a safety margin to cover uncertainties on the properties of fluids and even in the knowledge of the process, but the use of an excessively large fouling factor will result in an oversized heat exchanger with two or three times more area than is necessary. Although many tabulations based on the experiment are available and provide typical fouling factors such as the TEMA RGP-T-2.4 table [19], an acceptable assessment of the effects of fouling needs to be judged and evaluated for each particular application. Such tabulations can, however, serve as a guide in the absence of more specific information.

A number of semi empirical models have been proposed over the years for the prediction of the rate of fouling in heat exchangers or for estimating a fouling factor to be used in heat transfer calculations.

The first work on this subject began in the late 1950s with Kern and Seaton [17].

The modeling resulting from this work is based on the assumption that two processes act simultaneously. The first process is that of particle deposition characterized by a deposition flux that is constant if the concentration is also constant. The second process is that of the re-entrainment of particles characterized by a re-entrain flow $\phi_r$ dependent on the mass of particles ($m_p$) deposited. The particle balance of the deposit is expressed according to the following equation:

$$\frac{dm_p}{dt} = \phi_d - \phi_r \tag{2}$$

The deposition process is designed as the serialization of particle transport and adhesion mechanisms. The following assumptions are made:

- consideration of a single type of fouling;
- homogeneity of the deposit;
- not taken into account of the phase of initiation of the deposit and the state of surface;
- constancy of the properties and thermo-physical characteristics of the fluid and the deposit.

The particle wall transport phase controls the deposition process while the shear stress controls the re-entrain phase of the particles. Thus, considering the proportionality of $\phi_{\mathrm{d}}$ as a function of the deposited mass of particles, we can write the following equations:

$$\phi_d = k_p * (C_b - C_w) \tag{3}$$

$$\phi_r = C_1 * \tau_w * m_p \tag{4}$$

Or
$k_p$ is the transport coefficient.
$C_b$ the particle concentration in the fluid.
$C_w$ the particle concentration at the wall.
$C_1$ a dimensional constant.
$\tau_w$ the shear stress exerted by the fluid on the deposit.
Equation (2) thus becomes:

$$\frac{dm_p}{dt} = k_p * (C_b - C_w) - C_1 * \tau_w * m_p \quad (5)$$

The solution of Eq. (5) is thus:

$$m_p = \frac{K_p * (C_b - C_w)}{C_1 * \tau_w} * [1 - \exp(-C_1 * \tau_w * t)] \quad (6)$$

Assuming that $\tau = \frac{1}{C_1 * \tau_w}$ and $m_p^* = \frac{K_p * (C_b - C_w)}{C_1 * \tau_w}$

We can thus express the equation as follows:

$$m_p = m_p^* * \left[1 - exp\left(-\frac{t}{\tau}\right)\right] \quad (7)$$

Considering that the initial fouling flow is equal to the deposition flow and that the thermo physical properties of the deposit (conductivity and density) are constant, it is thus possible to express Eq. (7) in the form of a thermal fouling resistance:

$$Rf(t) = Rf^* * \left[1 - exp\left(-\frac{t}{\tau}\right)\right] \quad (8)$$

With $Rf(t)$, the evolution of the fouling resistance as a function of time expressed in [$m^2$-K/W] $Rf^*$, the asymptotic value of the fouling resistance expressed in [$m^2$-K/W] (this value characterizes the situation where the deposition rate equals the breakout speed). $t$, the time expressed in [$s$] $\tau$, the characteristic time expressed in [$s$] and whose value is generally attributed to the time required for the fouling resistance to reach its asymptotic value if the evolution of this kinetics was linear.

The Kern and Seaton [17] model therefore provides a mathematical description of the concept of simple fouling. This equation verifies the asymptotic behavior of the formation of a particulate deposit on the exchange surface of a heat exchanger. All models and theory of fouling are based on this model.

An apparent weakness of the Kern and Seaton [17] model is that the re-entrain flow depends on the thickness of the deposition layer. As a result, it is only once a significant deposit thickness has accumulated that the role of the re-entrain term becomes significant [20]. In the case of high speed flow, the deposit would be completely removed.

We also note that this model requires to go back on the values of $Rf^*$ and $\tau$. In general, there is no way to predict these values unless detailed experimental work has been done [21]. These values are thus established on a given installation and especially under given operating conditions. Thus, a modification, even minor, of the operating conditions (for example, water quality, flow modification) significantly modifies the parameters of the model and leads to a bad modeling of the fouling [22].

However, we note various works that make it possible to know the impact of the flow velocity ($Um$) on the value of the asymptotic fouling resistance $Rf^*$, or even to determine the value $Rf^*$.

Different authors thus propose a relationship of proportionality of type:

$$Rf^* \sim Um_i \tag{9}$$

With regard to the tube exchangers: for Kern and Seaton [17] the value of $i$ is −1. For Watkinson [23], this constant takes the value of −1.2 to −2.

As far as plate heat exchangers are concerned, Muller-Steinhagen [24] has in its study demonstrated a relation of proportionality between the asymptotic resistance of fouling $Rf^*$ and the inverse of the speed squared (i.e., an exponent $i - 2$ in Eq. (9)), without providing a general relationship.

In the same context, Grandgeorge [25] proposes an empirical relation resulting from several experiments on different industrial size plate heat exchangers. In this context, Grandgeorge [25] established that the use of the initial pressure drop in the heat exchanger ($\Delta Po$) in place of the flow velocity makes it possible to correlate with the aid of a single relationship the value of the asymptotic resistance $Rf^*$. The relationship is as follows:

$$Rf^* = \frac{1}{4 * \Delta Po} \tag{10}$$

Based on these observations, this model has been revised and modified by various researchers with various descriptions of the term deposition and re-entrain: Only empirical parameters were added and derived solely from the experimental study [20, 25].

## 6. Experimental procedure

The phosphoric acid concentration loop is allowed to concentrate—by evaporation—the phosphoric acid from 28 to 54% $P_2O_5$ in a forced-circulation evaporator closed loop, functioning under vacuum feeded by a barometric condenser. The system used for concentration composed of a stainless steel tubular heat exchanger, a centrifugal pump, a boiler or expansion chamber, a barometric condenser and a basket filter [26].

The inclusion of the dilute acid is done at the basket filter where it mixes with the circulating acid in order to protect the pump from abrasion and to limit the heat exchanger fouling, which reduces the stop frequency for washing. The circulation pump then aspirates the mixture formed and sends it to the inlet of the heat exchanger at a temperature in the order of 70°C. The heat exchanger allows heating the phosphoric acid at a temperature in the order of 80°C. The steam undergoes a condensation at a temperature of around 120°C at the level of the exchanger. The condensate will be sent to a storage tank before being returned to the utility center.

The overheated mixture of the acid outgoing the exchanger then passes into the boiler where an amount of water evaporates and the production of concentrated acid is done by overflow in inner tube of the boiler and the rest will be recycled. The condenser also ensures the re-entrain of incondensable outgoing of the boiler by the effect of water tube created by the waterfall. At the foot of the barometric guard, the seawater is gathered in a guard tank before being rejected to the sea (**Figure 3**).

Our experimental study is based on the following hypotheses.

1. The flow of two fluids (Phosphoric acid and steam) is at counter current.

2. Values of the thermo-physical properties of the fluids were considered constant.

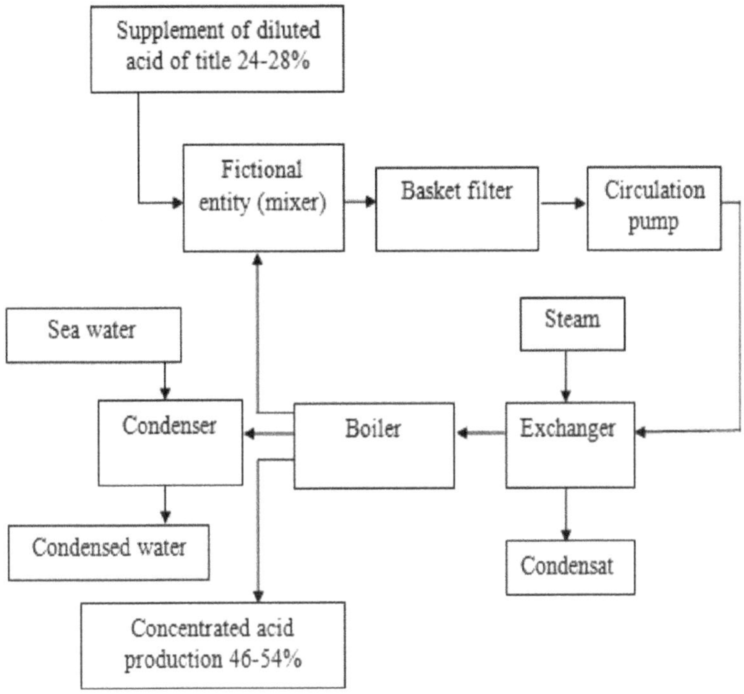


**Figure 3.**
*Simplified diagram of the phosphoric acid concentration unit.*

3. The thermal losses were neglected.

4. The inlet and outlet temperatures of the two fluids are determined at the extremities of the heat exchanger.

5. Pump suction and discharge pressure measurements are taken at the extremities of the circulation pump.

## 7. Calculation method

The experimental data was collected out during 1 year. The method that we used to follow the fouling evolution consists in carrying out a heat balance at the boundaries of the heat exchanger by the intermediary of measurements of the inlet and outlet temperatures pump suction and discharge pressure measurements and acid density measurement (**Figure 4**). The latter was taken each 2 h during all the day.

This method, albeit indirect, makes it possible to detect the necessary moment to shut down the installation for cleaning. In the current study, the temporal evolution of the fouling resistance of the phosphoric acid was studied.

The calculation of the fouling resistance was done using the following relation:

$$Rf(t) = \frac{1}{\mathrm{U_s}} - \frac{1}{\mathrm{U_p}} = \frac{1}{U(t)} - \frac{1}{U(t=0)} \tag{11}$$

### 7.1 Calculation of $U_s$

The overall heat transfer coefficient at the dirty state was given in the time course, via the expression:

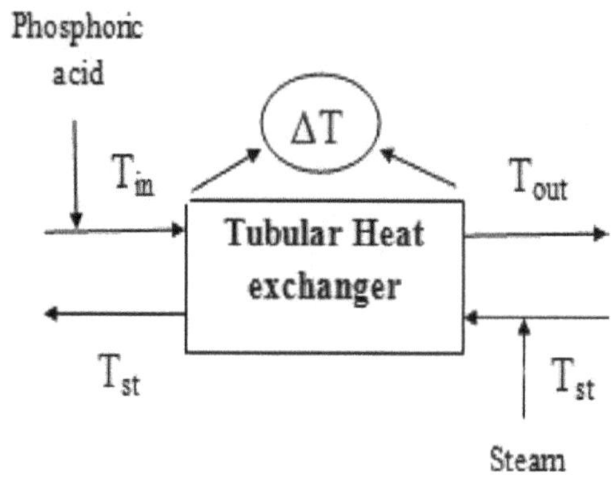


**Figure 4.**
*The measurement method at the boundaries of the heat exchanger.*

$$U_s = U(t) = \frac{\dot{m}_{ac,cir} * Cp_{ac} * (T_{out,ac} - T_{in,ac})}{A * F * \Delta T_{ml}} \quad (12)$$

This relation is taken by the evaluation of energy on the heat exchanger by supposing the isolated system and the physical properties of the two fluids, as well as, the heat transfer coefficients stay constant along the exchanger.

### 7.2 Calculation of $U_p$

In the phosphoric acid concentration unit, the operating conditions at the limits of the heat exchanger unstable, it is necessary to disclose the heat exchange coefficients in the proper conditions $U_p$ corresponding to the new operating conditions. Assuming that the cleaning between operational runs is perfect and that the heat exchangers are totally free of fouling at the beginning of a new run. The initial value of the overall heat transfer coefficient at the beginning of every cycle is considered as the value of the overall heat transfer coefficient in the clean state.

$$U_p = U(t = 0) = \frac{\dot{m}_{ac,cir} * Cp_{ac} * (T_{out,ac} - T_{in,ac})}{A * F * \Delta T_{ml}} \quad (13)$$

## 8. Results and discussion

### 8.1 Temporal evolution of the fouling resistance

The evolution of the fouling resistance in the phosphoric acid concentration process heat exchanger was followed for a study period quoted previously. This heat exchanger is already in service for 100 days before the be-ginning of the present study. However, it has carried out a stop that lasted 12 hours then its return to service. This stop is for the heat exchangers cleaning.

All the results from the fouling resistance are presented on **Figure 5**.

According to the values of these resistances, which are the majority higher than zero, this experimental data presents a fouling state. This is evident since, as mentioned before, this exchanger is in service for more than 3 months. The curves presented show that the temporal evolution of the fouling resistance, seems to follow an asymptotic evolution, which conforms to the model of Kern and Seaton [17], with the absence of the induction period. That can be explained to the rapid evolution of this phenomenon associated in particular with the characteristics of the

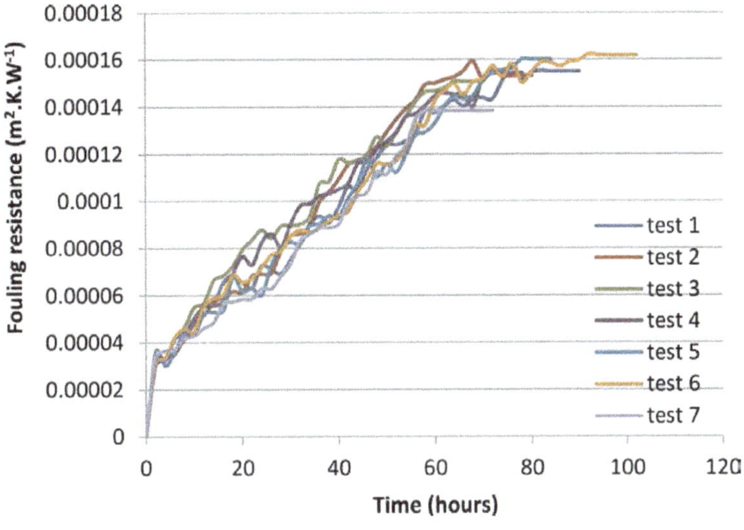


**Figure 5.**
*Variation of the fouling resistance as a function of time for the stainless-steel-tubular heat exchanger.*

treated phosphoric acid. As it appears clearly as the fouling resistance increases with the time until reaching a maximum value varied from $1.38 * 10^{-4}$ to $1.61 * 10^{-4}$ $m^2.K.W^{-1}$.

The series functioned for more than 5 days, a sufficient period so that the resistance asymptotic value is reached. The fluctuation observed on these curves are due to the variation of flow, which, acting on the shear stress to the wall, causes re-entrain deposit particles or their deposition according to the sent flow value.

## 8.2 Temporal evolution of the overall heat transfer coefficient

From Eq. (11), we notice that the overall heat transfer coefficient is inversely proportional to the fouling resistance.

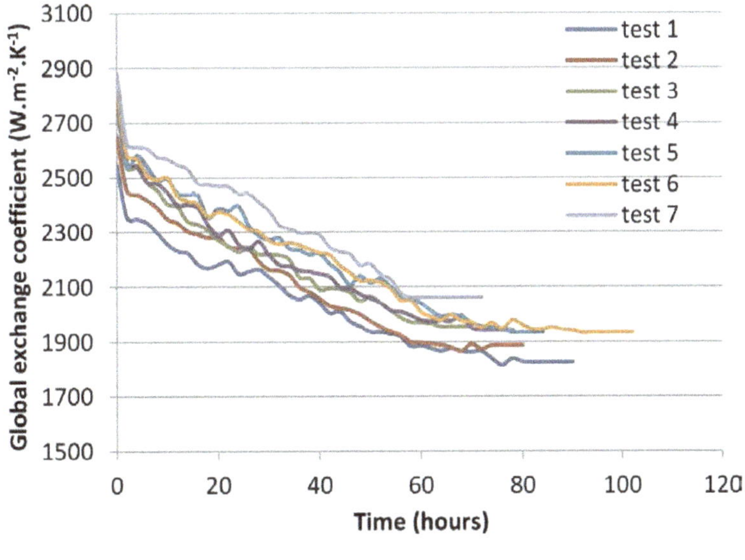


**Figure 6.**
*Variation of the overall heat transfer coefficient as a function of time for the stainless-steel-tubular heat exchanger.*

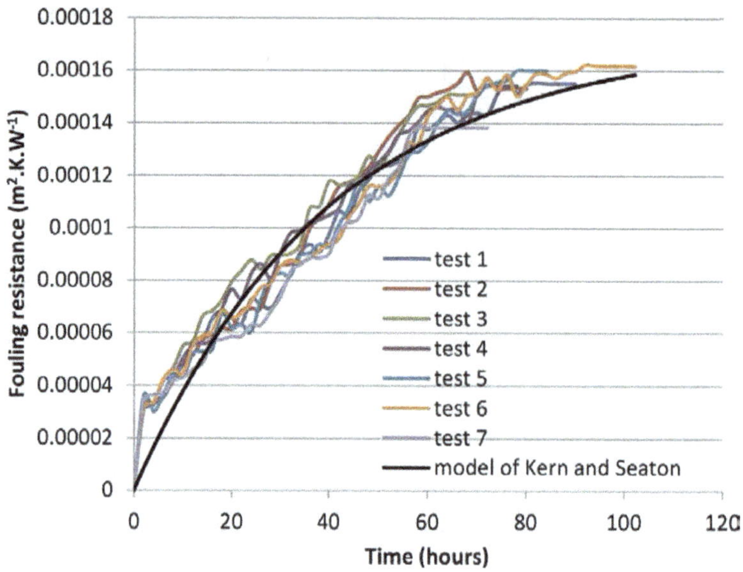


**Figure 7.**
*Kinetics of fouling of the stainless-steel-tubular heat exchanger.*

The fouling resistance increases over time, which leads to a decrease in the flow of heat exchanged between the phosphoric acid and the steam, and subse-quently the decrease in the overall heat transfer coefficient. As it appears clearly on **Figures 5** and **6**, when the fouling resistance increases with the time, the overall heat transfer coefficient decreases until reaching a minimum value varied from 1821 to 2078 $W.m^{-2}.K^{-1}$.

## 8.3 Temporal evolution of the fouling resistance obtained from both measurement and the Kern and Seaton model

One of the earliest correlative models for the characterization of the asymptotic kinetics of fouling, we distinguish Kern and Seaton [17], whose general expression is as follows:

$$Rf(t) = Rf^* * \left[1 - exp\left(-\frac{t}{\tau}\right)\right] \quad (14)$$

This model gives rather satisfactory results, provided that the asymptotic value of the thermal resistance $Rf^*$ as well as the time constant $\tau$ are evaluated, which strongly conditions the accuracy of the model.

The analysis of the experimental data which makes it possible to carry out the plots of **Figure 7** gives us the results of the two greatness $Rf^*$ and $\tau$ for the stainless steel tubular heat exchanger. The asymptotic model is fairly faithful to the experimental data with determination coefficient $R^2$ close to 1 (**Table 1**).

| $Rf^*$ [$m^2.K.W^{-1}$] | $\tau$ [h] | $R^2$ |
|---|---|---|
| $1.72*10^{-4}$ | 40.32 | 0.975 |

**Table 1.**
*Values of the asymptotic fouling resistance, the time constant and the determination coefficient for the stainless-steel tubular heat exchanger.*

## 9. Conclusion

The monitoring of heat exchangers provides the sound knowledge of the fouling evolution in the specific conditions of the process. Deposit formation is a thermal resistance which leads important economic penalties.

The aim of this work was the study of the heat exchanger fouling phenomenon in the concentration phosphoric acid process. Secondly, the study concerned the temporal evolution of the fouling resistance and the overall heat transfer coefficient.

The results achieved revealed an asymptotic evolution of the fouling resistance, compliant with the model of Kern and Seaton with the lack of the induction period, which is explained by the consequence of an improper cleaning, or a deviation between the present study and the beginning of the heat exchangers functioning after the last stop.

## Nomenclature

| | |
|---|---|
| A | area, $m^2$ |
| Cp | specific heat capacity, $J.Kg^{-1}.K^{-1}$ |
| F | correction factor (=1 for a steam condenser) |
| $\dot{m}$ | mass flow rate, $kg.s^{-1}$ |
| P | pressure, bar |
| Q | thermal power, W |
| Rf | fouling resistance, $m^2.K.W^{-1}$ |
| T | temperature, K |
| t | time, h |
| U | overall heat transfer coefficient, $W.m^{-2}.K^{-1}$ |
| Greek letters | |
| $\Delta$ | difference of greatness between two points |
| $\tau$ | time required to reach 63.2% of $Rf^*$, h |
| Subscripts | |
| ac | acid |
| cir | circulation |
| ex | exchange |
| in | input |
| ml | logarithmic mean |
| 0 | clean |
| out | output |
| p | clean state |
| s | dirty state |
| st | steam |
| * | asymptotic value |

## Author details

Rania Jradi[1*], Ali Fguiri[1], Christophe Marvillet[2] and Mohamed Razak Jeday[1]

1 Research Laboratory "Process, Energy, Environment and Electrical Systems", National Engineering School of Gabes (ENIG), Gabes, Tunisia

2 CMGPCE Laboratory, French Institute of Refrigeration (IFFI), National Conservatory of Arts and Crafts (CNAM), Paris, France

*Address all correspondence to: raniajradi@yahoo.fr

## References


[1] Demasles H, Mercier P, Tochon P, Thonon B. Guide de L'encrassement des Echangeurs de Chaleur. France: Editions-GRETh; 2007

[2] Farajollahi B, Gh Etemad S, Hojjat M. Heat transfer of nanofluids in a shell and tube heat exchanger. International Journal of Heat and Mass Transfer. 2010;**53**:12-17

[3] Lalot S, Pálsson H. Detection of fouling in a cross-flow heat exchanger using a neural network based technique. International Journal of Thermal Sciences. 2010;**49**: 675-679

[4] Pogiatzis T, Ishiyama EM, Paterson WR, Vassiliadis VS, Wilson DI. Identifying optimal cleaning cycles for heat exchangers subject to fouling and ageing. Applied Energy. 2012;**89**: 60-66

[5] Mohanty DK, Singru PM. Use of C-factor for monitoring of fouling in a shell and tube heat exchanger. Energy. 2011;**36**:2899-2904

[6] Yeap BL, Wilson DI, Polley GT, Pugh SJ. Mitigation of crude oil refinery heat exchanger fouling through retrofits based on thermo-hydraulic fouling models. Chemical Engineering Research and Design. 2004;**82**:53-71

[7] Radhakrishnan VR, Ramasamy M, Zabiri H, Thanh V Do, Tahir NM, Mukhtar H, et al. Heat exchanger fouling model and preventive maintenance scheduling tool. Applied Thermal Engineering. 2007;**27**: 2791-2802

[8] Behbahani RM, Muller-Steinhagen H, Jamialahmadi M. Investigation of scale formation in heat exchangers of phosphoric acid evaporator plants. The Canadian Journal of Chemical Engineering. 2008;**84**:189-197

[9] Jamialahmadi M, Müller-Steinhagen H. Heat exchanger fouling and cleaning in the dihydrate process for the production of phosphoric acid. Chemical Engineering Research and Design. 2007;**85**:245-255

[10] Amjad Z. Calcium sulfate dihydrate (gypsum) scale formation on heat exchanger surfaces: The influence of scale inhibitors. Journal of Colloid and Interfacial Science. 1988;**123**: 523-536

[11] Geddert T, Bialuch I, Augustin W, Scholl S. Extending the induction period of crystallization fouling through surface coating. Heat Transfer Engineering. 2009;**30**:868-875

[12] Shih WY, Albrecht K, Glater J, Cohen Y. A dual-probe approach for evaluation of gypsum crystallization in response to antiscalant treatment. Desalination. 2004;**169**:213-221

[13] Behbahani RM, Muller-Steinhagen H, Jamialahmadi M. Heat exchanger fouling in phosphoric acid evaporators-evaluation of field data-. In: Proceedings EIC on Heat Exchanger Fouling and Cleaning: Fundamentals and Applications. Mexique: Santa Fé; 2003

[14] Behbahani RM. Heat Transfer and Heat Transfer Fouling in Phosphoric Acid Evaporators. European Union: Lap Lambert Academic Publishing; 2010

[15] Taler D. Experimental determination of correlations for average heat transfer coefficients in heat exchangers on both fluid sides. Heat and Mass Transfer. 2013;**49**: 1125-1139

[16] Ebert WA, Panchal CB. Analysis of Exxon crude-oil slip stream coking data. In: Fouling Mitigation of Industrial Heat-Exchange Equipment. New York: Begell House; 1996:451-460

[17] Kern D , Seaton R. A theoretical analysis of thermal surface fouling. British Chemical Engineering. 1959;**4**: 258-262

[18] Riverol C, Napolitano V. Estimation of the overall heat transfer coefficient in a tubular heat exchanger under fouling using neural networks. Application in a flash pasteurizer. International Communications in Heat and Mass Transfer. 2002;**29**:453-457

[19] Standards of Tubular Exchanger Manufacturers Association. Tubular Exchanger Manufacturers Association edition. 9th ed. New York, Tarrytown; 2007

[20] Wang Y, Yuan Z, Liang Y, Xie Y, Chen X, Li X. A review of experimental measurement and prediction models of crude oil fouling rate in crude refinery preheat trains. Asia-Pacific Journal of Chemical Engineering. 2015;**10**:607-625

[21] Bott TR. Fouling of Heat Exchangers. Science B.V edition. Amsterdam: Elsevier; 1995

[22] Weber C, Tremeac B, Marvillet C, Castelain C. Analysis of different fouling predictive models in a heat exchanger from experimental data. In: Proceedings of ECOS 2016—The 29th International Conference on Efficiency, Cost, Optimization, Simulation and Environmental Impact of Energy Systems. Slovenia: Portoroz; 2016

[23] Watkinson AP. Particulate fouling of sensible heat exchangers. Vancouver: The University of British Columbia; 1968

[24] Muller-Steinhagen H. Particulate fouling for low finned tubes and for plate heat exchangers. In: International Conference on Fouling in Process Plant. England: St Catherine College of Oxford; 1988

[25] Grandgeorge S. Contribution à l'étude de l'encrassement particulaire des échangeurs à plaques corruguées en milieu liquide. L'Institut National Polytechnique de Grenoble; 1996

[26] Becker P. Phosphates and Phosphoric Acid: Raw Materials, Technology and Economics of the Wet Process. 2nd ed. New York: CRC Press; 1989

# 7

# Molecular Dynamics Simulation-Based Study on Enhancing Thermal Properties of Graphene-Reinforced Thermoplastic Polyurethane Nanocomposite for Heat Exchanger Materials

*Animesh Talapatra and Debasis Datta*

**Abstract**

Molecular dynamics (MD) simulation-based development of heat resistance nanocomposite materials for nanoheat transfer devices (like nanoheat exchanger) and applications have been studied. In this study, MD software (Materials Studio) has been used to know the heat transport behaviors of the graphene-reinforced thermo-plastic polyurethane (Gr/TPU) nanocomposite. The effect of graphene weight per-centage (wt%) on thermal properties (e.g., glass transition temperature, coefficient of thermal expansion, heat capacity, thermal conductivity, and interface thermal con-ductance) of Gr/TPU nanocomposites has been studied. Condensed-phase optimized molecular potentials for atomistic simulation studies (COMPASS) force field which is incorporated in both amorphous and forcite plus atomistic simulation modules within the software are used for this present study. Layer models have been developed to characterize thermal properties of the Gr/TPU nanocomposites. It is seen from the simulation results that glass transition temperature (Tg) of the Gr/TPU nanocomposites is higher than that of pure TPU. MD simulation results indicate that addition of graphene into TPU matrix enhances thermal conductivity. The present study provides effective guidance and understanding of the thermal mechanism of graphene/TPU nanocomposites for improving their thermal properties. Finally, the revealed enhanced thermal properties of nanocomposites, the interfacial interaction energy, and the free volume of polymer nanocomposites are examined and discussed.

**Keywords:** molecular dynamics (MD) simulations, thermo-mechanical properties, glass transition temperature, coefficient of thermal expansion, thermal conductivity

## 1. Introduction

Optimum design of heat exchanger using nanotechnology is a burning field to reduce the energy consumption. Recently, application of nanosolids, nanofluids, and nanogases is the promising nanolevel research area of interest for energy savings in heat exchanger. Investigation of the nanolevel heat transfer using molecular dynamics (MD) simulation is only a new pioneer concept for the last few years [1–5]. Here investigations are based on atomic movement within a nanosystem during MD simulation. Experimental study on the heat transfer of nano-size devices or equipments are very time-consuming and expensive for the existing testing capabilities [6–9]. Many studies have been done for enhancing the energy consumption of heat exchanger by improving thermal properties. This thermal behavior of nanocomposite can bring a huge transformation and innovation in the heat transfer. The ultrathin thermal polyurethane heat transfer material can be applied to a number of different fabrics in heat exchanger. The use of graphene (Gr) in thermoplastic polyurethane (TPU), that is, Gr/TPU nanocomposite in place of traditional material in heat exchanger, increases the heat transfer rate in a significant manner. Hussein studied to calculate thermal properties of metals and nonmetals at room temperature for applications in heat exchanger and found that metallic materials are preferably suitable for heat transfer application [10]. But there is some limitation for application of metals due to corrosion. To overcome the situation, implementation of polymer heat exchanger technology for the past decades is a pioneering innovation for heat exchanger materials [11]. The major limitation of polymer for application in heat exchanger is very low thermal conductivity. To improve that property graphene-reinforced polymer nanocomposite is a suitable candidate material in evaporation and condensation applications within heat exchanger [12]. Thermal performance of polymer nanocomposite heat exchangers mainly deals with on shell and tube heat exchangers, plate heat exchangers, finned tube heat exchangers, immersed tube heat exchangers, and hollow fiber heat exchangers [13]. Currently, thermoplastic elastomer is used in heat exchanger applications. Thermoplastics elastomer can be repeatedly softened by heating and solidified by cooling as long as the material is not thermally damaged by overheating [14]. The thermal expansion of thermoplastic polymer can be beneficial with regard to fouling because repeated expansion and contraction of the polymer channels can lead to scale detachment [15]. It is seen from previous studies that there are less number of simulation-based studies like MD simulation on enhanced heat transfer in heat exchanger materials. Detail results of MD simulation are obtained by solving Newton's equation of motion of every atom within nanoscopic system. The basic dynamics parameters of all atoms, that is, position, velocity, and interaction force, play a vital rule during MD simulation. Nanoheat transfer problems of nanocomposites are related to thermo-mechanical properties of nanomaterials. For the design of nanodevices like nanoheat exchanger (NHE), the concepts of the nanothermal properties with temperature variation and dimension of the nanodevice is very much promising ideas. Determination of the thermal properties of nanocomposites by MD simulations is very time-consuming and a challenging task. The objective of the chapter is to characterize thermal properties like thermal conductivity and thermal expansion coefficient of graphene-reinforced polyurethane nanocomposite using molecular dynamics (MD) modeling for nanoheat exchanger material application.

MD simulation consists of many parts which are mainly (a) molecular interactions, (b) molecular minimization, (c) algorithms, (d) ensemble, (e) boundary conditions, (f) atomistic stress calculation, etc. MD simulation helps to determine the position ($r_i$) and velocity ($v_i$) vectors of atom *i* with the time integration method (e.g., the velocity Verlet algorithm):

$$r_i(t+\Delta t) = r_i(t) + v_i(t)\Delta t + \frac{1}{2}a_i(t)\Delta t^2 \tag{1}$$

$$v_i(t+\Delta t) = v_i(t) + \frac{a_i(t) + a_i(t + t\Delta t)}{2}\Delta t \quad (2)$$

where ($\boldsymbol{a}_i$) is the acceleration of atom $i$, $t$ is the current time, and $\Delta t$ is the time step.

$$a_i = \frac{F_i}{m_i} \quad (3)$$

where $m_i$ is mass of atom $i$th and $\boldsymbol{F}_i$ is the force vector of atom $i$th obtain from the gradient of the total potential energy ($E$) on atom $i$

$$F_i = \frac{dE}{dr_i} \quad (4)$$

In MD simulation, a model system is built at the atomistic level with prescribed potentials (also known as the force field) acting between the atoms. The potential energy is dependent on the force field that is applied to the system. The potential energies of the system are determined from both bonded and nonbonded energies. The total potential energy combines all energetic contributions shown in the following equation:

$$\mathrm{E} = \mathrm{E_{bonded}} + \mathrm{E_{non\text{-}bonded}} \quad (5)$$

$$\mathrm{E} = (\mathrm{E_{valance}} + \mathrm{E_{cross\text{-}term}}) + \mathrm{E_{non\text{-}bonded}} \quad (6)$$

$$\mathrm{E} = (\mathrm{E_{bond}} + \mathrm{E_{angle}} + \mathrm{E_{torsion}} + \mathrm{E_{inversion}}) + \mathrm{E_{cross\text{-}term}} + (\mathrm{E_{vdW}} + \mathrm{E_{coulomb}}) \quad (7)$$

The different types of bond and nonbond energy equations have been shown below:

$$E_{bond} = \frac{1}{2}k_b(b - b_0)^2 \quad (8)$$

where $k_b$ is the stretching force constant, $b_0$ is the equilibrium bond length, and b is the actual bond length.

$$E_{bend} = \frac{1}{2}k_\theta(\theta - \theta_0)^2 \quad (9)$$

where $k_\theta$ is the angle-bending force constant, $\theta_0$ is the equilibrium bond angle, and $\theta$ is the actual bond angle.

$$E_{torsion} = \frac{1}{2}k_\phi(1 + \cos(n\phi - \phi_0)) \quad (10)$$

where $k\varphi$ is the torsional barrier, $\varphi$ is the actual torsion angle, n is the periodicity, and $\varphi_0$ is the reference torsional angle.

$$E_{inversion} = \frac{1}{2}k_\omega(\omega - \omega_0)^2 \quad (11)$$

where $K_\omega$ is the force constant and $\omega$ is the angle between the axis and the plane.

$$E_{vdw} = \sum \frac{A_{ij}}{r_{ij}^{12}} - \frac{B_{ij}}{r_{ij}^{6}} \quad (12)$$

where $A_{ij}$ and $B_{ij}$ are the repulsive and attractive term coefficients, respectively, and $r_{ij}$ is the distance between the two atoms.

$$E_{coulomb} = \frac{1}{\varepsilon} \frac{q_1 q_2}{r_{ij}} \tag{13}$$

where $q_1$ and $q_2$ are the charges on the interacting atoms, ε is the dielectric constant, and $r_{ij}$ is the interatomic distance.

The Condensed-phase Optimized Molecular Potentials for Atomistic Simulation Studies (COMPASS) [16] which is incorporated in both amorphous and forcite plus atomistic simulation modules in the Material Studio is used for this present study. COMPASS functional form has 11 valence terms (including diagonal and off-diagonal cross coupling terms) and 2 nonbond interaction terms (the Coulombic and Lennard-Jones functions for electrostatic and van der Waals (vdW) interactions, respectively). During all the simulations, the temperature and pressure are maintained by Andersen and Berendsen method. The calculation of nonbonded interactions is simulated by applying a cutoff distance of 12.5 Å. The spline and buffer widths are 1 and 0.5 Å, respectively.

Experimental methods for prediction of the enhanced thermal properties of graphene-reinforced nanocomposites are limited because nanometer scale measurements are difficult and costly. Thus, MD simulation techniques are only an economical path to characterize nanomaterial and nanocomposites for heat exchanger material within small length and small time.

## 2. MD simulation models and methods for thermal property calculation

Heat can transfer through electrons and phonons, by excitations and by scattering in the nanocomposites [17]. This different ways heat transfer methods help to understand the mechanism of thermal properties enhancement in graphene-based nanocomposites. In order to study the enhanced thermal properties of graphene-reinforced thermoplastic polyurethane nanocomposites, the wide range thermal parameters of the nanocomposite material (like thermal conductivity, coefficient of the thermal expansion, glass transition temperature, etc.) have to be calculated. To calculate these parameters, different simulation models are constructed using molecular modeling software package by Materials Studio 2017. In MD simulation study, there are three types of models which are developed, namely concentrated model (CM), layer model (LM), and interfacial model (IM). Different models have been used to study different properties of the nanocomposite with respect to different parameters. In this study, we have focused mainly on layer models to characterize enhanced thermal properties. Both graphene and polyurethane models are simulated separately before constructing the graphene-reinforced thermoplastic polyurethane (Gr/TPU) nanocomposites. The condensed-phase optimized molecular potential for atomistic simulation studies (COMPASS) force field has been selected to describe the atomistic behavior for the simulation models. In MD simulation, each atom is modeled as a point mass and interacts with other atoms through force field. The position and momentum of atoms are updated based on Newton's equation of motion.

After the construction all of the graphene and TPU model using amorphous cell module within the Material Studio, build layer option is used to construct the layer models of the Gr/TPU nanocomposites. It is seen that with high weight fraction condition of graphene, nanocomposites behave likely more brittle than polymer matrix due to growth of void and chain disentanglement. So, 1% weight fraction of

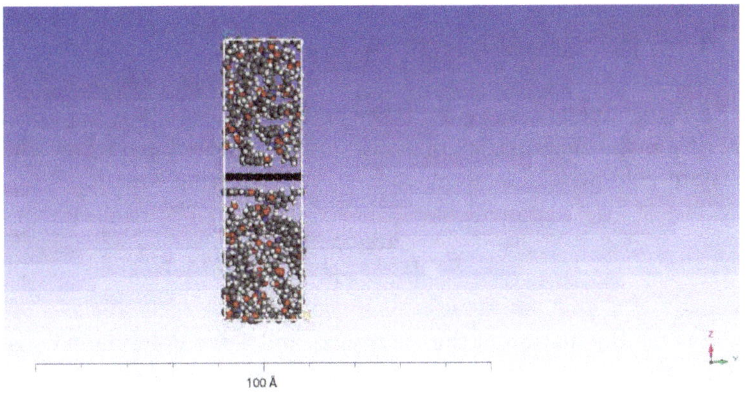


**Figure 1.**
*Developed 1% Gr/TPU nanocomposite model.*

graphene-reinforced nanocomposites is considered in this study. The constructed nanocomposite lattice parameters with dimensions a = 22 Å, b = 22 Å, and c = 95 Å are constrained in such a way that after dynamic equilibrium process density of Gr/TPU, nanocomposite is 1.38 g/cc (nearly experimental value). MD simulation run is divided into two parts, namely equilibration run and production run. Equilibration run helps to develop the molecular structure under the condition of the desired thermodynamic state, while the production run helps to calculate different thermal parameters, namely specific heat, thermal expansion, and thermal conductivity. In equilibration run, two important conditions have to be fulfilled. One is the minimum energy stabilized condition at a prescribed temperature, and another is initial stress-free structure within periodic boundary condition. So, first steps nanocomposite models are moved through energy minimization, canonical ensemble (constant number of atoms, volume, and temperature) (NVT)) dynamic simulations, and temperature annealing cycle, respectively. The duration of the dynamic run is considered 200 ps with an integration time step of 1 fs (femto-second). This process is followed by graphene as a rigid structure so that the lattice dimension (c) in the z-direction will be changed. Temperature annealing cycle involves temperature up and down from 300 to 600 K to get the minimization of energy in the structure. The annealing time is set for 500 ps, during which the temperature is raised from 300 to 600 K with a rate of 6 K/ps and cool down to 300 K with the same rate. In the second step, the non-constrained parts (TPU) within lattices are compressed in such a way so that the final density of nanocomposite will be 1.38 g/cc (nearly experimental value) after using isothermal-isobaric (NPT) ensembles. The lowest energy structure models are fully relaxed under an isothermal-isobaric NPT ensemble (i.e., constant numbers of atoms, pres-sure, and temperature) at 300 K and 1 atm for 500 ps. The isothermal-isobaric (NPT) ensembles help to relax the lattices parameters and angles in order to obtain a final reasonable equilibrated structure. These steps generate various curves of various parameters such as energy, pressure, volume, and temperature versus simulation run time. These curves are very important to study thermal properties of the nanocomposite. **Figure 1** shows the developed 1% Gr/TPU nanocomposite model by MD simulation.

After dynamic equilibrium process density of 1% Gr/TPU, the nanocomposite is 1.38 g/cc which is shown in **Figure 2**.

Heat capacity is one of the important thermal properties for the nanocomposite system. In this work, MD simulation is applied to calculate the isobaric heat capacity ($C_p$), and the value of Cp can be determined according to the following equation:

$$Cp = \frac{(KE + PE + PV)^2}{K_B\,T} \tag{14}$$

where KE is the kinetic energy, PE is the potential energy, P is the pressure, V is the volume, $K_B$ is the Boltzmann constant, and T is the temperature. The specific heat at constant volume ($C_v$) is obtained from the following equation:

$$Cv = \frac{(\delta E^2)_{NVT}}{k_B T^2} \tag{15}$$

where δE is the fluctuation of the energy, $k_B$ and T are Boltzmann constant, and absolute temperature, respectively.

In order to study the glass transition temperature and coefficient of thermal expansion (CTE), a high-temperature annealing protocol is followed. At each temperature, the system is equilibrated by isothermal-isobaric (NPT) ensemble in MD simulation at atmospheric pressure for 500 ps. The temperature is raised up to 600 K and equilibrated for 500 ps using an NPT ensemble under atmospheric pressure and then dropped by 20 K each time until it reached 300 K. The cooling down method is applied after the heating up method by decreasing the temperature with the same settings and simulation time. Since each temperature drop is only 20 K, the structure is re-equilibrated very quickly every time its temperature is decreased. For each temperature, the volume of the simulation box V is examined over the duration time of the MD simulation, and the average value is calculated. From the volume versus temperature relationship curve, it is seen that there is a discontinuity in the volume versus temperature slope, which gives the glass transi-tion temperature (Tg) of nanocomposite. The volume versus temperature (V-T) results are important to know two factors; first, this provides a means of determin-ing the quality of the force field used in the simulations, and second, prediction of the volumetric glass transition temperature (Tg) [18]. The simulation result is in good agreement with experiment and demonstrates the accuracy of COMPASS force field. The volumetric coefficient of thermal expansion (VCTE) is defined by (α) [19]:

$$\alpha = \frac{1}{V_0}\frac{\Delta V}{\Delta T} \tag{16}$$

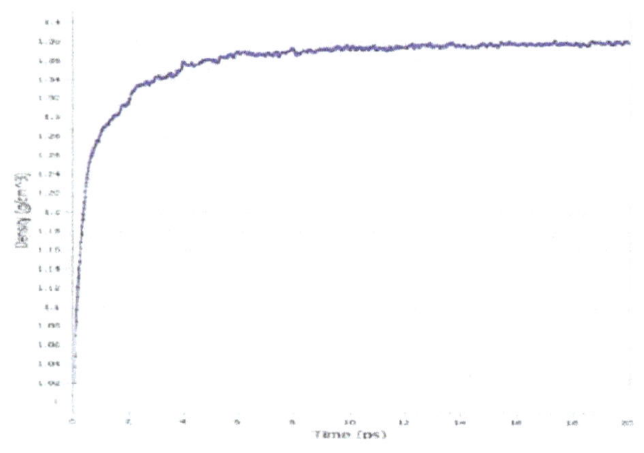

**Figure 2.**
*Density (g/cc) versus time (ps) in dynamic equilibrium run.*

where $V_0$ is the equilibrated system volume before the cooling simulation starts. The fractional change of volume with respect to temperature ($\Delta V/\Delta T$) is obtainable from volume versus temperature relationship curve. It is seen that change of volume with respect to temperature ($\Delta V/\Delta T$) is a different value above glass transition temperature (Tg) for graphene-reinforced thermoplastic polyurethane nanocomposite. So, volumetric coefficient of thermal expansion (VCTE) has two values due to glass transition temperature (Tg) [20]. The glass transition temperature and volumetric coefficient of thermal expansion (VCTE) of nanocomposite can also be obtained by calculating the free volume as a function of temperature, since the free volume undergoes an abrupt change when the material goes through glass transition. By probing the lattice cell with a spherical probe, using the "atom volume and surfaces" tool of the Materials Studio (MS), the free volume in the nanocomposite is calculated as a function of temperature during the annealing process as shown in **Figure 3**.

Free volume is the volume that is not occupied by either the graphene or the TPU chains. The free volume fraction (FVF) can be obtained by the following equation [21]:

$$FVF = \frac{V_f}{V_f + V_o} \tag{17}$$

where $V_f$ is the free volume and $V_0$ is the occupied volume of the polymer chains.

Thermal conductivity is the sum of the phonon contribution and the electronic contribution. Therefore, total thermal conductivity (K)

$$K = K_e + K_p \tag{18}$$

where $K_p$ and $K_e$ are the phonon contribution and the electronic contribution to the thermal conductivity, respectively. Though, electrons contributed thermal conductivity is neglected in most graphene-reinforced nanocomposite. Thermal conductivity is an important thermal property relevant to thermal management applications. Thermal conductivity is generally calculated using equilibrium or nonequilibrium MD approaches. Equilibrium molecular dynamic (EMD) facilitates thermal conductivity prediction in all directions using one simulation, whereas nonequilibrium molecular dynamic (NEMD) requires the use of a thermal gradient and therefore only enables the calculation of thermal conductivity in one

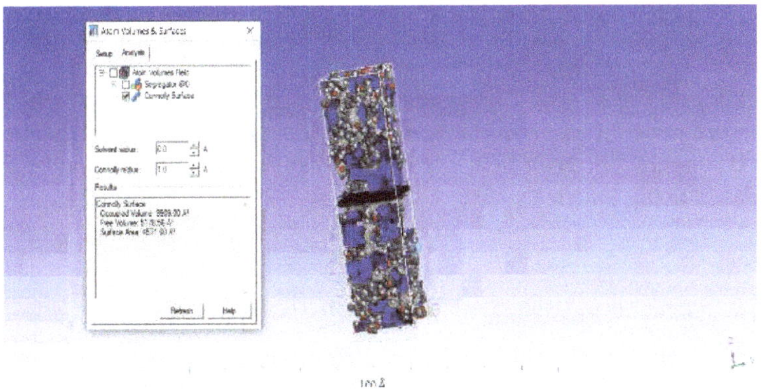

**Figure 3.**
*Free volume calculation using atom volume and surfaces tool in MS.*

direction. The indirect method is an equilibrium molecular dynamic (EMD) method which is derived from Green-Kubo approach [22, 23], where current fluctuations are used to compute the thermal conductivity via the fluctuation-dissipation theorem [24]:

$$K_{\alpha\beta} = \frac{V}{k_B T^2} \int_0^\infty \langle J_\alpha(t) J_\beta(0) \rangle dt \tag{19}$$

where $k_B$ is the Boltzmann constant, V and T denote the volume and temperature of the system, $J_\alpha$ is the heat flux in the $\alpha$ direction, and the angular brackets denote the ensemble average. The heat flux vector can be written as

$$J(t) = \frac{1}{V} \frac{d}{dt} \sum_{i=1}^{N} r_i E_i \tag{20}$$

where $r_i$ and $E_i$ are the position and total energy of the ith atom, respectively. EMD is particularly useful for geometries where periodic boundary conditions can be applied. EMD is often computationally more expensive, and the results are more sensitive to the simulation parameters. In EMD, the system is set to the desired temperature, and then a constant energy scheme is used with the well-known Green-Kubo relations to calculate the thermal conductivity tensor. The direct method is a NEMD method in which a temperature difference is introduced into the simulation domain and the thermal conductivity is computed according to Fourier's l a w a s $K = -J/\Delta T$, where J and $\Delta T$ are heat flux and temperature gradient across the system, respectively. For nonequilibrium MD (NEMD) methods, a long slab of polymer nanocomposite is constructed, and a difference in temperature is established between a heat source and a sink at the ends of the slab, and the flux is calculated. Equilibrium systems are simulated by nonequilibrium MD (NEMD) based on our homemade PERL script to compute their thermal conductivities. The nonequilibrium state can be established either by applying two thermostats at different temperatures to maintain a constant temperature at the two ends of the system or by artificially swapping atom velocities in different regions to impose a constant heat flux also known as the reverse nonequilibrium MD (RNEMD) method based on Muller-Plathe's approach [25]. In the reverse nonequilibrium MD (RNEMD) method, the energy exchange is carried out by exchanging the kinetic energy of two particles: the hottest particle in the cold layer and the coldest particle in the hot layer. The energy E is therefore variable and needs averaging over many exchanges. In the RNEMD method, the simula-tion box was divided into a number of slabs in the heat flux (z) direction with the same thickness. The heat flux was generated by exchanging the kinetic energy between the highest kinetic (the hottest) atom in the heat sink and the lowest kinetic energy (the coldest) atom in the heat source. The larger momentum exchange rate in RNEMD method suggests higher energy exchange frequency between heat source and heat sink. The thermal conductivity was calculated using Fourier's law:

$$K = -\frac{J_q}{\nabla T} \tag{21}$$

where $\nabla T$ denoted the time-integrated temperature gradient from least squares approximation of the discrete temperatures according to the heat flow direction. The temperature of each slab was calculated using the virial theorem, and $J_q$ is the heat flux given as

$$J_q = \frac{E}{2A_c \Delta t} \tag{22}$$

where E is the subtracted energy from the heat sink. $A_c$ is cross-sectional area and $\Delta t$ are the time step.

Interface thermal conductance (ITC) is developed between graphene and polymer matrix within nanocomposite due to their weak interactions. However, the thermal conductivity (TC) of graphene-reinforced TPU nanocomposites are far below the thermal conductivity (TC) of graphene. Such a low filler efficiency is most likely attributed to the low interface thermal conductance (ITC) between graphene and polymer. The simulating method (pump-probe transient thermo-reflectance method) is used to calculate the interface thermal conductance (ITC) between the filler (graphene) and matrix (TPU). The graphene-TPU interfacial thermal resistance (R) is then calculated by following equation:

$$R = \frac{\Delta T}{J_q} \tag{23}$$

## 3. Results and discussion

In the equilibration stage within MD simulation, NPT dynamic run is carried out for 500 ps at room temperature and pressure to generate curves of energy, density, pressure, and temperature versus time. These curves are used to decide the cutoff between equilibration and production runs. It is observed that all the models are equilibrated at 50 ps, that is, no fluctuations after 50 ps. At the end of the equilibrations, the density of the nanocomposites stabilized at an average density of 1.3 g/cc with a standard deviation of 0.02 g/cc. The reason behind for the density differs from experimental value because MD simulation deals with material defect-free and impurities. It is seen in the volume versus temperature (V-T) curve during NPT dynamic run that free volume change affects the thermal properties of the nanocomposites. Further it is also observed that free volume within lattice increases according to the strain application. The glass transition mainly depends on two factors: (a) free volume and (b) the mobility of chain segments. Initially there is no graphene in TPU chains for that free volume is zero in the simulated cell and polymers are free to move within this cell volume. As a result, high values for the entropy and low values for the bulk, shear, and Young's moduli. Further, the incorporation of the graphene into the simulated cells increases free volume and decrease the entropy and increase the values of the bulk, shear, and Young's moduli. Since the entropy of the system is related to the free volume and the Connolly surface of the TPU nanocomposites, the prediction of these parameters may give a concept on the enhancement of the nanocomposite thermal properties. The simulation results show that the TPU matrix reinforced with graphene have a tendency to increase glass transition temperature (Tg) for stronger interlocking between the graphene and TPU molecules. The Connolly surface to volume ratio is small values for the neat TPU and increase with the increase in the graphene loading. The free volume is defined as the volume on the side of the Connolly surface without atoms. This simulation study reveals that Tg of TPU is in the range of 285–305°C and for the Gr/TPU nanocomposite is 350 K (experimental 223–323 K). From the V-T curve, volumetric coefficient of thermal expansion (α) is evaluated from the slope above and below Tg, which is between $2.6 \times 10^{-5}$ and $2.4 \times 10^{-4}/K^{-1}$ (experimental $3.15 \times 10^{-}/K^{-4}$). The free volume change rapidly when the material goes through glass transition, according to Fox and Flory's theory of glass transition [20].

Further simulation study reveals that graphene/TPU nanocomposite thermal conductivity is 1.5 W/mK, whereas TPU thermal conductivity is 0.2 W/mK. NEMD simulations are used to calculate the thermal conductivity either by imposing a thermal gradient into the system of particles or by introducing a heat flux flow in the reverse nonequilibrium MD (RNEMD) method. In the present study, heat flux flow method is used to calculate thermal conductivity in the longitudinal direction. The total number of layers is 40. Two types of exchange method are used in the present study, namely VARIABLE and FIXED. About 1 kcal/mol energy exchange is taken in FIXED method. The number of exchanges is taken as 500 for equilibrium stage under NVT and 1000 for production stage under NVE. The time steps in between two exchanges are fixed at 100. Due to the presence of the graphene-TPU interface, there exists a temperature jump $\Delta T$ at the interface. The present study obtained values are in good agreement with previous values obtained from simulations and experimental measurements in the literatures [26–31]. The present study contributes some novel procedures during MD simulation work which will not be done in previous researchers.

## 4. Conclusions

After all earlier studies, it can be concluded that the development of new technologies are giving a new attention on to investigate nanoscale phenomena (including nanoscale heat transfer). Therefore, MD simulation is the only nanoscale tool to investigate the enhancement of thermal properties of graphene-reinforced nanocomposites for heat exchanger material. Based on the current simulation results, it is found that graphene-reinforced TPU nanocomposites demonstrate higher moduli, higher glass transition temperature, and lower values of CTE than pure TPU, that is, without reinforcements. This provides useful information to understand the nanoheat transport behaviors within TPU nanocomposites for the future development of thermal nanodevice. By taking advantage of low-cost simulations to establish material designs, overall materials development costs can be dramatically reduced, and development times can be expedited.

## Acknowledgements

The authors would like to thank the organizer committee members of TEQIP which funded short-term course on mechanics of composite using Material Studio in NIT, Durgapur, and further license software support to conduct MD simulation. Thanks to nanoHUB Pro for instructions and online helps. Also thanks to Accelrys (recently BIOVIA) Materials Studio community members help for simulation study.

## Nomenclature

| | |
|---|---|
| $E$ | total potential energy |
| $k_b$ | the stretching force constant |
| $k_\theta$ | the angle-bending force constant |
| $k_\varphi$ | the torsional barrier |
| $K_\omega$ | the force constant |
| $T$ | thermodynamic temperature |
| $V$ | volume of nanocomposites |

| | |
|---|---|
| $C_p$ | specific heat of nanoparticle bulk material |
| k | thermal conductivity |
| $k_B$ | Boltzmann constant |
| $K_p$ | the phonon contribution |
| $K_e$ | the electronic contribution |
| $J_q$ | the heat flux |
| R | interfacial thermal resistance |
| $\alpha$ | the volumetric coefficient of thermal expansion |
| $\rho$ | density of nanomaterial |
| $r_i$ | the position of the ith atom |
| $E_i$ | total energy of the ith atom |

## Author details

Animesh Talapatra[1*] and Debasis Datta[2]

1 MCKVIE, Howrah, West Bengal, India

2 IIEST, Howrah, West Bengal, India

*Address all correspondence to: animesh_talapatra@yahoo.co.in

## References

[1] Komarneni S. Feature article. Nanocomposites. Journal of Materials Chemistry. 1992;**2**:1219-1230

[2] Jordan J, Jacob KI, Tannenbaum R, Sharaf MA, Jasiuk IM. Experimental trends in polymer nanocomposites—A review. Materials Science and Engineering A. 2005;**393**(1–2):1-11. DOI: 10.1016/j.msea.2004.09.044

[3] Podsiadlo P, Tang Z, Shim BS, Kotov NA. Counterintuitive effect of molecular strength and role of molecular rigidity on mechanical properties of layer-by-layer assembled nanocomposites. Nano Letters. 2007; **7**(5):1224-1231

[4] Chan L, Zuo X, Wang L, Wang E, Song S, Wang J, et al. Flexible carbon nanotube-polymer composite films with high conductivity and super hydrophobicity made by solution process. Nano Letters. 2008;**8**(12): 4454-4458

[5] Lehn JM, Fendler JH, Meldrum F. The colloid chemical approach to nanostructured materials. Advanced Materials. 1995;**7**(7):607-632

[6] Seol JH, Jo I, Moore AL, Lindsay L, Aitken ZH, Pettes MT, et al. Two-dimensional phonon transport in supported graphene. Science. 2010; **328**(5975):213-216

[7] Klemens PG, Pedraza DF. Thermal conductivity of graphite in the basal plane. Carbon. 1994;**32**(4):735-741

[8] Allen PB, Feldman JL. Thermal conductivity of glasses: Theory and application to amorphous Si. Physical Review Letters. 1989;**62**(6):645

[9] Allen PB, Feldman JL, Fabian J, Wooten F. Diffusons, locons and propagons: Character of atomic vibrations' in amorphous Si. Philosophical Magazine B. 1999;**79** (11–12):1715-1731

[10] Hussein AM, Bakar RA, Kadirgama K, Sharma KV. Experimental measurement of nanofluids thermal properties. International Journal of Automotive and Mechanical Engineering. 2013;**7**:850-863

[11] Cevallos JG, Bergles AE, Bar-Cohen A, Rodgers P, Gupta SK. Polymer heat exchangers-history, opportunities, and challenges. Heat Transfer Engineering. 2012;**33**(13):1075-1093

[12] Chen H, Ginzburg VV, Yang J, Yang Y, Liu W, Huang Y, et al. Thermal conductivity of polymer-based composites: Fundamentals and applications. Progress of Polymer Science. 2016;**59**:41-85

[13] Chen X, Su Y, Reay D, Riffat S. Recent research in polymer heat exchangers—A review. Renewable and Sustainable Energy Reviews. 2016;**60**: 1367-1386

[14] T'Joen C, Park Y, Wang Q, Sommers A, Han X, Jacobi A. A review on polymer heat exchangers for HVAC&R applications. International Journal of Refrigeration. 2009;**32**: 763-779

[15] Zarkadas DM, Sirkar KK. Polymeric hollow fiber heat exchangers: An alternative for lower temperature applications. Industrial and Engineering Chemistry Research. 2004;**43**: 8093-8106

[16] Sun H. COMPASS: An ab initio force-field optimized for condensed-phase applications overview with details on alkane and benzene compounds. The Journal of Physical Chemistry B. 1998; **102**(38):7338-7364

[17] Xu Z. Heat transport in low-dimensional materials: A review and perspective. Theoretical and Applied Mechanics Letters. 2016;**6**(3):113-121

[18] Choi J, Yu S, Yang S, Cho M. The glass transition and thermo elastic behavior of epoxy-based nanocomposites: A molecular dynamics study. Polymer. 2011;**52**(22):5197-5203

[19] Shiu SC, Tsai JL. Characterizing thermal and mechanical properties of graphene/epoxy nanocomposites. Composites Part B: Engineering. 2014; **56**:691-697

[20] Yang S, Qu J. Computing thermo mechanical properties of cross-linked epoxy by molecular dynamic simulations. Polymer. 2012;**53**(21): 4806-4817

[21] Schmidtke E, Günther-Schade K, Hofmann D, Faupel F. The distribution of the unoccupied volume in glassy polymers. Journal of Molecular Graphics & Modelling. 2004;**22**(4):309-316

[22] Green MS. Mark off random processes and the statistical mechanics of time-dependent phenomena. II. Irreversible processes in fluids. The Journal of Chemical Physics. 1954;**22**(3): 398-413

[23] Kubo R. Statistical-mechanical theory of irreversible processes. I. General theory and simple applications to magnetic and conduction problems. Journal of the Physical Society of Japan. 1957;**12**(6):570-586

[24] Sellan DP, Landry ES, Turney JE, Mc Gaughey AJH, Amon CH. Size effects in molecular dynamics thermal conductivity predictions. Physical Review B. 2010;**81**:214-305

[25] Müller-Plathe F. A simple nonequilibrium molecular dynamics method for calculating the thermal conductivity. The Journal of chemical physics. 1997;**106**(14):6082-6085

[26] Fox TG, Loshaek S. Influence of molecular weight and degree of crosslinking on the specific volume and glass temperature of polymers. Journal of Polymer Science. 1955;**15**(80):371-390

[27] Wu SL, Shi TJ, Zhang LY. Latex co-coagulation approach to fabrication of polyurethane/graphene nanocomposites with improved electrical conductivity, thermal conductivity, and barrier property. Journal of Applied Polymer Science. 2016;**133**(11):13

[28] Lee SH, Jung JH, Oh IK. 3D networked graphene-ferromagnetic hybrids for fast shape memory polymers with enhanced mechanical stiffness and thermal conductivity. Small. 2014; **10**(19):3880-3886

[29] Li A, Zhang C, Zhang YF. Thermal conductivities of PU composites with graphene aerogels reduced by different methods. Composites Part A: Applied Science and Manufacturing. 2017;**103**: 161-167

[30] Yadav SK, Cho JW. Functionalized graphene nanoplatelets for enhanced mechanical and thermal properties of polyurethane nanocomposites. Applied Surface Science. 2013;**266**:360-367

[31] Materials Studio. User's Manual, Version 1.2. San Diego, CA: Accelrys, Inc.; 2001

# 8

# Applications of Compound Nanotechnology and Twisted Inserts for Enhanced Heat Transfer

*Hussain H. Al-Kayiem, Muna S. Kassim and Saud T. Taher*

**Abstract**

Nanoadditives are a type of heat transfer enhancement techniques adopted in heat exchangers to improve the performance of industrial plants through improvement of the thermal properties of base fluids. Recently, various types of inserts with nanofluids are adopted to enhance the thermal performance of double pipe heat exchangers. In the current article, $TiO_2$/water nanofluid with multiple twisted tape inserts was investigated as a hybrid enhancement technique of heat transfer in straight pipes. The investigations were carried out experimentally and numerically at Reynolds numbers varied from 5000 to 20,000. Using nanofluid with 0.1% $TiO_2$ nanoparticles volume fraction demonstrated enhanced heat transfer with slight increase in pressure drop. Results are showing a maximum increase of 110.8% in Nusselt number in a tube fitted with quintuple twisted tape inserts with 25.2% increase in the pressure drop. However, as the article is representing a part of specified book on heat exchangers, the literature has been extended to provide sufficient background to the reader on the use of nanotech, twisted inserts, and hybrid of compound nanofluids and inserts to enhance heat transfer processes.

**Keywords:** enhanced heat transfer, hybrid heat transfer enhancement, multi-longitudinal vortices, multiple twisted tapes, $TiO_2$ nanofluid

## 1. Introduction

Heat exchangers (HEXs) are typical thermal systems in industrial and engineering applications. They are adopted as means of heat dissipation in a wide range of thermal processes ranging from huge scale to microscale. HEXs are involved in the power production process, chemical and food industries, electronics cooling, environmental production engineering, waste heat recovery, manufacturing industry, air conditioning, and refrigeration applications. HEXs constitute a multibillion-dollar industry in the United States alone, and there are over 300 companies engaged in the manufacture of a wide array of heat exchangers [1].

The performance of heat exchanger can significantly increase by the heat transfer augmentation techniques that lead to the reduction of heat exchanger size, as well as operating cost reduction. Enhancement techniques can be classified either as passive or active techniques. A passive technique does not need any external power input, and the additional power needed to enhance heat transfer is taken from the available power in the system such as extended surfaces, treated surfaces, and twisted tape. Active techniques require external power such as mechanical aids and surface vibration [2].

Various methods of heat transfer enhancement are used in HEXs. These techniques are expressive to be manufactured and adopted to increase the thermal system efficiency by increasing the rate of heat transfer process and/or to reduce the size of thermal systems. They can be classified into two main categories; (i) active methods: which use an external power source; (ii) passive methods: which use several techniques without a power source such as turbulence generators (as propeller, spiral fin, twisted tapes, ribs ... etc.), or by using additives like the nanoadditives. However, in their review paper on the tape enhanced heat transfer [2], classified third enhancement techniques. They proposed compound methods, when two passive methods are used simultaneously. The method adopted in this paper is compound enhancement as twisted tapes and nanoadditives. As such, below are some backgrounds on the nanoadditives and twisted tapes to enhance heat transfer.

### 1.1 Nanoenhanced heat transfer

Nanoadditives are a type of heat transfer enhancement method through enhancement of the thermal properties of base materials. They may be synthesized with base fluids to produce nanofluids for thermal transport or synthesized with phase change materials (PCM) to produce nanocomposites for thermal energy storage as in [3].

Numerous works and review papers have been published in the field of nanoenhanced fluids, such as the paper of [4] who presented and compared the preparation, stability and thermophysical properties of nanofluids. It was concluded that nanofluids have, in general, better thermo-physical properties even at a very low particle concentration (typically 1.0% or less) than conventional heat transfer fluids. The only drawback is increment in the viscosity which leads to a higher pressure drop. Similar article by [5], published in 2018, has also reviewed the fabrication, stability, and thermophysical properties of nanofluids. After presenting the progress of studies on nanofluid thermophysical characterization, they identified some possible opportunities for future research that can bridge the gap between in-lab research and commercialization of nanofluids as this nanofluids are receiving large attention due to their potential usage. Reference [6], who reviewed the current scenario and future prospective of nanofluids, who also supported the conclusion of [5], ended with that the study of nanofluids has been materialized as a new field of scientific interest and innovative application. Reference [3] investigated nanofluid characteristics concerning thermal behavior in a plain tube, which showed significant enhancement rate in heat transfer with increase in nanoparticles diameter and volume concentration. Reference [7] experimented the thermophysical properties of $Al_2O_3$ and $TiO_2$ nanoparticles with distilled water and proposed parabolic equations for dynamic viscosity and thermal conductivity at specific conditions.

### 1.2 Enhanced heat transfer by twisted inserts

Twisted tape heat transfer enhancement mechanism is attributed to producing swirl flow in form of secondary recirculation act on the axial flow. This fluid mechanism increases the tangential and radial turbulent components leading to higher mix fluid layers and reduces the thermal and hydrodynamic boundary layer thickness due to increase in turbulent fluctuation and combination occurring between core region and near wall region in fluid flow, which cause higher temperature gradient near the wall leading to increased heat transfer rate.

Reference [8] reported for single twisted tape insert in circular tube in which heat transfer rate increased significantly than tube without inserts, and as twist ratio decreases, the heat transfer rate and friction factor become larger. Subsequently, [9] experimented the single twisted tape insert with many different twist ratios to describe the mechanism of heat transfer enhancement and proposed correlations for Nusselt number and friction factor.

A recognized contribution to the field of heat transfer enhancement is done by the thermofluid team in University of Pretoria—South Africa. They have started their investigation on heat transfer enhancement since 2010 and since then, they published a large number of experimental and numerical investigation on the enhanced heat transfer by inserts. In 2013, [10] investigated the heat transfer enhancement in laminar flow regime in circular tubes using rib roughening and twisted tape inserts. The work is carried out experimentally, and Nusselt number and the friction factor were measured. The work concluded major finding that the center-cleared twisted tapes in combination with transverse ribs perform significantly better than the individual enhancement technique acting alone for laminar flow through a circular duct up to a certain amount of center-clearance. The reported results are useful for the design of solar thermal heaters and heat exchangers. In 2017, another two papers have been published by the same team [11, 12]. Both papers have been carried out numerically. In [11], heat transfer behavior in a tube with inserted twisted tape swirl generator is investigated numerically, for different values of the twist ratio and diameter ratio and for Reynolds numbers within the range of 100–20,000. Results have shown that the tube use of twist tape enhances heat transfer generally, but, accompanied with a higher pressure drop. Improvement of the thermal-hydraulic performance can only be observed for certain configurations and Reynolds numbers. In [12], simulations were conducted for laminar, transitional, and turbulent flow regimes for four different rib angle of attack values and for a plain tube without ribs, as benchmark case. Within the investigated range, the larger thermal performance factors are observed to occur for the intermediate Reynolds numbers. Maximum values in the range of 2.0–2.5 are predicted for the Reynolds number of 2000, where a subsequent drop to values within the range of 1.0–1.5 is found to occur for Reynolds numbers around 3000–4000, which may be attributed to the transitional effects. In 2018, a paper [13] is published by the same team, which reported results of experimental investigation of heat transfer performance of corrugated tube with spring tape inserts in turbulent forced convection in Re range of 10,000–5000. Air is adopted as working fluid. Results show that Nusselt numbers can be increased considerably, depending on pitch and spring ratios. The heat transfer enhancement to the pressure drop penalty is realized to be larger than unity for all cases. Values around 2.8 occur for cases with the smallest pitch and spring ratios. Predictive Nusselt number and friction factor correlations are proposed.

### 1.3 Compound nanoadditive/twisted tape enhanced heat transfer

By combining the nanofluid enhancement with inserts, [14] found that using $Al_2O_3$ nanoparticles with distilled water and full-length twisted tape gave a superior performance with a noticeable increment in friction loses than plain tube. Reference [15], achieved 20 and 2.5% increase in overall heat transfer coefficient and friction factor, respectively, by using nanofluid with single twisted tape insert in double pipe heat exchanger than that without insert. Reference [16] presented an experimental analysis of the turbulent flow in tube fitted with (single, dual, triple, and quadruple) twisted tapes and nanofluid under turbulent flow conditions. The results shown that Nusselt number and friction factors increased as the number of tapes and volumetric concentration increased. Also, the increment in heat transfer rate by increase in nanofluid volumetric concentration only was higher than that of increase in twisted tape number only. It must be mentioned that the volumetric efficiency of the tube was not taken into consideration.

Reference [17] has carried out experimental and numerical investigations on similar tube flow and similar nanofluid with plain tube, tube fitted with dual plain tape inserts, tube fitted with dual twisted tapes inserts and tube fitted with dual helical screw twisted tape inserts. He concluded that, a maximum enhancement of 82.2% is achieved in the Nusselt number by using tube fitted with dual helical screw twisted tapes inserts and $TiO_2$/water nanofluid flow than that observed with the plain tube and distilled water flow. And the maximum friction factor observed for the same model of the tube fitted with dual helical screw twisted tape inserts and nanofluid are up to 17.34% than that of the plain tube. Reference [18] carried out a wide range of experimental study on the convective heat transfer enhancement using combined techniques. One of these techniques is the use of twisted tape along the whole tube length of a micro-fin tube that effectively combined the features of extended surfaces, turbulators, and artificial roughness. Nanofluids are used for improving the thermo-physical properties of the fluid. Ag-water nanofluid in a micro-fin tube with nonuniform twisted tapes insert is examined under turbulent flow. The effects of the twist ratios of nonuniform twisted-tapes of $3.0 > 2.8 > 2.6$, $3.0 > 2.6 > 2.2$, and $3.0 > 2.4 > 1.8$, in counter and co-current flow arrangements and nanofluid concentrations of 0.007, 0.016, and 0.03% vol. are investigated. They claimed that heat transfer, friction loss, and thermal performance factors are increasing as the twist ratios are decreasing for nonuniform twisted tapes and increasing nanofluid concentrations. The optimum condition are achieved in using the micro-fin tube with a nonuniform twisted-tape in a counter-current-arrangement with twist ratios in a series of $3.0 > 2.4 > 1.8$ with Ag-water nanofluid at a concentration of 0.3% vol. The enhancements are up to 112.5% for the heat transfer rate and 1.62 for thermal performance.

On other combined enhancement attempt in microchannel heat exchangers, [19] carried out experimental investigation on heat transfer for pulsating flow of GOP-water nanofluid. The effects of mass fraction of graphene oxide (GOPs) and flow pulsating frequency on heat transfer and pressure drop in a microchannel with arrayed pin-fins have been investigated. Five different mass fractions of graphene oxide nanofluids were prepared and used as working fluids. Experiments were performed under the condition that the pulsating frequency was from 1 to 5 Hz, the mass fraction was from 0.02 to 0.2%, and the average Reynolds numbers were 272, 407, and 544. The results show that the heat transfer is enhanced significantly when the frequency is in the range of 2–5 Hz. For the frequency of 1 Hz, the pulsating flow shows a negative effect on temperature uniformity. With the increase of mass fraction, the heat transfer performance is improved, while no significant change is found in pressure drop. The pulsating flow leads to a significant enhancement of pressure drop for frequency at 2 Hz. The combination of pulsating and nanofluid can obtain higher heat transfer efficiency under limited size of microchannel heat sink and low inlet Reynolds numbers. The results provided good guide for the design of microchannel heat exchangers.

Since conventional fluids, such as water, have a relatively poor heat transfer characteristic, the nanoenhancing technique opens the door to gain more benefits from these conventional fluids especially in heat transfer intensification field.

As pointed above and mentioned by many other researchers [20–23], the field of nanoenhanced heat transfer with multiple inserts is still virgin. Further experimental data are essential to support the literature, and to enhance the understanding of hydrothermal behavior in thermal systems. So far, few studies have been carried out on multiple twisted tapes with nanofluid effect on thermo-hydraulic characteristics in thermal systems.

The objective of this chapter is to scrutinize the effect of compound multiple twisted tape (TT) inserts with $TiO_2$-water nanofluid on heat exchange enhancement in double pipe heat exchangers. Three cases of tubes fitted with single, triple, and quintuple plain twisted tapes have been investigated experimentally and simulated numerically. In addition, two bench-mark cases have been investigated: first case is with plain tube with pure water flow, and the second case is with plain tube with nanofluid flow. Results have been manipulated and presented in terms of Nusselt number for heat transfer and friction factor for pressure drop.

## 2. Problem formulation

The basic geometry adopted in this investigation is a straight tube with 1000 mm length, $L$ and 50 mm internal diameter, $D$. Pure tube, with pure water flow and no inserts, was considered as the benchmark case to compare the thermal enhancement and the pressure drop. The other cases were investigated with a different number of inserts and with 0.1 vol.% $TiO_2$/water nanofluid.

It is worth mentioning that this research meant to investigate the flow characteristics only in compound thermal and hydrodynamic in fully developed region where the entrance effect becomes insignificant beyond a pipe length of 8 times the diameter for turbulent flow [8, 14]. The fully developed region was calculated to be accomplished at 500 mm from the tube inlet.

### 2.1 Twisted tape inserts

Twisted tape inserts (TT) are heat transfer enhancement devices which are dividing the flow within the tube resulting in higher velocity near the tube surface. They, also, creating spiral flow creates swirl or secondary flow in the main flow which increases local velocities and promotes mixing. They are widely used over decades to generate the swirl flow in the thermal fluid resulting in increased heat transfer coefficient, with a penalty of increased pressure drop across the flow passage. Thus, reduction in the thermal system, like the heat exchangers, can be achieved. Types of TT are shown in **Figure 1a**.

Main parameters that are commonly adopted to characterize the TT are the empty tube Reynolds number (Re), half-pitch ($y$), and twist ratio ($Y$). The main geometrical TT identifiers are shown in **Figure 1b**.

The half-pitch ($y$) is defined as the distance between two points on the edge of a TT, which lies down on the same plane as the TT completes 180° of revolution.

The twist ratio, $Y$, is defined as the ratio of the half-pitch to the internal tube diameter, $Y = y/D_p$.

Tube fitted with single twisted tape (STT), tube fitted with triple twisted tapes (TTT), and tube fitted with quintuple twisted tapes (QTT) are considered in the present experimental and numerical investigations. The schematics of cross sectional view of these models and twisted tapes and the geometries are illustrated in **Figure 2**.

The twisted tapes have the same length of the tube and has a width; $L$ = 1000 mm, width, $w$ (mm), thickness, $\delta$ (mm) and pitch of 180° twist, $y$ (mm) as explained in **Table 1**. The swirl direction corresponding to tape arrangement was co-swirl flow, and all tapes were aligned to be twisted in the same direction.

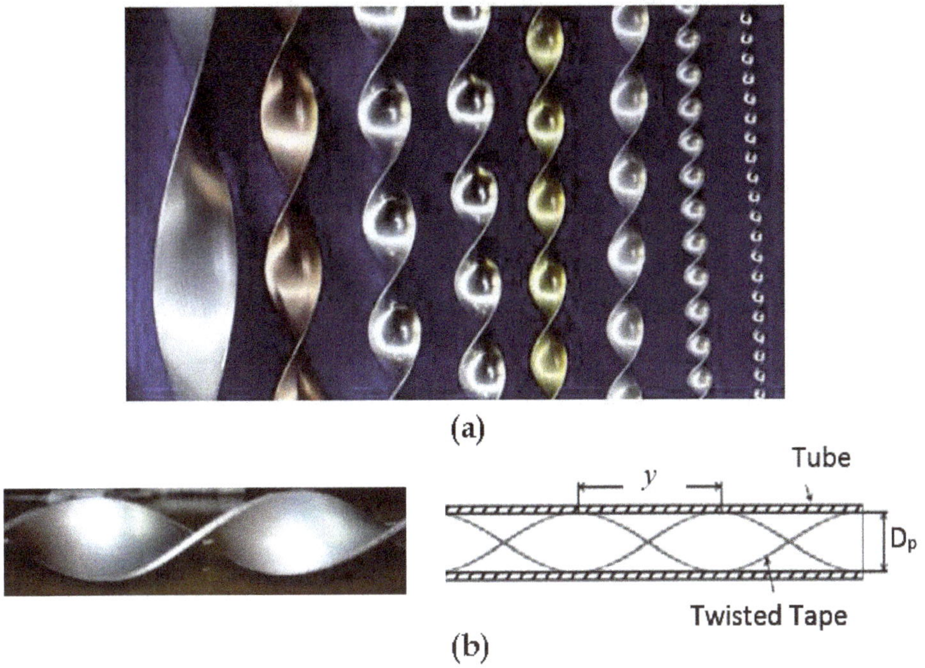


**Figure 1.**
*Twisted tape inserts; (a) different types of twisted tape inserts (courtesy of visual capitalist [24], with permission); (b) topologies of twisted tape insert.*

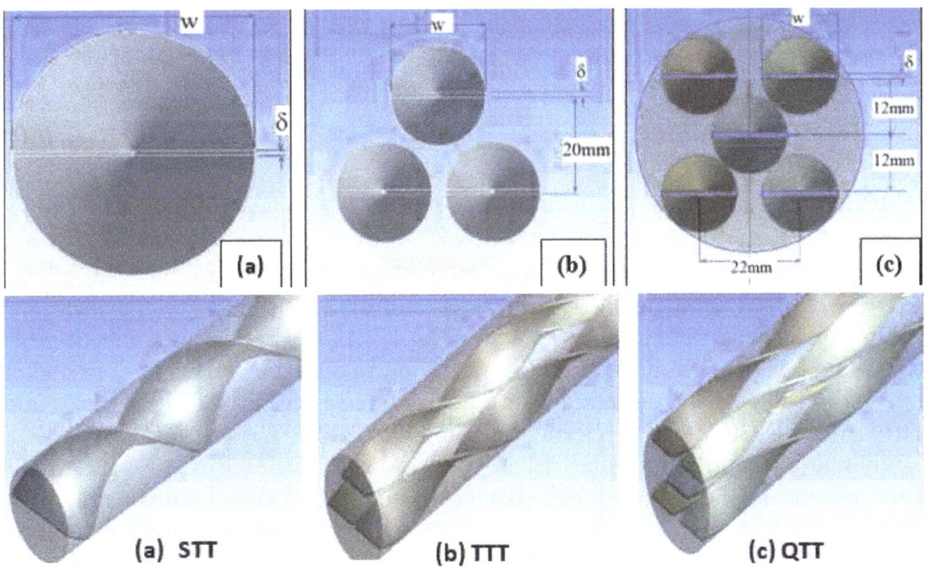


**Figure 2.**
*Schematic views for tubes fitted with; (a) single twisted tape (STT), (b) triple twisted tapes (TTT), and (c) quintuple twisted tapes (QTT).*

## 2.2 Working fluids

Water was used in the primary tests to investigate the effect of inserts only on the thermo-hydraulic performance of tubes. Latterly, nanofluid of $TiO_2$ pre-dispersed in water was used to obtain the enhancement by both nanotechnology

| Model | No. of tapes | Tape width (mm) | Tape pitch ($y$) (mm) | Tape length (mm) | Tape thickness ($\delta$) (mm) | Tape material |
|---|---|---|---|---|---|---|
| STT | 1 | 48 | 100 | 1000 | 1 | Aluminum |
| TTT | 3 | 21 | 100 | 1000 | 1 | Aluminum |
| QTT | 5 | 16 | 100 | 1000 | 1 | Aluminum |

**Table 1.**
*Geometries and details of TT inserts.*

| | Density ρ ($kg/m^3$) | Specific heat ($C_p$) (J/kg·K) | Thermal conductivity, $k$ (W/m·K) | Dynamic viscosity, $\mu$ (kg/m·s) |
|---|---|---|---|---|
| $TiO_2$ | 4230 | 692 | 8.4 | — |
| Water | 997 | 4179 | 0.613 | $855 \times 10^{-6}$ |

**Table 2.**
*Properties of titanium oxide particles and water.*

| Property | Formula | Relation no. | Recommended by | Value |
|---|---|---|---|---|
| Density, $\rho$ | $\rho_{nf} = \varphi\rho_{np} + (1-\varphi)\rho_{bf}$ | (1) | Pak and Cho [25] | 1000.2 $kg/m^3$ |
| Specific heat, $C_p$ | $(C_P)_{nf} = \varphi(C_P)_{np} + (1-\varphi)(C_P)_{bf}$ | (2) | Maddah et al. [15] | 4175.5 J/kg·K |
| Thermal conductivity, $k$ | $k_{nf} = [1 + 3\varphi])k_{bf}$ | (3) | Said et al. [7] | 0.6148 W/m·K |
| Dynamic viscosity, $\mu$ | $\mu_{nf} = [1 + 2.5\,\varphi + 6.2\,\varphi^2])\mu_{bf}$ | (4) | Batchelor [26] | $857.143 \times 10^{-6}$ kg/m·s |

*Note: φ is the percentage of volume fraction of nanoparticles.*

**Table 3.**
*Correlations of 0.1% volume of $TiO_2$-in-water nanofluid properties and predicted values.*

and twisted tape inserts. In the present study, 0.1 vol.% $TiO_2$ nanoparticles of size (<50 nm) mixed with distilled water was sonicated continuously by ultrasonic vibrator generating pulses of 240 W at 40 + 4 kHz to break down any possible nanoparticle agglomeration. Nanofluid was found to be stable during tests period of tests and no intermediate mixing process considered necessary. $TiO_2$ particles and water have the considered properties illustrated in **Table 2**.

Volume concentration φ desired in this study was 0.1% for $TiO_2$/water. The thermophysical parameters are calculated for nanofluid; **Table 3** contains the popular and valid models, which were used to evaluate property values and shows values.

## 3. Experimental implementation

### 3.1 Apparatus

Experimental setup was designed and fabricated to investigate the thermo-hydraulic characteristics of the double pipe heat exchanger, as shown in **Figure 3**. The test section comprised of copper tube of 1000 mm length with inside and

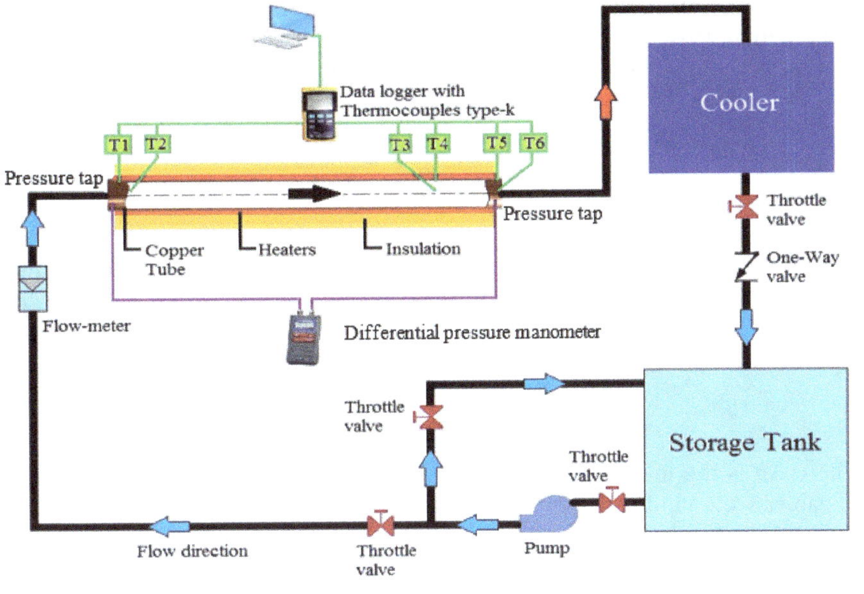


**Figure 3.**
*Schematic diagram of the experimental setup.*

outside diameters of 50 and 53 mm, respectively. An electric heater with 5 kW capacity was wounded on the outer surface of the copper tube and insulated by glass-wool insulation. Thermocouples, referred to as (T1–6) in **Figure 3**, were installed to measure wall and fluid temperature at inlet, fully developed, and outlet regions. All thermocouples were calibrated before being fixed, and all were connected to the data logger to record temperature readings. A differential pressure manometer was used to measure pressure drop along the test section. The working fluid was heated inside the test section and allowed to cool by passing through a cooler (evaporative cooling system). By recirculation, the working fluid returns to the storage tank in the flow loop and then is pumped again to the test section. Flow meter was placed at the entrance of the test section to measure the working fluid flow rate. Throttle valves were incorporated to allow controlling the working fluid flow rate and for maintenance and emergency, if any. To ensure steady state condition for each run, period of 30 minutes was permitted prior to starting the measurement and data acquisition.

### 3.2 Verification of experimental measurements

Heat transfer in convective flow is commonly evaluated in terms of Nusselt number, Nu, as in Eq. (5).

$$\mathrm{Nu} = \frac{h.D_p}{k} = \frac{\left[\bar{\bar{q}}\,/(T_s - T_b)\right]D_p}{k} \tag{5}$$

Nusselt number values were validated with the experimental results of previous researches shown in **Table 4**. Values of Reynolds numbers, Re, were evaluated as per Eq. (6), as follows:

$$\mathrm{Re} = \frac{VD_p}{v} \tag{6}$$

where $v$ is the mean velocity (=measured flow rate/pipe inlet cross section area), $\nu$ is the kinematic viscosity of the working fluid, and $D_p$ is the pipe or tube diameter.

Prandtl number is considered 7.56 for water. For nanofluid, Pr is evaluated using Eq. (7), as follows:

$$\mathrm{Pr} = \frac{Cp_{nf} \times \mu_{nf}}{k_{nf}} \tag{7}$$

The friction factor was predicted using the known Darcy-Weisbach equation (Eq. 11), for both, water and nanofluid flow, as follows:

$$f = \frac{\Delta p}{0.5\,\rho\,{V_{in}}^2\left(\frac{L}{D_h}\right)} \tag{11}$$

where $\Delta p$ is the measured pressure drop over the 1.0 m pipe length. The measured values were vindicated with results gained from well-established correlations, illustrated in **Table 5**.

| Correlation | Eq. No. | Suggested by | Case of HEX |
|---|---|---|---|
| $Nu = \frac{\left(\frac{f}{8}\right)\left(Re_{D_h} - 1000\right)Pr}{1 + 12.7\left(\frac{f}{8}\right)^{\frac{1}{2}}\left(Pr^{\frac{2}{3}} - 1\right)}$ | (8) | Gnielinski, in [27] | Tube with water flow |
| $Nu = 0.074\,Re_{nf}^{0.707}Pr_{nf}^{0.385}\varphi^{0.074}$ | (9) | Duangthongsuk and Wongwises [21] | Tube with nanofluid flow |
| $Nu = 0.024\ Re^{0.8}pr^{0.4} \times \left[1 + \frac{0.769}{\frac{y}{w}}\right]^{1.1}\left[\frac{T_b}{T_s}\right]^{0.45}\left[\frac{\pi + 2 - \frac{2\delta}{D}}{\pi - \frac{4\delta}{D}}\right]^{0.2}\left[\frac{\pi}{\pi - \frac{4\delta}{D}}\right]^{0.8}$ | (10) | Bergles and Manglik [9] | Tube with twisted tape(s) and water flow |

**Table 4.**
*Adopted correlations from the literature to predict Nusselt number used to verify the experimental results.*

| Correlation | Eq. No. | Suggested by | Case of HEX |
|---|---|---|---|
| $f = 0.318\,{Re_{D_h}}^{-0.25}$ | (12) | Blasius, in [27] | Tube with water flow |
| $f = \left(0.79\ln Re_{D_h} - 1.64\right)^{-2}$ | (13) | Petukhov, in [27] | Tube with water flow |
| $f = 0.961\varphi^{0.052}\,{Re_{D_h}}^{-0.375}$ | (14) | Duangthongsuk and Wongwises [21] | Tube with nanofluid flow |
| $f = \left[0.079\,Re^{-0.25}\left[\frac{\pi + 2 - \frac{2\delta}{D}}{\pi - \frac{4\delta}{D}}\right]^{1.25}\left[\frac{\pi}{\pi - \frac{4\delta}{D}}\right]^{1.75}\right]\left[1 + 2.06\left(1 + \left(\frac{2\frac{y}{w}}{\pi}\right)^2\right)^{-0.74}\right]$ | (15) | Bergles and Manglik [9] | Tube with triple twisted tapes and water flow |

**Table 5.**
*Adopted correlations from the literature to predict friction factor used to verify the experimental results.*

### 3.3 Experimental uncertainty and errors

The used instruments were well calibrated, and the margin of error and uncertainty of each instrument were evaluated.

- Thermocouple type-K used in this study to measure temperature of surfaces. Thermocouple type-K probes were used to measure the fluid temperature. All thermocouples and probes were calibrated and certified by the Central Organization for Standardization and Quality Control (COSQC)—Baghdad—Iraq [certificate number 734/2016 dated 30 Oct. 2016]. Mean uncertainty within the temperature measurement in the experiment was $\pm 0.63°C$.

- Flow meters type LZS-25 used to measure the flow rate of heat transfer fluids were calibrated using a scaled container and time recording technique, which is well identified as a standard procedure for flow meter calibration. The relative uncertainty of flow measurement was $\pm 2.34\%$.

- Differential manometer was calibrated against another certified manometer type Lutron (PM-9100). The relative uncertainty was $\pm 2.1\%$.

- The two pressure gauges were calibrated and certified by the Central Organization for Standardization and Quality Control (COSQC)—Baghdad—Iraq [certificate number PRE/918/2016 dated 06 Oct. 2016]. Max uncertainty was $\pm 0.012$ bar.

The possible error in the prediction of Re and Nu was estimated. The uncertainty in Re values was estimated by Eqs. (16)–(18) using the uncertainties of the measuring instruments used for measurements of relevant variables in Re:

$$Re = \frac{4\,\dot{m}}{\pi \mu d} = \frac{4\rho \dot{V}}{\pi \mu d} \tag{16}$$

$$\Delta Re = \left[ \left[ \left( \frac{\sigma Re}{\sigma \dot{V}} \right) \bullet \dot{V} \right]^2 + \left[ \left( \frac{\sigma Re}{\sigma d} \right) \bullet d \right]^2 \right]^{0.5} \tag{17}$$

Getting,

$$\Delta Re / Re = \left[ \left[ \left( \frac{\Delta \dot{V}}{\dot{V}} \right) \right]^2 + \left[ \left( \frac{\Delta d}{d} \right) \right]^2 \right]^{0.5} \tag{18}$$

= 0.0665 or 6.65%.

Uncertainty in Nu values was estimated by Eq. (19) using the uncertainties of the measuring instruments used for measurements of relevant variables in Nu,as

$$\Delta Nu / Nu = \left[ \left[ \left( \frac{\Delta h}{h} \right) \right]^2 + \left[ \left( \frac{\Delta d}{d} \right) \right]^2 \right]^{0.5} \tag{19}$$

= 0.0625 or 6.25%

## 4. Computational methods

### 4.1 Numerical simulation and conditions

The numerical analyses included single phase flow based on steady and three dimensional continuity; momentum and energy equations were performed using the ANSYS-Fluent commercial code. RNG $k$-$\varepsilon$ model was chosen to solve the considered cases because the effect of swirl on turbulence is included in this model, which enhances the accuracy of swirling flows.

### 4.2 Boundary conditions

The thermo-fluid process in the current simulated was solved as steady, incompressible, and 3D flow. The following was adopted to set the boundary conditions for the simulation:

- The flow was considered as internal flow under uniform heat flux condition for all cases with $\overline{\overline{q}} = 30{,}000\ W/m^2$.
- No slip condition was also applied to the tube wall.
- Results are considered only at the fully developed region ($L > 10D_p$).
- Water and 0.1 vol.% $TiO_2$/water are considered as the working fluids, and calculations for the thermal properties are done at the inlet temperature of 300 K, as in **Tables 2** and **3**.
- The nanofluid was considered as a single-phase fluid with changed physical parameters as density, thermal capability, thermal conductivity, and viscosity.
- All cases were investigated within the turbulent flow regime studied with a range for Reynolds number of 5000–20,000.
- The hydraulic diameter at each inlet and outlet has the same value.
- At the outlet, a pressure outlet condition was used, and the gauge pressure is set to zero.
- Turbulence intensity obtained from the expression recommended by [28].

$$T.I. = 0.16\,Re_{D_p}^{-0.125} \tag{20}$$

Other flow quantities are extrapolated from the interior domain by the solver in Fluent software. The SIMPLE (Semi-Implicit Pressure Linked Equations) algorithm were chosen as solver method. In addition, a convergence criterion of $10^{-5}$ was used for energy and mass conservation of the calculated parameters.

### 4.3 Grid generation

To avoid blurred curved areas, tetrahedral cells are used for meshing the computational domains, as shown in **Figure 4**. Mesh, with around 3,000,000 elements, was decided to represent the domains for the current simulation.

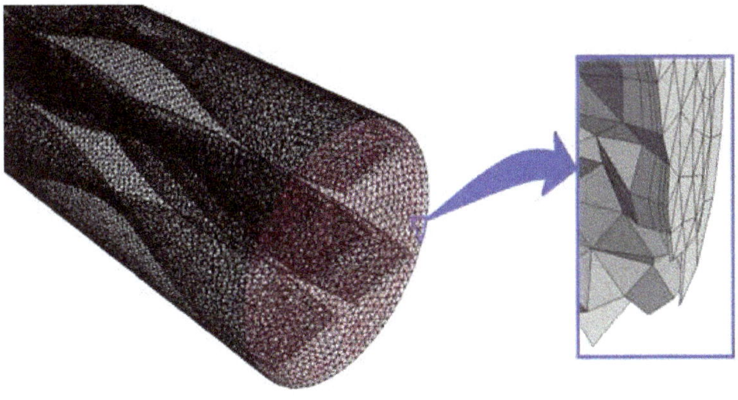

**Figure 4.**
*Tetrahedral mesh cells for the case of QTT with enlarged small cut for clarification.*

**Orthogonal Quality** mesh metrics

| Unacceptable | Bad | Acceptable | Good | Very good | Excellent |
|---|---|---|---|---|---|
| 0-0.001 | 0.001-0.14 | 0.15-0.20 | 0.20-0.69 | 0.70-0.95 | 0.95-1.00 |

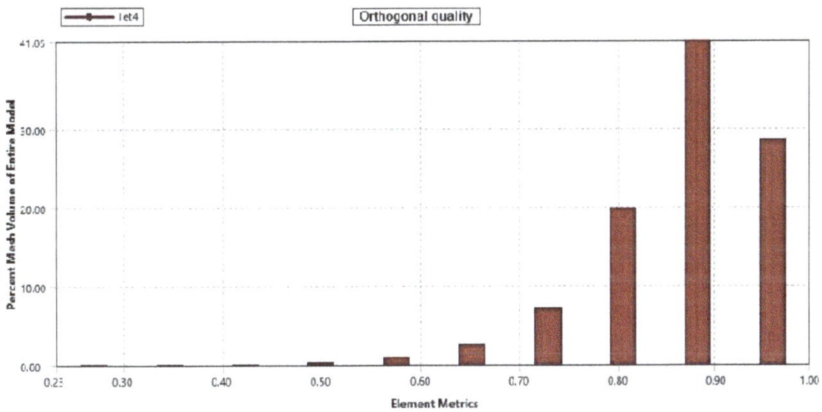


**Figure 5.**
*Mesh cell orthogonal quality metrics.*

**Skewness** mesh metrics spectrum

| Excellent | Very good | Good | Acceptable | Bad | Unacceptable |
|---|---|---|---|---|---|
| 0-0.25 | 0.25-0.50 | 0.50-0.80 | 0.80-0.94 | 0.95-0.97 | 0.98-1.00 |

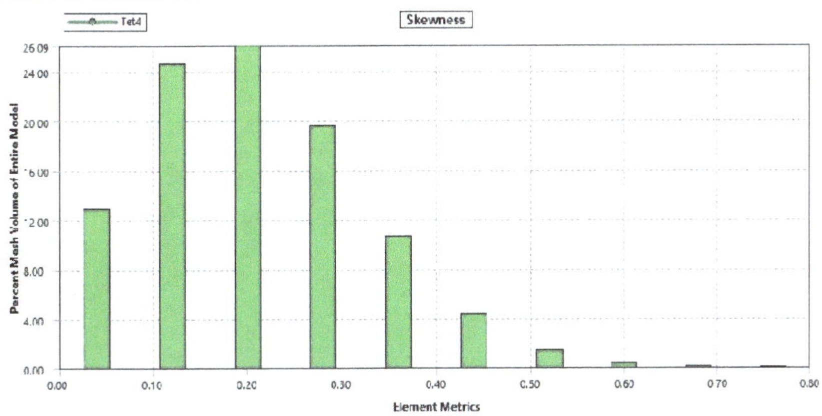


**Figure 6.**
*Mesh cell skewness quality metrics.*

The same way that element numbers in the mesh are important, mesh quality has a remarkable role in the numerical solution accuracy. Grid quality is commonly identified through orthogonal quality and skewness. Orthogonal quality describes how much the mesh criteria are within the correct range that is valid for physical value prediction. Orthogonal quality is presented in **Figure 5** for the grids performed to represent computational domains. The average orthogonal quality value gained in the present numerical procedure, of 0.8563, is within a very good quality range. Skewness determines how the generated cells are close to the ideal configuration and it governs solution ability to converge, as illustrated in **Figure 6**. The average skew-ness value of depended grids of 0.22288 is within excellent simulation range.

## 5. Results and discussion

The presentation and discussion of results are not based on sub-sectioning the numerical and experimental results. Instead, the results are sub-sectioned based on the hydrothermal parameters, like the velocity, pressure, and thermal performance.

### 5.1 Hydrodynamics analysis

The hydrodynamics analysis of the flow within internal conduits includes the velocity and pressure structures.

#### *5.1.1 Velocity field analysis*

The flow structure in the pipe is characterized by analyzing the velocity field and the pressure distribution in the flow domain. In the current work, there are four different geometrical configurations including plane pipe flow and another three cases with single, triple, and quintuple twisted tape inserts inside the pipe. CFD is a powerful tool to provide flow visualization and assist in the analysis of the flow field structure. Fields of velocity, predicted by computational simulation, in case of water flowing are depicted in **Figure 7**. The velocity contours shown are taken for

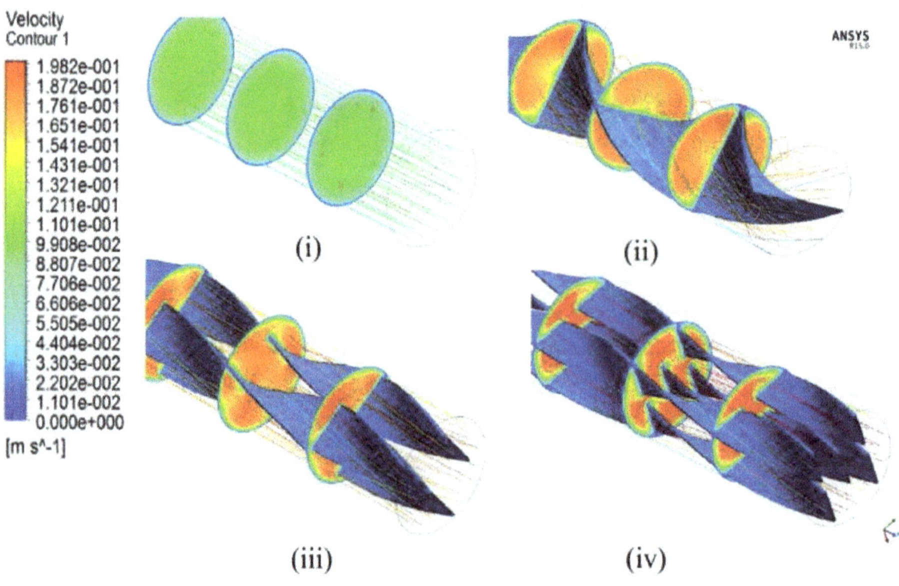


**Figure 7.**
*Velocity contours and streamlines of water flow for; (i) PT, (ii) STT, (iii) TTT, and (iv) QTT.*

Re = 5000 and at different axial locations in the fully developed region. As can be seen in **Figure 7**, frame ii, the velocity is increasing by 50.0% for STT higher than PT. This happens due to secondary flow caused by twisted tape geometry, which increases mean velocity by changing flow type from linear motion to swirl motion and also due to reduction in hydraulic diameter which leads to increase in velocity at constant Reynolds number and the reduction in hydraulic diameter causing a decrease in flow cross sectional area, which resulted in an increment in mean velocity value to satisfy equation of continuity (the rate of mass enters a system is equal to the rate of mass leaves the system).

For TTT and QTT, the velocity increased by 13.3 and 27.4% higher than STT due to narrowing the flow passages. Swirl motion and turbulence fluctuation are also increasing by increases in tape number due to multi-passage flow interactions. In addition, the tapes are breaking the flow field uniformity, and mix fluid flow layers between near wall region and core region lead to the appearance of many regions of high velocity; high velocity region in plain tube appears only at the core of it, which increases as the number of twisted tape increases, and that fission leads to increase in average velocity of fluid flow.

Longitudinal vortices in flow fields are shown in **Figure 8**; it was found that the number of vortices generated in the flow equals the number of twisted tapes and formed around it.

Apparently, the velocities of the nanofluid are nearly the same as those of water under the considerable nanoparticle volume fraction, which discloses that nanofluid will not require an added disadvantage over pumping power.

### *5.1.2 Analysis of pressure field and pressure drop*

As a fluid flows through the tube, there will be a pressure drop due to the shear drag at the contact wall in addition to the pressure required to pump the fluid inside the tube which is in tube with inserts higher than those without. The main determinants of pressure drop are fluid viscosity and fluid velocity. Pressure contours are illustrated in **Figure 9** for the computational domains considered at Reynolds number 5000 for water flow and on longitudinal revolution surface along the axial direction.

In the case of STT, **Figure 9**, frame a, pressure drop is 98.7% higher than in PT due to the fact that twisted tape insert increases frictional shear forces within the surface area of the inserts.

**Figure 10** presents the predicted pressure drop for all simulated cases including pure tube and tube with single, triple, and quintuple twisted tape inserts with water

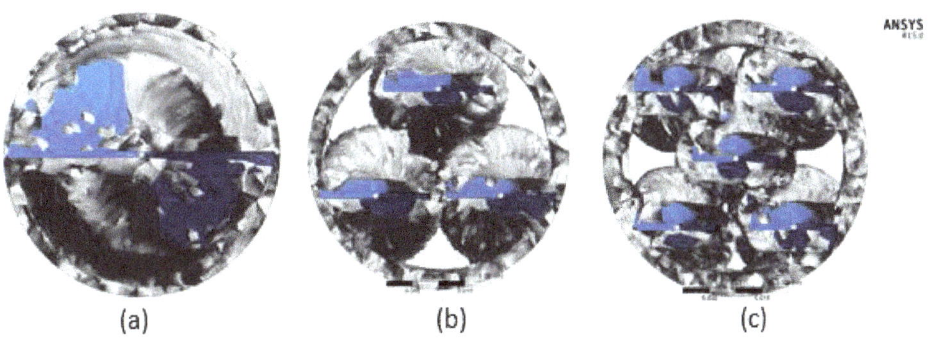


**Figure 8.**
*Longitudinal water vortices for tube fitted with; (a) single twisted tape, (b) triple twisted tapes, and (c) quintuple twisted tapes.*

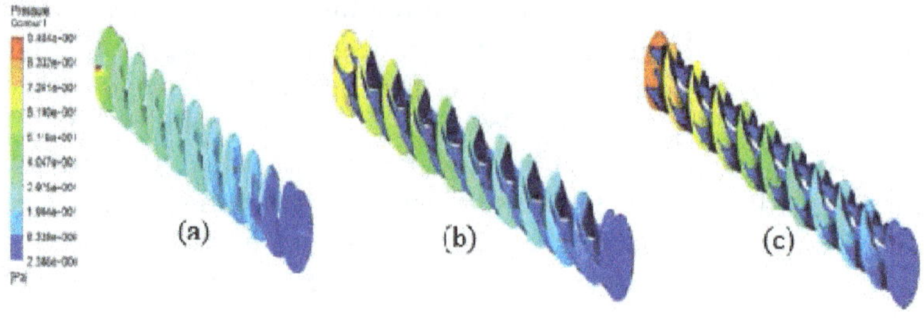


**Figure 9.**
*Pressure contours for a longitudinal revolution surface along the computational domains of water flow; (a) STT, (b) TTT, and (c) QTT.*

flow and with nanofluid flow. The pressure drop in TTT and QTT is 31.2 and 64.5% higher than STT, respectively. This is due to swirl motion achieved by each one of them, where the secondary motion generated by twisted tapes have an effect on velocity proportionally, where velocity gradient effect on shear forces acting on fluid flow causes pressure drop.

Pressure drop increases slightly by using nanofluid at 0.1 vol.% $TiO_2$. The numerical results show a percentage difference between water and nanofluid up to 3% for the same model investigated. However, experimental results for pressure agree well with those calculated numerically with a maximum divergence of 6.8%. It is obvious, from **Figure 10**, that pressure drop of water and for nanofluid increases with increasing Reynolds number. The small increase in pressure drop of nanofluid than water illustrates that using nanofluids with higher particle volume fraction may cause small penalty in pressure drop.

### *5.1.3 Friction factor*

The friction factor is influenced by velocity variation, pressure drop, and contact surface topologies with the fluid flow. The models were examined numerically and experimentally in Reynolds number range varying from 5000 to 20,000.

The experimental results of the friction factor are reasonably matching with the results obtained from correlations 12 and 13 for the plain tube with water flow with

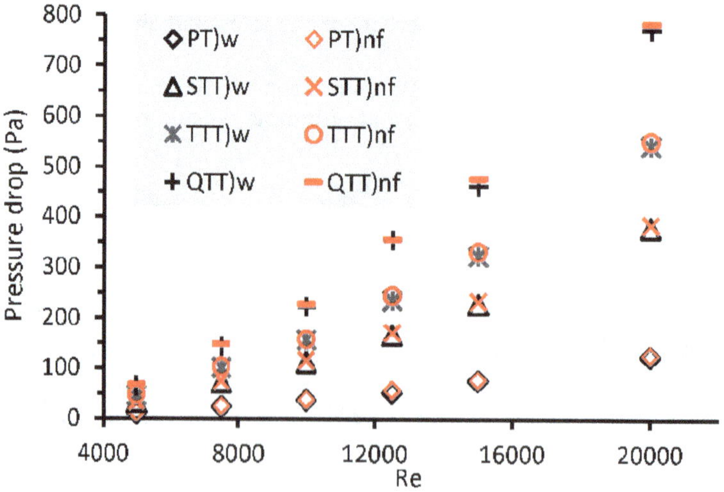


**Figure 10.**
*Simulation results of pressure drop obtained from water (w) and 0.1 vol.% $TiO_2$/water nanofluid (nf) for different inserts at various Reynolds numbers.*

a maximum deviation of 4.1%. Further verification was carried out by comparing the experimental results of the friction factor of single TT insert in the tube with water flow by comparison with the results gained by a correlation developed by Bergles, as in Eq. (15). The maximum deviation was 3.0%. The verification results are shown in **Figure 11**.

Measured friction factors coincide well with the calculated values from correlations of validation. As the fluid velocity increases, the friction factor decreases. Therefore, friction factor decreases with Reynolds number increasing. This is because Reynolds number increases the momentum, overcomes the viscous force of the fluid, and consequently lowers the shear between the fluid and the tube wall.

Reference [16] developed a correlation for the friction factor prediction for nanofluid flow in a plain tube, shown as Eq. (14). Comparison between the predicted results by Eq. (14) and the experimental results in the current investigations are shown in **Figure 12**. Good match between the experimental and empirical

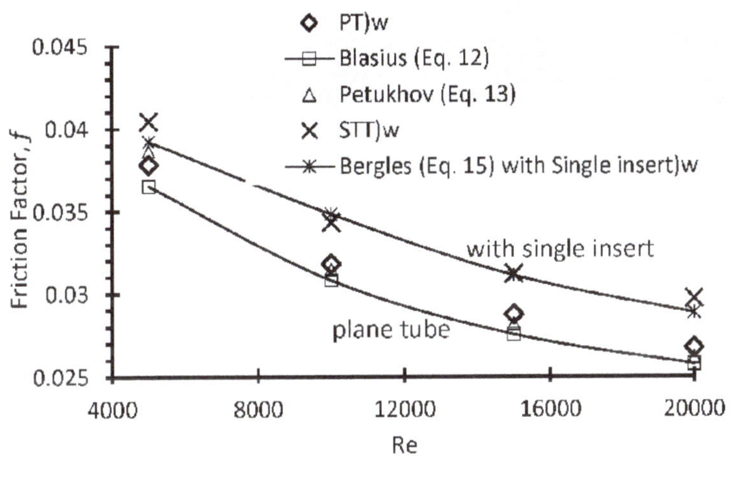


**Figure 11.**
*Verification of experimentally measured friction factors and those predicted from; (a) PT with water flow and prediction by Blasius (Eq. 12) and Petukhov (Eq. 13); (b) STT with water flow and prediction by Bergles (Eq. 15).*

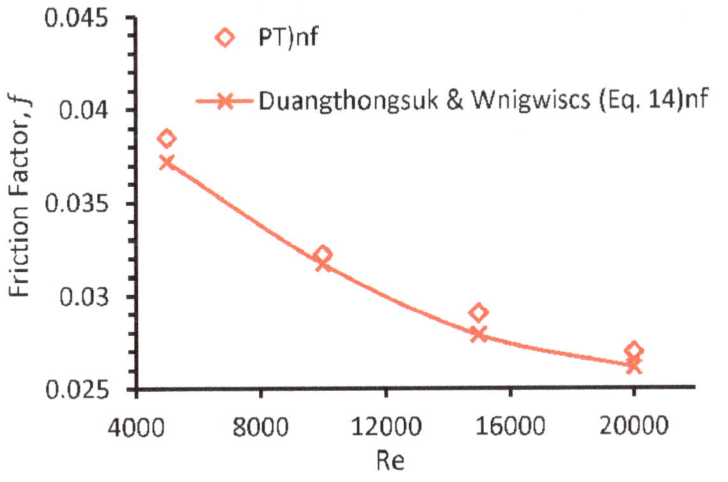


**Figure 12.**
*Verification of experimentally measured friction factors and predicted by Duangthongsuk and Wongwises (Eq. 14) for 0.1 vol.% $TiO_2$/water nanofluid flow in plain tube.*

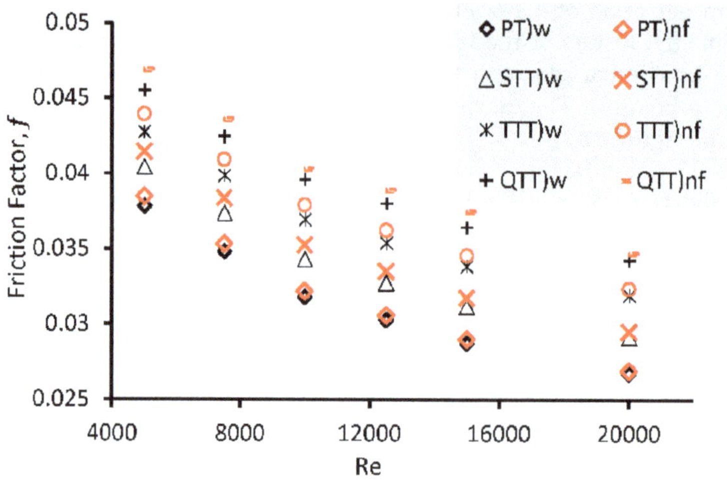


**Figure 13.**
*Experimental measurement results for the friction factor verses Reynolds number for plain tube and tubes with single, triple, and quintuple twisted tape inserts of flow cases of water flow and 0.1 vol.% $TiO_2$/water nanofluid flow.*

values has been achieved. Within the tested range of 5000–20,000 Re, the correlation shows overprediction of around 2.0–3.0% in the friction factor values compared to the experimental results.

The variations of the friction factor with Reynolds number for the tubes with different twisted tape inserts, with flows of water and nanofluid, are compared in **Figure 13**. The friction factors of the nanofluids are slightly higher than those of the base liquid. The tube fitted with plain twisted tape inserts (STT) when water flow has friction factor of 6.6–8.7% higher than plain tube. This is attributed to the flow blockage and swirl flow due to tape insert; for the same case, friction factor increases by 1.2–2.4% using nanofluid.

The additional dissipation of pressure of the fluid caused by the fluid disturbance due to increase of tape number results in an increase in pressure drop, which causes increase of friction factor. As the number of inserts increases, the pressure drop significantly increases. For water flow, the friction factors of TTT and QTT are 12.2–17.74% and 18.428–24.65% higher than that in plain tube. For nanofluid flow, the friction factors of TTT are 13.3–18.4% and for QTT are 19.6–25.2% higher than that in plain tube.

Results of predicted friction factor from the numerical simulation show the same trend as of the experimental ones, where differentiations between results within 6.04% are considered as an acceptable limit. Numerical simulation results of the friction factors for cases of inserts using water and nanofluid are shown in **Figure 14**. All cases demonstrated a slight increase in the numerically predicted friction factor values when 0.1 vol.% $TiO_2$ is used as working fluid. This is attributed to the slight increase in the viscosity of the nanofluid compared to the viscosity of pure water.

### 5.2 Thermal analysis

The heat transfer enhancement, in terms of Nusselt number, is influenced by velocity variation, friction factor, nanoparticles volume fraction, twisted tape dimensions, and other parameters. The four models were examined experimentally and numerically within Reynolds number ranging from 5000 to 20,000. The

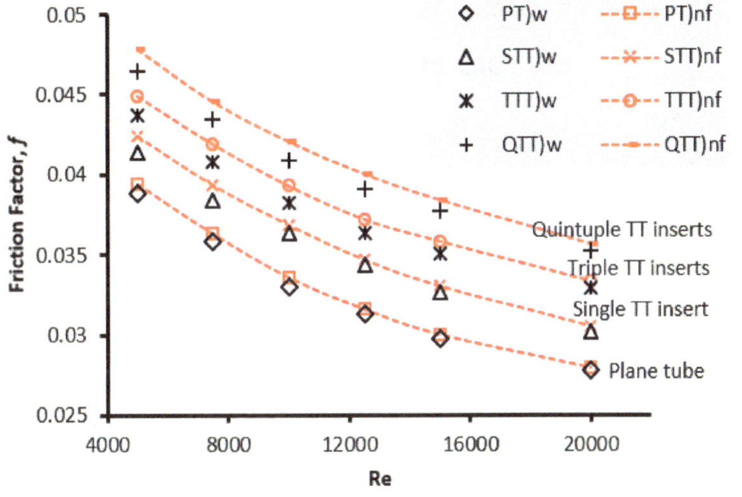


**Figure 14.**
*Numerical simulation results of the friction factor versus Reynolds number for plain tube and tubes with single, triple, and quintuple twisted tape inserts of flow cases of water flow and 0.1 vol.% $TiO_2$/water nanofluid flow.*

measured thermal parameters were verified by comparing the measurement results with well-established correlations to predict Nu. The verification results are shown in **Figure 15a** and **b**. The experimental results are matching those results obtained from correlations 8 and 15 with a deviation of 2.6–7.4%.

Further verification was carried out for the case of nanofluid flow in the pipe by comparing the experimental measurement Nu results with correlation 9 prediction Nu results, as shown in **Figure 16**. Very good agreement between the experimental and correlation results was demonstrated. The predicted results of Nu by the correlation are higher than the experimental results of Nu. As Re increased, a slight increase in the margin of error was observed.

The variations of Nusselt number with Reynolds number for the tubes with different twisted tape inserts are compared in **Figure 17** for both nanofluid and base fluid. It can generally be observed that the Nusselt number increases as the Reynolds number increases. This arises as a result of the momentum that overcomes the viscous force of the fluid as the Reynolds number increases and in effect diminishes the shear between the fluid and the tube wall.

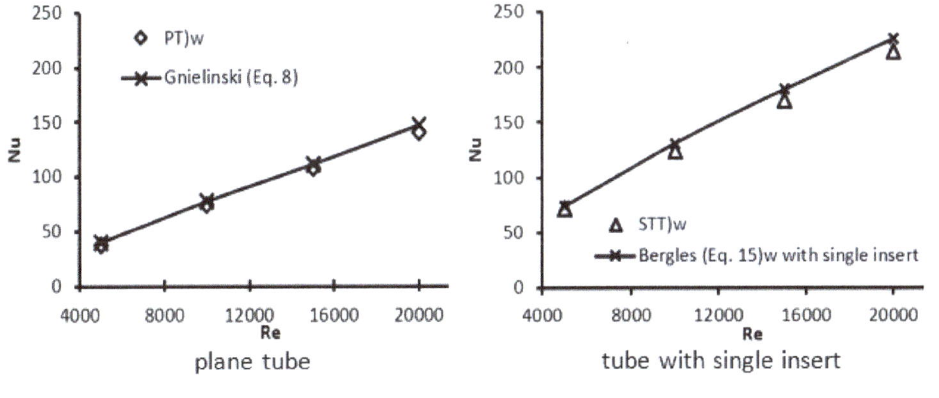


**Figure15.**
*Verification of experimental results for heat exchange performance by: (a) comparison with Gnielinski correlation for water flow in plain tube. (b) Comparison with Bergles correlation for water flow in tube with single TT insert.*

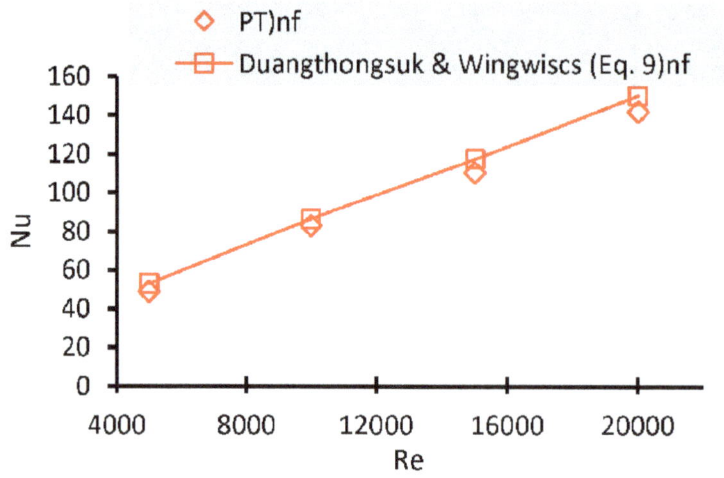


**Figure 16.**
*Validation of experimental results for Nusselt number to the plain tube with Duangthongsuk and Wingwiscs correlation for 0.1 vol.% $TiO_2$/water nanofluid flow.*

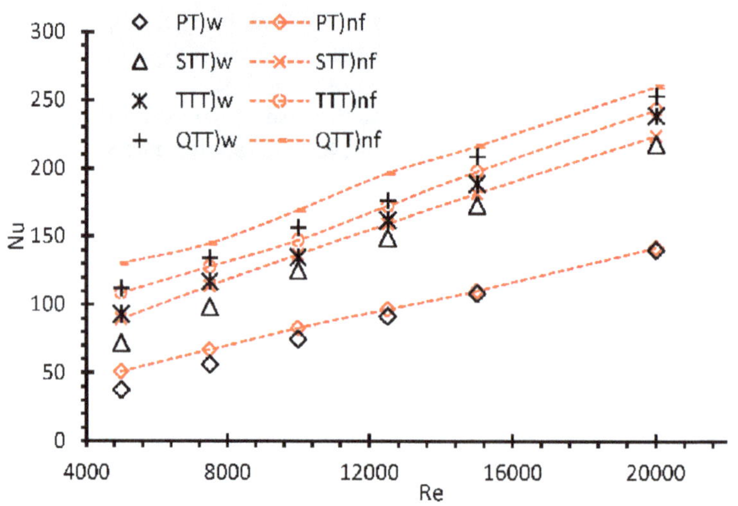


**Figure17.**
*Variation of Nusselt number with Reynolds number for plain tubes and tubes with single, triple, and quintuple twisted tape inserts operating with water flow and 0.1 vol.% $TiO_2$/water nanofluid flow.*

For water flow, the STT has Nusselt numbers of 43.4–63.2% higher than plain tube. This enhancement in heat transfer rate returns to the act of twisted tape that generates swirl motion, which leads to the better temperature distribution at the core region and increases turbulence intensity at near wall region that results in higher temperature gradient there and enhances heat transfer coefficient. Also, secondary flow with greater enhancement was realized at higher Reynolds numbers.

In case of water flow, Nusselt numbers of TTT and QTT are 50.12–83.74% and 57.48–100.066% higher than that in measured in PT. It is obvious that Nusselt number increases with increasing twisted tape number due to increase in secondary motion violence, which disperses the high temperature region near wall to uniformly distribute all over flow cross sectional area, which is recognized as the key factor of heat transfer enhancement.

For 0.1 vol.% $TiO_2$/water nanofluid flow, PT, STT, TTT, and QTT have Nusselt numbers of 1.3–30.4%, 46.1–83.2%, 53.8–97.5%, and 59.9–110.8%, respectively, higher than that in PT with water flow. This behavior is due to the fact that nanoparticles presented in the base liquid increase the thermal conductivity, which leads to an increase in heat transfer performance.

## 6. Conclusions

Heat transfer and fluid flow characteristics through tubes induced with multiple plain twisted tape inserts using water and 0.1 vol.% $TiO_2$/water nanofluid were investigated experimentally and numerically. The investigations were carried out for turbulent flow with Reynolds number ranging between 5000 and 20,000. The findings may be summarized as follows:

- It is found that tube fitted with quintuple twisted tape inserts performed highest in augmenting heat transfer. Nusselt numbers obtained by using quintuple twisted tapes inserts are up to 110.8% for nanofluid flow and 100.0% for water flow higher than in plain tube with water flow.

- As the number of twisted tapes increases, higher Nusselt number is obtained, compared to plain tube.

- The pressure drop and friction factor increase significantly by using twisted tape inserts and slightly by nanofluid.

- Using nanofluids with twisted tape inserts are resulting in more heat transfer augmentation than using each one individually.

- Using twisted tape inserts need to increase fluid pumping power due to high pressure drop produced. When using nanofluid without inserts, pumping power is not affected as the increase in the pressure drop is negligible compared to the case of pure water flow.

Accordingly, the combined technique of twisted tape inserts and nanofluid is recommended for heat exchanger application, as the heat transfer is considerably enhancing, with reasonable penalty in pumping power.

## Acknowledgements

The authors acknowledge the supports provided by both, Al-Mustansiriyah University, Baghdad, Iraq for providing all the logistic and technical supports to produce this research using the relevant labs in the mechanical engineering department and Universiti Teknologi PETRONAS—Malaysia for providing the financial support to publish the research results using the budget under grant YUTP —FRG, CS: 015LC0-026.

## Abbreviations

| | |
|---|---|
| $D$ | tube diameter (mm) |
| $f$ | friction factor |

| | |
|---|---|
| $I$ | turbulence intensity |
| $K$ | thermal conductivity (W/m•K) |
| $L$ | tube length (mm) |
| $Nu$ | Nusselt number |
| $\Delta p$ | pressure drop (Pa) |
| $Pr$ | Prandtl number |
| $\bar{\bar{q}}$ | heat flux (W/m$^2$) |
| $Re$ | Reynolds number |
| $V$ | mean velocity (m/s) |
| w | tape width (mm) |
| $y$ | tape pitch 180° (mm) |
| $\delta$ | tape thickness (mm) |
| $v$ | kinematic viscosity (m$^2$/s) |
| $\varphi$ | nanoparticles volume fraction (%) |
| $\rho$ | fluid density (kg/m$^3$) |
| $\mu$ | fluid dynamic viscosity (kg/m s) |
| $c_p$ | fluid-specific heat (J/kg•K) |

### Subscripts

| | |
|---|---|
| b | bulk |
| h | hydraulic |
| in | inlet |
| nf | nanofluid |
| s | surface |
| w | water |

### Acronyms

| | |
|---|---|
| PT | plain tube |
| STT | tube with single twisted tape |
| TTT | tube with triple twisted tapes |
| QTT | tube with quintuple twisted tapes |

## Author details

Hussain H. Al-Kayiem[1]*, Muna S. Kassim[2] and Saud T. Taher[2]

1 Mechanical Engineering Department, Universiti Teknologi PETRONAS, Perak, Malaysia

2 Mechanical Engineering Department, Al-Mustansiriyah University, Baghdad, Iraq

*Address all correspondence to: hussain_kayiem@utp.edu.my

## References

[1] Kakaç S, Liu H. Heat Exchangers: Selection, Rating and Thermal Design. 2nd ed. USA, New York: CRC Press LLC, Taylor & Francis Inc.; 2002

[2] Kumar CN, Murugesan P. Review on twisted tapes heat transfer enhancement. International Journal of Science & Engineering Research. 2012; **3**(4):1-9

[3] Al-Kayiem HH, Saw CL, Afolabi L. Review on nanomaterials for thermal energy storage technologies. Journal of Nanoscience and Nanotechnology. 2013;**3**:60-71. DOI: 10.2174/ 22113525113119990011

[4] Dey D, Kumar P, Samantaray S. A review of nanofluid preparation, stability, and thermo-physical properties. Heat Transfer - Asian Research. 2017;**46**(8):1413-1442. DOI: 10.1002/htj.21282

[5] Ali N, Teixeira JA, Addali A. A review on nanofluids: Fabrication, stability, and thermophysical properties. Journal of Nanomaterials. 2018;**2018**: 6978130. DOI: 10.1155/2018/6978130

[6] Narayanan MV, Rakesh SG, Nanofluids SG. A review on current scenario and future prospective. IOP Conference Series: Materials Science and Engineering. 2018;**377**:012084. DOI: 10.1088/1757-899X/377/1/012084

[7] Said Z, Saidur R, Hepbasli A, Rahim NA. New thermophysical properties of water based $TiO_2$ nanofluid-the hysteresis phenomenon revisited. International Communications in Heat and Mass Transfer. 2014;**58**:85-95. DOI: 10.1016/j. icheatmasstransfer.2014.08.034

[8] He Y, Men Y, Zhao Y, Lu H, Ding Y. Numerical investigation into the convective heat transfer of $TiO_2$ nanofluids flowing through a straight tube under the laminar flow conditions. Applied Thermal Engineering. 2009;**29**: 1965-1972. DOI: 10.1016/j. applthermaleng.2008.09.020

[9] Bergles AE, Manglik RM. Heat transfer and pressure drop correlations for twisted tape inserts in isothermal tube: Part II-transition and turbulent flows. Transaction of ASME Journal of Heat Transfer. 1993;**115**:890-896. DOI: 10.1115/1.2911384

[10] Bhattacharyya S, Saha S, Saha SK. Laminar flow heat transfer enhancement in a circular tube having integral transverse rib roughness and fitted with centre-cleared twisted-tape. Experimental Thermal and Fluid Science. 2013;**44**:727-735. DOI: 10.1016/ j.expthermflusci.2012.09.016

[11] Bhattacharyya S, Chattopadhyay H, Benim AC. Simulation of heat transfer enhancement in tube flow with twisted tape insert. Progress in Computational Fluid Dynamics An International Journal. 2017;**17**(3):193-197. DOI: 10.1504/PCFD.2017.084356

[12] Bhattacharyya S, Chattopadhyay H, Benim AC. Computational investigation of heat transfer enhancement by alternating inclined ribs in tubular heat exchanger. Progress in Computational Fluid Dynamics, An International Journal. 2017;**17**(6):390-396. DOI: 10.1504/PCFD.2017.088818

[13] Bhattacharyya S, Benim AC, Chattopadhyay H, Banerjee A. Experimental investigation of heat transfer performance of corrugated tube with spring tape inserts. Experimental Heat Transfer. 2018;**32**(5):411-425. DOI: 10.1080/08916152.2018.1531955

[14] Pourrajabi M, Nezhad AR, Pourrajabi MR. Numerical study of turbulent flow and heat transfer of a nanofluid in a circular tube with twisted

tape insert. In: International Symposium on Advances in Science and Technology. Iran: 7th SASTech; 2013. DOI: 10.1007/s10973-017-6900-5

[15] Maddah H, Farokhi M, Aghayari R, Jahanizadeh S, Ashtary K. Effect of twisted-tape turbulator and nanofluid on heat transfer in a double pipe heat exchanger. Journal of Engineering. 2014;**2014**:920970. DOI: 10.1155/2014/920970

[16] Safikhani H, Eiamsa-ard S. Multi-objective optimization of $TiO_2$-water nanofluid flow in tubes fitted with multiple twisted tape inserts in different arrangement. Journal of Transport Phenomena in Nano-Micro Scale. 2015; **3**(2):89-99. DOI: 10.7508/tpnms.2015.02.003

[17] Saleh FA. Computational and experimental investigations on thermal and fluid flow characteristics for different models of tapes inserts with $TiO_2$/water nanofluid under turbulent conditions. International Journal of Mechanical & Mechatronics Engineering. 2018;**19**(2):1-14

[18] Eiamsa-ard S, Wongcharee K. Convective heat transfer enhancement using Ag-water nanofluid in a micro-fin tube combined with non-uniform twisted tape. International Journal of Mechanical Sciences. 2018;**146–147**: 337-335. DOI: 10.1016/j.ijmecsci.2018.07.040

[19] Xu C, Xu S, Wei S, Chen P. Experimental investigation of heat transfer for pulsating flow of GOPs-water nanofluid in a microchannel. International Communications in Heat and Mass Transfer. 2020;**110**:104403. DOI: 10.1016/j.icheatmasstransfer.2019.104403

[20] Chun BH, Kang HU, Kim SH. Effect of alumina nanoparticles in the fluid on heat transfer in double-pipe heat exchanger system. Korean Journal of Chemical Engineering. 2008;**25**(5): 966-971. DOI: 10.1007/s11814-008-0156-5

[21] Duangthongsuk W, Wongwises S. Effect of thermophysical properties models on the predicting of the convective heat transfer coefficient for low concentration nanofluid. International Communications in Heat and Mass Transfer. 2008;**35**(10): 1320-1326. DOI: 10.1016/j.icheatmasstransfer.2008.07.015

[22] Santra AK, Sen S, Chakraborty N. Study of heat transfer due to laminar flow of copper-water nanofluid through two isothermally heated parallel plates. International Journal of Thermal Sciences. 2009;**48**(2): 391-400. DOI: 10.1016/j. ijthermalsci.2008.10.004

[23] Namburu PK, Das DK, Tanguturi KM, Vajjha RS. Numerical study of turbulent flow and heat transfer characteristics of nanofluids considering variable properties. International Journal of Thermal Sciences. 2009;**48**(2):290-302. DOI: 10.1016/j.ijthermalsci.2008.01.001

[24] Available from: https://www.visualcapitalist.com/ [Retrieved on: 26 May 2020]

[25] Pak BC, Cho YI. Hydrodynamic and heat transfer study of dispersed fluids with submicron metallic oxide particles. Experimental Heat Transfer 1998; **11**(2):151-170. DOI:10.1080/08916159808946559

[26] Batchelor GK. The effect of Brownian motion on the bulk stress in the suspension of spherical particles. Journal of Fluid Mechanics. 1977;**83**(1):97-117. DOI: 10.1017/S0022112077001062

[27] Incropera FP, DeWitt PD, Bergman TL, Lavine AS. Fundamentals of Heat and Mass Transfer. USA, New York: John-Wiley & Sons Inc.; 2006

# 9

# Plate Heat Exchangers: Artificial Neural Networks for their Design

*Bayram Kılıç*

**Abstract**

Heat exchangers are installation components in which two fluids with different temperatures are separated from each other with the help of plates and these plates make heat transfer between two fluids. The biggest advantage of plate heat exchangers over other type of heat exchangers is their heat transfer efficiency. The thinness of the plates separating the two fluids compared to other material alternatives increases the amount of heat transfer and thus reduces the heat losses that may occur during heat transfer. Plate heat exchangers are not only efficient but also prevent the formation of residue and dirt that can accumulate over time in the used applications. It also protects the system against excessive pressure that may occur in the installation. In this study, some information about plate heat exchangers is given such as classification, plate geometry, pressure losses, and thermal calculations. Also, the data obtained from the experimental work were used to obtain some relativity in order to use it in plate heat exchangers and artificial neural networks (ANN) method was used for this purpose. Artificial neural network method is used in many engineering applications. The most important advantages of this method are rapid formation, simple formation and high learning capacity.

**Keywords:** plate heat exchanger, heat transfer, artificial neural networks, thermodynamic

## 1. Introduction

One of the most common and used processes in engineering practice is the heat exchange between two or more fluids at different temperatures. The devices in which these changes are made are generally called heat exchangers (HEX) and in practice they are used in power plants, chemical industries, heating, air conditioning, cooling, vehicles, electronic devices, use of alternative energy sources, can be found in many places. As can be seen from the above written, HEXs that is used in various applica-tions in practice, can be used in different structures, capacities, sizes and types according to the intended use. HEXs are the most important heat transfer equipments of the industry and they can be seen in different kinds and capacities at almost every stage of chemistry, power plants, cooling, heating and air conditioning processes under different names such as evaporator, condenser, heater and cooler. From the point of view of machine and chemical engineering education, the HEXs are a very good application for this branch of science which contains all of the basic subjects of these engineering branches: materials, strength, thermodynamics and heat transfer science. As it can be understood, HEXs are always used in daily life. Thus, the design ought to be followed up to the best detail, and the nearest investigation results ought to be gotten by utilizing related projects, and studies ought to be directed to improve the designs. Decreasing the amount of heat transferred in the HEX causes the performance of the HEX to reduce. This implies loss of performance in the energy system utilizing HEXs. The improvement of heat transfer permits the system dimensions to be kept at the proper values, thus reducing system cost and operating costs.

In the case of plate heat exchangers (PHEX), the surfaces with the basic heat transfer are made of thin metal plates. These metal surfaces might be level or wavy. They can be examined in three groups: sealed plate, spiral plate and lamellar. Heating, cooling and ventilation applications achieve the high efficiency, affordabil-ity and compact design they require thanks to PHEXs. By replacing tubular HEXs with daytime HEXs with PHEXs, PHEXs have gained a rapidly increasing market share in the entire industry. The wide selection range of plates in various sizes and materials provides superior flexibility to PHEXs. This flexibility is a great advantage for many thermal process HEXs [1].

Some correlations can be obtained using the data obtained from experimental studies and then these correlations can be used in energy applications. Artificial neural networks (ANN) method is mostly used for this purpose in energy applications. Artificial neural networks estimate the output value corresponding to this data using, data that makes up the network. It is known that artificial intelligence methodologies are used in the analysis of PHEXs as in energy systems. However, an analysis using ANN methodology for heat transfer rate and effectiveness in PHEXs for different surface angles has been performed in a small number. This study focuses on the usability of ANN methodology for performance analysis of PHEXs.

## 2. Plate heat exchangers

In PHEXs, the surfaces with the basic heat transfer are made of thin metal plates. These metal surfaces might be level or wavy. They generally have a higher total heat transfer coefficient than shell-tube type HEXs. **Figure 1** shows the structure of a PHEX.

Plate heat exchangers; They can be examined in three groups as sealed plated, spiral plated and lamellar [1]:

a. Heat exchangers with sealed plates; Heat exchangers with sealed plates are made by packing the thin metal plates into a frame and packing them. On each side of each metal plate there are holes for fluid to pass through. When the plates are assembled and packed, using appropriate seals prevents the fluids from

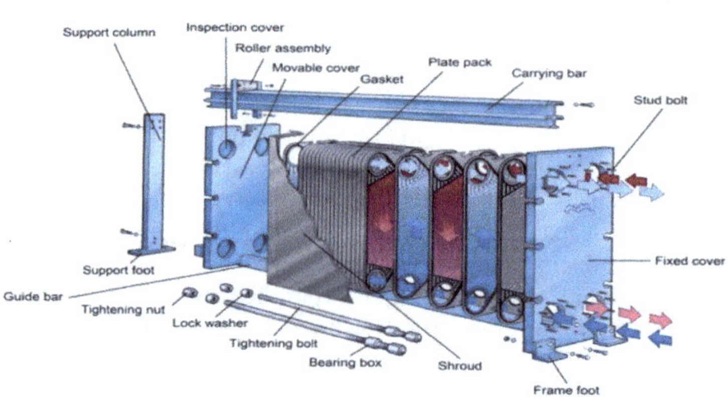


**Figure 1.**
*The structure of the plate heat exchanger [2].*

intermixing and leaking out. The hot and cold fluids flow through the spaces between the plates without mixing. The plates are made wavy to provide rigidity, to stabilize the distance between the plates and to improve heat transfer.

b. Spiral PHEXs; Plate type heat exchangers in which heat transfer surfaces are formed from plates and not formed by cylindrical pipes. Spiral plate heat exchangers are obtained by spirally wrapping two elongated thin metal plates 150–1800 mm wide each for a fluid, forming two spiral parallel edges. A uniform clearance can be maintained between the two plates. Both sides of the plates are covered with sealed covers. Various flow configurations are possible and different types of spiral heat exchangers can be manufactured depending on the flow configurations.

c. Lamellar heat exchangers; It is obtained by placing a bundle made of pipes (lamella) placed in a body. The lamella is usually held together by a point or an electric sewing source. One of the fluids flows through the lamellar tubes and flows through the other fluid lamellae. There are no surprise plates in the body. The flow is single-pass and the same direction or opposite flow arrangement can be used.

### 2.1 Design of plate heat exchangers

The heat exchangers are mainly regarded as pressure vessels. For this reason, the choice of design pressure, design temperature and appropriate material is crucial in the design of PHEXs. Once these criteria are established, the necessary calculations are made for the design. In the design of a PHEX, the surface A for heat transfer, the logarithmic mean temperature difference ($\Delta T_m$) and the total heat transfer coefficient (K), pressure drop, physical size and economics are significant factors.

### 2.2 The importance of material used in plate heat exchangers

One of the factors that enable PHEXs to work without problems for long years in the desired thermal conditions is that the heat exchanger material is of a certain quality. In order for materials to be subjected to ISO 9001 quality testing, it is necessary to mark each item for retrospective analysis.

### 2.3 Plate material

It is very important to choose the plate material according to the flow rate used and the maximum desired working strength. Generally, the plate materials given in **Table1** are used and the most commonly used material type is 1.440 / AISI 316 [2, 3].

| | |
|---|---|
| AISI 304 | NI 200/201 |
| AISI 316 | G-30 |
| AISI 316 L | C- 4 |
| 254 SMO | INCONEL 625 |
| 654 SMO | INCONEL 825 |
| TITANYUM | MONEL 400 |
| Ti – PD | TANTALUM |
| C-276 | C – 22 |

**Table 1.**
*Material types of heat exchangers.*

PHEXs are produced from different materials in today. These alternative materials are aluminum, carbon steel, stainless steel, nickel alloys, zirconium and titanium. For many chemical processes, Zirconium is a cost-effective material of construction in shielding process equipment systems from destructive corrosive leaks. With incredibly high resistance to corrosion, Zirconium HEXs can withstand some of the harshest situations. This translates to decreased maintenance expense, with downtime kept to a minimum. Copper has many preferred specifications for thermally efficient and durable HEXs. Above all else, copper is an perfect conductor of heat. This means that copper's high thermal conductivity allows heat to pass through it speedly. Other required specifications of copper in HEXs include its corrosion resistance, biofouling resistance, maximum allowable stress and internal pressure, creep rupture strength, fatigue strength, hardness, thermal expansion, specific heat, antimicrobial specifications, yield strength, high melting point, ease of fabrication, and ease of joining.

In chemical processes, the use of titanium HEXs has been found to be a cost-effective method of resisting leaks from corrosion on a process line. Titanium HEXs has superior corrosion resistance, high heat transfer efficiency, non-breaking property and provide an extended service life compared to other materials.

## 2.4 Gasket material

In plate heat exchangers it is the limit factor contour. Therefore, it is very important to choose the 'right' gasket material. Various sealing types according to application are listed below. One of the most important points that should not be forgotten is that the exchanger gaskets are produced in three qualities, normal, sulfur and peroxide. These three quality seal materials are peroxidized when compared with each other, and the material always ensures the best performance in terms of operating conditions [4, 5].

## 2.5 Plate geometries used in plate heat exchangers

Almost all plates used today are of the fishermen type. Traditionally, fish hatch plates are now made with two arrows, one is the "high-theta" with high resistance to flow and the other is the "low-theta" with low resistance against flow.

These two types of plates can be combined in three different ways, each with different characteristics in terms of heat transfer and pressure drop.

The symmetrical plates and the primary and secondary sides are geometrically identical and the plate deck can best be used on only one side. A later innovation is the asymmetrical plaque, these two teams have arrows, one high-theta and the other low-theta. Plates having different plate geometries are given in **Figures 2–6**

These two plates can be combined in the form of six different flow channels, each with different heat transfer characteristics. Gasket materials commonly used in plate heat exchangers are given **Table 2**.

The asymmetric plates also provide asymmetrical plate deck formation, where the geometry of the primary side is different from the geometry of the second-ary side. These two aspects can be used separately in the best way, thus reducing the required heat transfer surface and providing better utilization of the existing pressure reduction.

Plates are manufactured in different thicknesses up to 8 mm by 0.4 mm in accordance with the designed pressure rating and maintenance stability. Thin plates are not used in cleaning applications and other applications where the unit needs to be opened. The most commonly used materials in geothermal applications are AISI 316, 254 SMO and commercial pure (grade 1) titanium. Other high alloy materials are also used where geothermal fluids are very corrosive.

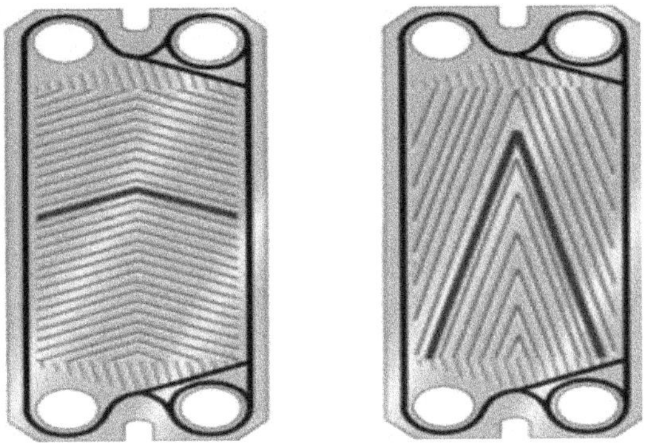

**Figure 2.**
*Two arrowhead chevron plates.*

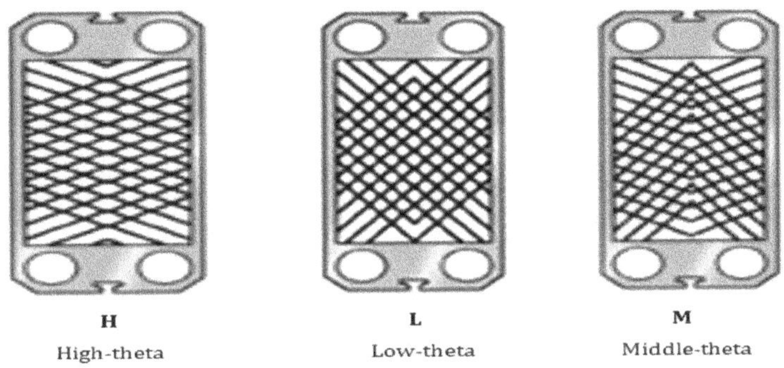


**Figure 3.**
*Three different channels by combining two symmetrical plates.*

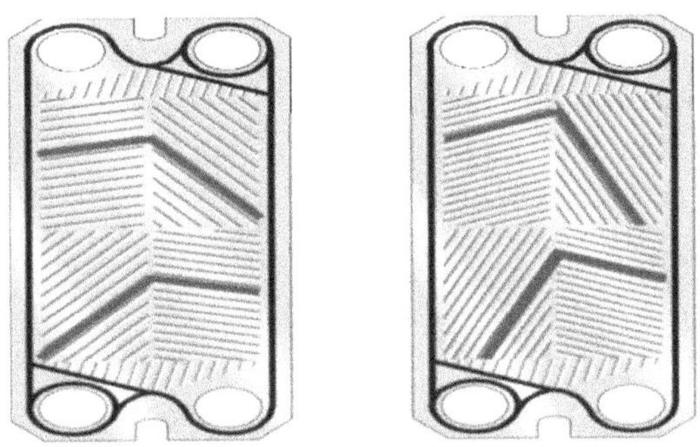

**Figure 4.**
*Asymmetrical high and low-theta plates.*

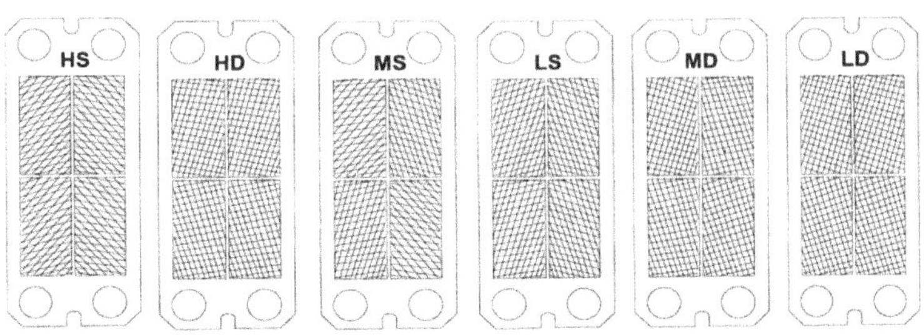


**Figure 5.**
*Six different flow channels combined from two different platters.*

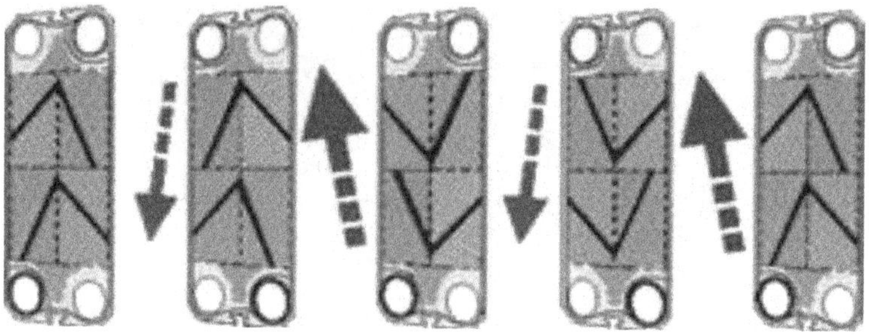

**Figure 6.**
*Asymmetric best use at PHE.*

| |
|---|
| Nitrile (Nbr) |
| Hnbr |
| Etilen Propilen (Epdm) |
| Florokarbon (Fpm) |
| Viton Gf |
| PTFE Encapsulated NBR |

**Table 2.**
*Gasket materials commonly used in plate heat exchangers.*

## 3. Thermodynamic analysis of plate heat exchangers

If the heat transfer in a HEX is assumed to be only between the fluids in it and the absence of a heat loss in the center, it can be written in the PHEX with the following relation [6–9]:

$\dot{Q}$ = Heat in the HEX (W).

= Heat given by the hot fluid (W).

= Heat received by the cold fluid (W).

$K$ = Total heat transfer coefficient (W/m$^2$°K)

$$\dot{Q} = K\,A\,\Delta T_m \tag{1}$$

The temperatures given and received during the cooling and heating of hot and cold fluids can be found by the mass fluxes of fluids and the enthalpy of entrances and exits and can be written as [10]:

$$\dot{Q} = \dot{m}\left(h_i - h_o\right) \tag{2}$$

If the temperatures of the fluids change when the heat is taken and given, the amount of heat that passes is [10]:

$$\dot{Q} = \dot{m}_h c_{ph}\left(T_{hi} - T_{ho}\right) = \dot{m}_c\, c_{pc}\left(T_{co} - T_{ci}\right) \tag{3}$$

After a certain period of operation, particles, metal salts or various chemical elements may accumulate in the fluids on the heat exchanger surfaces. Occasionally, due to corrosive effects, an oxidation layer may form on these surfaces. All these layers are brought to an additional thermal resistance during heat transfer. This contamination resistance (or factor), as indicated by the $R_f$ symbol, can be found as follows, in the sense that the thermal resistances of the heat transfer surfaces are dirty and clean [1]:

$$R_f = \frac{1}{K_{dirty}} - \frac{1}{K_{clean}} \tag{4}$$

Due to the roughness of the metal surfaces, there is a contact resistance between these two surfaces due to the poor contact between the two metals. The contact resistance on two surfaces causes a decrease in temperature on these surfaces. In order to take these situations into consideration, a resistance definition can be made as follows [1]:

$$R_t = \frac{(T_A - T_B)}{\frac{\dot{Q}}{A}} \tag{5}$$

Consequently, the total heat transfer coefficient at the surface of the HEX can be found by the following Eq. [1]:

$$\frac{1}{K} = \frac{1}{\alpha_1} + R_{f1} + \frac{\delta_1}{\lambda_1} + R_{t,1-2} + \frac{\delta_2}{\lambda_2} + R_{t,2-3} + \frac{\delta_3}{\lambda_3} + R_{f2} + \frac{1}{\alpha_2} \tag{6}$$

In the construction of the heat calculations of the heat exchangers, the expression of the mean logarithmic temperature difference ($\Delta T_m$) is required if Eq. (1) is used. The mean logarithmic temperature difference value is determined by the flow rate in the heat exchanger. **Figure 7** and **Figure 8** show temperature distributions along the length of the HEX when the flow is parallel and opposite.

The mean logarithmic temperature difference ($\Delta Tm$) can be expressed as:

$$\Delta T_m = \frac{\Delta T_1 - \Delta T_2}{\ln \frac{\Delta T_1}{\Delta T_2}} \tag{7}$$

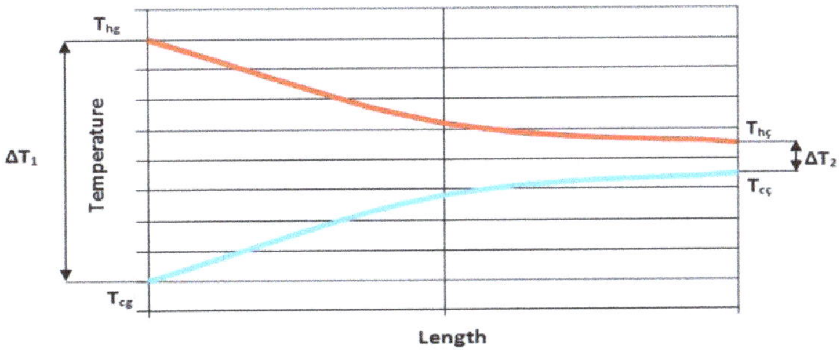


**Figure 7.**
*Temperature distribution in a parallel flow HEX [1].*

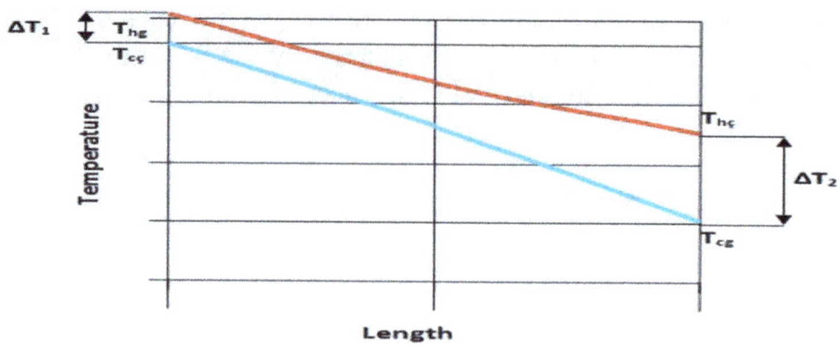


**Figure 8.**
*Temperature distribution in the reverse flow HEX [1].*

If the last equation is moved to Eq. (1):

$$\dot{Q} = \frac{KA(\Delta T_1 - \Delta T_2)}{\ln\dfrac{\Delta T_1}{\Delta T_2}} \tag{8}$$

expression is obtained.

The efficiency of HEXs can be calculated with the help of the following Equation [11]:

$$\varepsilon = \frac{\dot{Q}}{\dot{Q}_{\max}} \tag{9}$$

$C_h = \dot{m}_h\, c_{ph}$ and $C_c = \dot{m}_c\, c_{pc}$ are the thermal capacity values of hot and cold fluids, the actual heat transfer in the HEX can be written as [12, 13]:

$$\dot{Q} = C_h\left(T_{hi} - T_{ho}\right) = C_c\left(T_{co} - T_{ci}\right) \tag{10}$$

If the $Q_{max}$ value is defined as the maximum possible heat transition, it can be written as follows, provided that it is smaller than the $C_h$ or $C_c$ thermal capacity outputs [14, 15]:

$$\dot{Q}_{\max} = C_{\min}\left(T_{hi} - T_{ci}\right) \tag{11}$$

### 3.1 Determination of heat transfer rate and effectiveness in plate heat exchangers using alternative an approach

Artificial Neural Network (ANN) was designed for the generalizations of biological nervous systems' mathematical models. When the simplified neurons were introduced, the first steps were taken towards the neural networks, which are also known as connectionist models or parallel distributed processing. The artificial neurons are the main component of the process, and they are also known as simply neurons and nodes. In order to represent the effects of synapses in a simplified mathematical neuron model, the connection weights modulating the effects of associated input signals are utilized, while the nonlinear characteristics of neurons are represented by using the transfer function. Then, the impulse of neuron is calculated as the weighted sum of the input signals and transformed by the transfer function. The artificial neurons are provided with the capability of learning by adjusting the weights in accordance with the preferred learning algorithm.

The data obtained from the experimental work [16] were used to obtain some relativity in order to use it in PHEXs and ANN method was used for this purpose. ANN method is used in many engineering applications. The most important advantages of this method are rapid formation, simple formation and high learning capacity. Two methods are used in thermal calculations in PHEXs. These methods are mean logarithmic temperature difference and effectiveness-NTU methods. In this study, the data obtained from the experiments [16] (hot water inlet–outlet temperature, cold water inlet–outlet temperature, flow rate and plate surface angle) are quite suit-able for using the mean logarithmic temperature difference method. These experi-mental data are the input variables of the network created for the ANN method. The effectiveness and the heat transfer rate of the PHEX are output variables. All data and calculated results obtained from experimental data can be modeled with ANN method and can be used to obtain results depending on different variables.

MATLAB Toolbox was used for ANN methodology. In the training of the data, the number of neurons in the hidden layer was changed between 3 and 12. 80% of the 227 data obtained from the experiments were randomly selected for training and 20% for testing. The generated network has 4 input variables as hot water inlet temperature ($T_{hi}$), cold water inlet temperature ($T_{ci}$), fluid flow (m) and plate surface angle (β). Effectiveness (ε) is output variable. The generated artificial neural network is shown schematically in **Figure 9**. In the created network, LOGSIG is selected as the transfer function, Forward Back Prop for the network type, and TRAINLM and TRAINSCG as the training function. The number of epochs used was 1000 values. Each network is run 10 times in order to get the best value.

To obtain the best result from ANN method, different algorithms and hid-den neurons are used in different numbers. The values of root mean square error (RMSE), the coefficient of determination ($R^2$) and coefficient of variation (cov) obtained for heat transfer rate and effectiveness value are given in **Tables 3** and **4**. The performance evaluation criteria obtained from the ANN model are given in **Table 5**. The best $R^2$ value for heat transfer rate was obtained as 0.999636 for the TRAINLM-5 training function. For the best effectiveness value, $R^2$ value was 0.999565 for the TRAINLM-12 training function.

In determining the heat transfer rate and effectiveness of the PHEX, the equations obtained from the ANN method were used. In the above equations, $E_i$ is the neuron summation function and $F_i$ is the neuron activation function. In represents the input variables and bn represents the bias value. The coefficients used in the formulas represent the weight values of each neuron's summation function of the hidden layer of the training network. In the above equations, hot water inlet temperature ($T_{hi}$), cold water inlet temperature ($T_c$), fluid flow (m) and plate surface angle (β) are used in ANN as 4 input variables. The weight coefficients and bias values used for the determination of the heat transfer rate and effectiveness are given in **Tables 6** and 7 respectively.

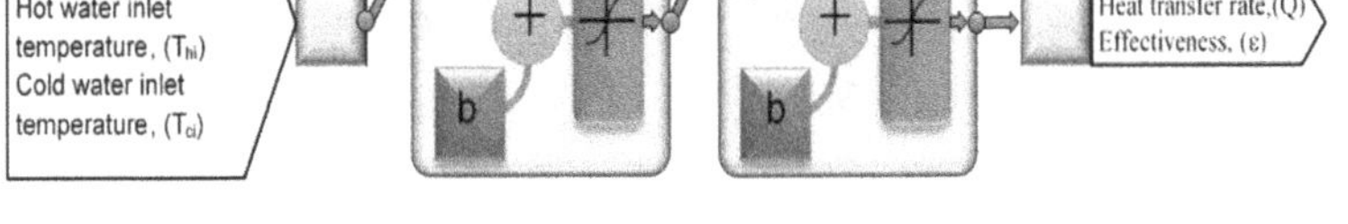


**Figure 9.**
*Schematic representation of artificial neural network [17].*

| Algorithm-Neuron | RMSE | cov | $R^2$ |
|---|---|---|---|
| Lm-3 | 53.5148 | 0.02003 | 0.999618 |
| Lm-4 | 52.7873 | 0.01975 | 0.999628 |
| **Lm-5** | **52.2255** | **0.01954** | **0.999636** |
| Lm-6 | 58.6761 | 0.02196 | 0.999540 |
| Lm-7 | 54.9293 | 0.02056 | 0.999597 |
| Lm-8 | 63.3763 | 0.02372 | 0.999464 |
| Lm-9 | 57.5448 | 0.02153 | 0.999558 |
| Lm-10 | 60.1653 | 0.02251 | 0.999517 |
| Lm-11 | 58.8812 | 0.02203 | 0.999537 |
| Lm-12 | 129.645 | 0.04852 | 0.997758 |
| SCG-3 | 95.9341 | 0.03590 | 0.998772 |
| SCG-4 | 76.4545 | 0.02861 | 0.999220 |
| SCG-5 | 75.8600 | 0.02839 | 0.999232 |
| SCG-6 | 61.0010 | 0.02283 | 0.999503 |
| SCG-7 | 71.3101 | 0.02669 | 0.999321 |
| SCG-8 | 61.6170 | 0.02306 | 0.999493 |
| SCG-9 | 63.3772 | 0.02372 | 0.999464 |
| SCG-10 | 54.6721 | 0.02046 | 0.999601 |
| SCG-11 | 62.3508 | 0.02333 | 0.999481 |
| SCG-12 | 59.8349 | 0.02239 | 0.999522 |

*The bold values are the best values of root mean square error (RMSE), the coefficient of determination (R2) and coefficient of variation (cov) obtained from the ANN model for to estimate of heat transfer rate and effectiveness.*

**Table 3.**
*The statistical values of the network for predicting the heat transfer rate.*

Heat transfer rate in the plate heat exchanger can be calculated by the following equations depending on the hot water inlet temperature ($T_{hi}$), cold water inlet temperature ($T_c$), fluid flow (m) and plate surface angle (β) [17].

$$\begin{aligned} E_6 = & -112.6039F_1 + 83.284F_2 - 89.934F_3 \\ & -70.4233F_4 + 229.4711F_5 - 107.9219 \end{aligned} \tag{12}$$

$$\dot{Q} = \left( \frac{1}{1 + e^{-E_6}} \right) 3920 \tag{13}$$

Similarly, the effectiveness in the PHEX can be calculated from the following equations depending on the hot water inlet temperature ($T_{hi}$), cold water inlet temperature ($T_c$), fluid flow (m) and plate surface angle (β) [17].

$$\begin{aligned} E_{13} = & \, 21.3112F_1 - 2.7185F_2 + 21.2966F_3 + 7.2653F_4 - 0.056092F_5 \\ & - 26.4603F_6 - 19.2568F_7 - 12.4354F_8 + 3.0543F_9 - 46.6099F_{10} \\ & + 58.215F_{11} + 0.28895F_{12} - 1.5743 \end{aligned} \tag{14}$$

$$\varepsilon = \frac{1}{1 + e^{-E_{13}}} \tag{15}$$

| Algorithm-Neuron | RMSE | cov | $R^2$ |
|---|---|---|---|
| Lm-3 | 0.0102575 | 0.022758 | 0.999475 |
| Lm-4 | 0.0096802 | 0.021477 | 0.999532 |
| Lm-5 | 0.0134191 | 0.029773 | 0.999101 |
| Lm-6 | 0.0097002 | 0.021522 | 0.999530 |
| Lm-7 | 0.0099292 | 0.022030 | 0.999508 |
| Lm-8 | 0.0097623 | 0.021660 | 0.999524 |
| Lm-9 | 0.0100210 | 0.022234 | 0.999498 |
| Lm-10 | 0.0096225 | 0.021350 | 0.999538 |
| Lm-11 | 0.0109986 | 0.024403 | 0.999396 |
| **Lm-12** | **0.0093323** | **0.020706** | **0.999565** |
| SCG-3 | 0.0124508 | 0.027625 | 0.999226 |
| SCG-4 | 0.0109349 | 0.024261 | 0.999403 |
| SCG-5 | 0.0129422 | 0.028715 | 0.999164 |
| SCG-6 | 0.0114423 | 0.025387 | 0.999346 |
| SCG-7 | 0.0121871 | 0.027040 | 0.999258 |
| SCG-8 | 0.0118729 | 0.026343 | 0.999296 |
| SCG-9 | 0.0112524 | 0.024966 | 0.999368 |
| SCG-10 | 0.0101688 | 0.022562 | 0.999484 |
| SCG-11 | 0.0109066 | 0.024199 | 0.999406 |
| SCG-12 | 0.0113416 | 0.025164 | 0.999358 |

*The bold values are the best values of root mean square error (RMSE), the coefficient of determination (R2) and coefficient of variation (cov) obtained from the ANN model for to estimate of heat transfer rate and effectiveness.*

**Table 4.**
*The statistical values of the network for predicting the effectiveness.*

| Thermodynamic Values | Method | Comparison Parameters | | |
|---|---|---|---|---|
| | | $R^2$ | RMSE | cov |
| Heat transfer rate | ANN | 0.999636 | 52.2255 | 0.01954 |
| Effectiveness | ANN | 0.999565 | 0.0093323 | 0.020706 |

**Table 5.**
*Performance evaluation criteria obtained from the ANN model.*

| Neuron position $(w_{ni})$ | $I_1(T_{hi})$ | $I_2(T_{ci})$ | $I_3(\dot{m})$ | $I_4(\beta)$ | $b_n$ |
|---|---|---|---|---|---|
| 1 | 1.8443 | 10,5531 | 2.1624 | −1.2641 | 18.7258 |
| 2 | 56.9616 | −52.142 | 27.1211 | −0.0728 | -20.0211 |
| 3 | −0.38653 | 0.42041 | −2.0398 | −5.9629 | 4.8897 |
| 4 | −2.6099 | 2.2846 | 3.6669 | 8.8609 | -7.3451 |
| 5 | −40.8109 | 37.1115 | −105.8138 | 33.7376 | 27.9695 |

**Table 6.**
*The weight coefficients and bias values used for the determination of the heat transfer rate.*

| Neuron position $(w_{ni})$ | $I_1(T_{hi})$ | $I_2(T_{ci})$ | $I_3(\dot{m})$ | $I_4(\beta)$ | $b_n$ |
|---|---|---|---|---|---|
| 1 | 11.1556 | 11.7598 | 39.362 | 9.2238 | 47.3869 |
| 2 | −37.921 | 46.173 | −7.0158 | −4.478 | −2.4622 |
| 3 | 16.3938 | −38.169 | −35.378 | −4.640 | 39.5329 |
| 4 | 13.9183 | −1.3878 | −10.187 | 13.3089 | 3.5955 |
| 5 | 92.1152 | 8.6141 | 28.7158 | −49.57 | −16.817 |
| 6 | 31.4252 | −30.344 | 3.3642 | 25.2838 | −23.704 |
| 7 | −14.906 | 11.1061 | −9.057 | −9.2724 | 10.1669 |
| 8 | 23.2713 | −4.3873 | −49.1203 | 20.2264 | 4.1424 |
| 9 | 68.8739 | −72.556 | 34.7909 | −44.281 | 13.6961 |
| 10 | −0.13074 | 3.2502 | 1.8366 | 11.4214 | −8.3915 |
| 11 | −10.7511 | 12.7919 | −3.6727 | 13.7598 | −7.913 |
| 12 | 35.966 | 58.1377 | 91.3142 | 3.7576 | −46.231 |

**Table 7.**
*Weight coefficients and bias values used to determine of the effectiveness.*

| Temperature $T_{hi}$(°C) | Temperature $T_{ci}$(°C) | Flow rate $\dot{m}$ (kg/s) | Plate surface angle β(°) | Heat transfer rate (W) | | Error (%) |
|---|---|---|---|---|---|---|
| | | | | Actual values | ANN | |
| 32.8 | 25 | 0.167 | 30 | 2167 | 2191 | 1.09 |
| 44.6 | 37.1 | 0.167 | 30 | 2167 | 2144 | 1.08 |
| 51,1 | 43.6 | 0.167 | 30 | 2167 | 2166 | 0.07 |
| 58.1 | 50.1 | 0.167 | 30 | 2377 | 2354 | 0.97 |
| 31.8 | 26.6 | 0.239 | 30 | 2195 | 2204 | 0.40 |
| 39.3 | 34 | 0.239 | 30 | 2295 | 2284 | 0.48 |
| 44.7 | 39.4 | 0.239 | 30 | 2295 | 2319 | 1.05 |
| 37.9 | 34.3 | 0.321 | 30 | 2144 | 2115 | 1.37 |
| 45.5 | 41.9 | 0.321 | 30 | 2144 | 2159 | 0.68 |
| 34.8 | 28.4 | 0.263 | 60 | 3077 | 3103 | 0.85 |
| 53.3 | 46.8 | 0.263 | 60 | 3407 | 3393 | 0.40 |
| 19.2 | 15.7 | 0.39 | 60 | 2608 | 2609 | 0.06 |
| 26.4 | 21.9 | 0.39 | 60 | 3259 | 3254 | 0.17 |
| 37.5 | 33.1 | 0.39 | 60 | 3422 | 3421 | 0.04 |
| 45.2 | 40.8 | 0.39 | 60 | 3585 | 3572 | 0.37 |
| 32.1 | 29 | 0.517 | 60 | 3023 | 3055 | 1.06 |
| 45.7 | 42.5 | 0.517 | 60 | 3455 | 3419 | 1.04 |

**Table 8.**
*Comparison of the actual values of the heat transfer rate with the estimated values obtained in the ANN model.*

The actual values of the heat transfer rate and the predicted values and error values obtained in the ANN model are given in **Table 8**. Eq. (16) is used in the calculation of error values [18].

$$Error = \frac{|A^e - A^p|}{A^e} x100 \quad (16)$$

Where, $A^e$ is the data obtained from experiments and $A^p$ is the estimates obtained from ANN. As seen in **Table 8**, the highest error value is 1.37. It has been determined that this value is an acceptable error value.

Similarly, the actual values of the effectiveness and the predicted values and error values obtained in the ANN model are given in **Table 9**. As seen in **Table 9**, the highest error value is 1.23. This value is also determined to be an acceptable error value.

## 3.2 Pressure drop in heat exchangers

In all HEXs there is a close physical and economic relationship between heat transfer and pressure drop. In a heat exchanger intended to be designed for a constant heat capacity, increasing fluid velocities increases the heat transfer coefficient and allows smaller heat exchangers (compact). In this way, a smaller size or more

| Temperature $T_{hi}$(°C) | Temperature $T_{ci}$(°C) | Flow rate $\dot{m}$ (kg/s) | Plate surface angle β(°) | Effectiveness | | Error (%) |
|---|---|---|---|---|---|---|
| | | | | Actual values | ANN | |
| 21.2 | 15.4 | 0.167 | 30 | 0.40 | 0,396 | 0.95 |
| 44.6 | 37.1 | 0.167 | 30 | 0.41 | 0,409 | 0.22 |
| 51.1 | 43.6 | 0.167 | 30 | 0.41 | 0,407 | 0.73 |
| 31.8 | 26.6 | 0.239 | 30 | 0.42 | 0,423 | 0,71 |
| 39.3 | 34 | 0.239 | 30 | 0.43 | 0,435 | 1.23 |
| 22.4 | 18.9 | 0.321 | 30 | 0.43 | 0,426 | 0.91 |
| 29.1 | 25.5 | 0.321 | 30 | 0.42 | 0,423 | 0.69 |
| 34.8 | 28.4 | 0.263 | 60 | 0.44 | 0,445 | 1.14 |
| 41.4 | 34.9 | 0.263 | 60 | 0.45 | 0,453 | 0.62 |
| 53.3 | 46.8 | 0.263 | 60 | 0.48 | 0,477 | 0.63 |
| 26.4 | 21.9 | 0.39 | 60 | 0.44 | 0,439 | 0.20 |
| 45.2 | 40.8 | 0.39 | 60 | 0.50 | 0,501 | 0.22 |
| 21.8 | 19.1 | 0.517 | 60 | 0.44 | 0,443 | 0.77 |
| 32.1 | 29 | 0.517 | 60 | 0.45 | 0,452 | 0.40 |
| 38.8 | 35.7 | 0.517 | 60 | 0.48 | 0,481 | 0.13 |
| 45.7 | 42.5 | 0.517 | 60 | 0.50 | 0,505 | 0.90 |
| 54.3 | 51 | 0.517 | 60 | 0.52 | 0,521 | 0,21 |

**Table 9.**
*Comparison of the actual values of the effectiveness with the estimated values obtained in the ANN model.*

compact heat exchanger design can be achieved at the same capacity with a lower investment cost. On the other hand, increasing the velocities of the fluids causes the pressure drop in the heat exchanger to increase. This increases the investment cost of the pump due to the operating costs of the system and the growth of the pump or fan, as it increases the power of the pump or fan. Therefore, in the design of a HEX, the heat transfer and pressure drop must be considered together and the most appropriate solution for the system should be sought.

Since the flow model is very complex even in the simplest heat exchangers, approximate solutions and experimental findings are utilized in the determination of pressure drop as well as theoretical analyzes. The total pressure drop in a HEX is considered in two ways, the pressure drop in the straight pipe and the local pressure drop. The pressure drop in the straight pipe indicates the pressure drop from the rubbing in the flowing fluid in the fixed section piping or conduits. The local pressure drop is the loss of flow and direction changes in the flow. Unlike fluids at the same temperature, the natural convection caused by the temperature distribution in the heat exchangers may cause an additional pressure loss (or sometimes gambling).

The total pressure loss in a heat exchanger is summed separately from the pressure losses in each step of the exchanger.

### 3.3 Local losses

Cross-sectioning, rotation, separation, or coupling of fluid as it flows through a channel also causes pressure losses. These are generally called local losses. Changes in the velocity and direction of the fluid create Eddy movements (eddies) that cause energy loss. Although local losses occur at very short distances, they remain effec-tive over a long period of time throughout the flow. These losses are generally [1];

$$\Delta P_y = \zeta \frac{\rho v^2}{2} \tag{17}$$

form. Where $\zeta$ is called the local loss coefficient and can be found from the relevant sources for various local loss factors either using formulas or diagrams.

### 3.4 Pressure loss caused by acceleration of the fluid and lifting force

The pressure loss during the acceleration of the fluid, in fixed cross-sections,

$$\Delta P_{iv} = \rho_o v_o^2 - \rho_i v_i^2 \tag{18}$$

Where $v_g$, $v_ç$ are the fluid velocity at inlet and outlet of the flow channel; $\rho_g$, $\rho_ç$, again indicate the density of the fluid at the inlet and outlet of the channel. The fluid is assumed to be incompressible, and in fluid fluids this value is the order of magnitude that other pressure losses can be neglected.

### 3.5 Pressure loss in sealed plate heat exchangers

The pressure loss in this type of heat exchanger is [19];

$$\Delta p_{gasketplate} = \lambda_{gasketplate} \frac{L_p \rho v^2}{d_h 2} \tag{19}$$

Turbulent flow;

$$\lambda_{gasketplate} = \mathbf{1{,}22}\,\mathrm{Re}^{-0{,}252} \tag{20}$$

Due to the protrusions on the plates, turbulence can pass through the Reynolds numbers, which are smaller than the values given in the flow flat exit channels. Therefore, in the case of sealed PHEXs, the flow at values such as Re > 100–400 is assumed to be turbulent.

### 3.6 The power required to maintain fluid motion

First, the pressure losses in the HEX and the pressure losses in the piping up to the heat exchanger are calculated. Later, the fan or pump power required to move the fluid in this system by calculating the sum of the pressure losses in the heat exchanger up to the HEX [20],

$$N = \frac{\dot{m}\sum(\Delta P)_t}{\rho\eta} \tag{21}$$

can be found in the equation.

## 4. Conclusions

PHEXs are the most frequently used heat transfer equipments in energy applications and can be under various names such as evaporators and condensers in almost every stage of chemistry, petrochemical industry, power plants, cooling, heating and air conditioning process in various types and capacities. From the point of view of machine and chemical engineering education, plate heat exchangers are a very good application for this branch of science which contains all of the basic subjects of these engineering branches: materials, strength, thermodynamics and heat transfer science. As can be seen, PHEXs are a commonly used construction in our daily lives. For this reason, its design should be done in detail, analysis results should be obtained with analysis programs and studies should be done to improve the designs. Decreased the amount of heat transferred in the PHEX causes the per-formance of the HEX to decrease. This means loss of capacity in energy system with plate heat exchanger. The regulation of heat transfer permits the system dimensions to be kept at the proper values, thus decreasing system cost and operating costs.

In this study, some equations were obtained to use in the plate heat exchangers by using the data obtained from the experimental work. ANN methodology was used for this purpose. As a result of the equations obtained for heat transfer and efficiency values in ANN application, approximate results were obtained at the value of 1.37 which is the highest error value for the real value heat transfer value and 1.23 for the efficiency value. When we look at the literature it is seen that these values are acceptable error values.

## Author details

Bayram Kılıç
Technical Sciences Vocational School, Burdur Mehmet Akif Ersoy University, Burdur, Turkey

*Address all correspondence to: bayramkilic@mehmetakif.edu.tr

## References


[1] Genceli O. Isı Değiştiricileri. Birsen Yayınevi, Istanbul, Türkiye, 1999. 424 p.

[2] https://ekinendustriyel.com/brazed-heat-exchanger/the-structure-of-plate-heat-exchanger/

[3] Camilleri R, Howey DA, McCulloch MD. Predicting the flow distribution in compact parallel flow heat exchangers. Applied Thermal Engineering.

[4] Çengel YA, Boles MA. Mühendislik Yaklaşımıyla Termodinamik. McGraw-Hill, 1994. 867 p.

[5] David K, Paul L. Evaluating of heat exchanger surface coatings. Applied Thermal Engineering.

[6] Dovic D, Palm B, Svaic S. Generalized correlations for predicting heat transfer and pressure drop in plate heat exchanger channels of arbitrary geometry. International Journal of Heat and Mass Transfer. 2009;52: 4553-4563.

[7] Dwivedi AK, Das SK. Dynamics of plate heat exchangers subject to flow variations. International Journal of Heat and Mass Transfer. 2007;50: 2733-2743.

[8] Faizal M, Ahmed MR. Experimental studies on a plate heat exchanger for small temperature different applications. Experimental Thermal and Fluid Science. 2012;36:242-248.

[9] Afonso IM, Cruz P, Maia JM, Melo LF. Simplified numerical simulation to obtain heat transfer correlations for stirred yoghurt in a plate heat exchanger. Food and Bioproducts Processing.

[10] Gherasim I, Taws M, Galanis N, Nguyen CT. Heat transfer and fluid flow in a plate heat exchanger part I. Experimental investigation. International Journal of Thermal Sciences. 2011;50:1492-1498.

[11] Gut JAW, Fernandes R, Pinto JM, Tadini CC. Thermal model validation of plate heat exchangers with generalized configurations. Chemical Engineering Science.

[12] Gyuwan H, Sangkwon J. Pressure loss effect on recuperative heat exchanger and its thermal performance. Cryogenics.

[13] Hajabdollahia H, Hajabdollahib Z. Investigating the effect of properties variation inoptimum design of compact heat exchanger using segmented method. Chemical Engineering Re-search and Design. 2016;112:46-55.

[14] Idario P, Nascimento E, Garcia C. Heat transfer performance enhancement in compact heat exchangers by using shallow square dimples in flat tubes. Applied Thermal Engineering.

[15] Karthik P, Kumaresan V, Velraj R. Experimental and parametric studies of a louvered fin and flat tube compact heat exchanger using computational fluid dynamics. Alexandria Engineering Journal.

[16] Kılıç B. Experimental investigation of effects to heat transfer of plate geometry with dynamic and thermal parameters in the plate heat exchangers. Ph.D. Thesis, Süleyman Demirel University, The Graduate School of Natural and Applied Sci-ences, Isparta, Turkey. 2013. (in Turkish).

[17] Kılıç B, İpek O, Şencan Şahin A. A comparative computational intelligence approach for heat transfer analysis of corrugated plate heat exchangers. Environmental Engineering and Management Journal.

[18] Selbas R, Sencan A, Kılıç B. Alternative approach in thermal analysis of plate heat exchanger. Heat Mass Transfer. 2009;45:323-329.

[19] Pandey SD, Nema VK. An experimental invstigation of exergy loss reduction in corrugated plate heat exchanger. Energy. 2011;36:997-3001.

[20] Reppich M. Use of high performance plate heat exchangers in chemical and process industries. International Journal of Thermal Science. 1999;38:999-1008.

10

# Evolutionary Design of Heat Exchangers in Thermal Energy Storage

*Miguel Rosa Oliveira Panão*

**Abstract**

The efficiency and ability to control the energy exchanges in thermal energy storage systems using the sensible and latent heat thermodynamic processes depends on the best configuration in the heat exchanger's design. In 1996, Adrian Bejan introduced the Constructal Theory, which design tools have since been explored to predict the evolution of the architecture in flow systems. This chapter reviews the fundamental knowledge developed by the application of the constructal principle to the energy flows in the design of heat exchangers of thermal energy storage systems. It introduces the Svelteness and scale analysis, as two constructal tools in the evolutionary design of engineering flow systems. It also includes the analysis on essential scales of several configurations, or energy flow architectures, toward establishing the main guidelines in the design of heat exchangers for storing thermal energy.

**Keywords:** thermal energy storage, heat exchangers, constructal theory, phase-change materials, flow architecture

## 1. Introduction

Engineering systems capture a fraction of the total amount of thermal energy available from renewable sources, and to increase the energy system reliability, the research and development of the flow architecture in thermal energy storage systems is of paramount importance.

One of the crucial issues is the characteristic fluctuations in the availability of energy from renewable resources and wasted energy in industrial processes. The design of efficient thermal energy storage systems is an essential step toward meeting the consumption demands of electricity and heat [1]. Therefore, there is growing attention in the development of thermal energy storage systems to produce adequate energy savings and utilization, with a relevant impact on numerous and diverse applications [2, 3].

There are two basic approaches to thermal energy storage. One using the sensible heat without phase-change (SHS - Sensible Heat Storage), and another using the sensible heat *and* phase-change (LHS - Latent Heat Storage), as depicted in **Figure 1**. The thermal balance describing each approach is given by

$$Q_{st}^{SHS} = V_{SHS}\rho_l c_{p,l}\left(T_f - T_i\right) \qquad (1)$$

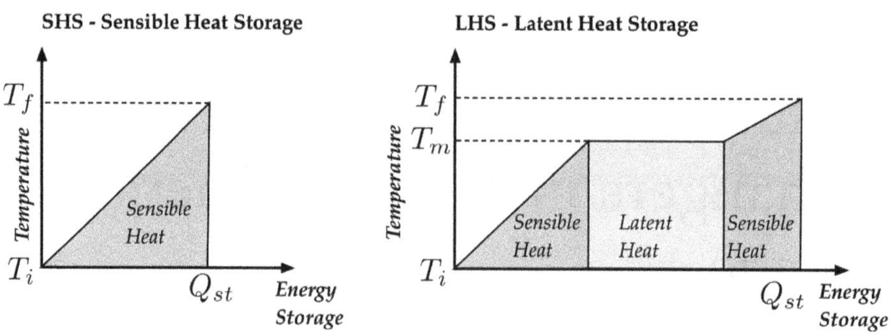


**Figure 1.**
*Thermal energy storage modes based on the sensible heat (SHS - left) and latent heat (LHS - right).*

$$Q_{st}^{LHS} = V_{LHS}\left[\rho_s c_{p,s}(T_m - T_i) + h_{sl} + \rho_l c_{p,l}(T_f - T_m)\right] \tag{2}$$

where $V$ is the storage material volume, $\rho_l, c_{p,l}$ are the fluid or melted material's density and specific heat, respectively, $h_{sl}$ is the latent heat of fusion of the Phase-Change Material (PCM), and $\rho_s$, $c_{p,s}$ are the PCM density and specific heat in its solid state, $T_i$ and $T_f$ are the initial and final temperatures of the energy storage process and $T_m$ corresponds to the melting temperature of the PCM in the LHS case.

In both approaches, the upper limit for the final temperature is the saturation value associated with the vaporization of the liquid. Because of its high heat capacity, water is the most used fluid for SHS. However, theoretically, Huang et al. [4] showed the energy stored in a water-based system is one order of magnitude lower than the energy stored in a PCM. And, experimentally, Kaygusuz [5] showed evidence of a PCM system able to store up to 60% of its theoretical maximum, which represents almost the double value of the theoretical storage capacity in water-based systems. These results were the motivation for further investment in the development of LHS systems. Compared to single-phase heat storage systems, LHS systems store the same amount of energy using more compact systems, reducing production and maintenance costs.

In LHS systems, **Figure 1** on the right represents the three theoretical stages of the storage process. Initially, the Thermal Storage Material (TSM) is in its solid-state, and it stores (charges) energy through:

Step 1) the sensible heat until the melting temperature;

Step 2) the latent heat component until a complete phase-change of the TSM from solid to liquid. In this stage, the time of the melting process depends on the advancement of the solid–liquid interface (melting front), according to the configuration of the thermal fluid circuit of the heat exchanger (HE) immersed in the TSM;

Step 3) after the total liquefaction of the TSM, energy storage continues until the liquid reaches the saturation temperature of the next phase-change without compromising the volume of the Thermal Storage facility.

The storage of energy in these steps depends on the thermal properties of TSM, and the thermodynamics of the storage process. Still, the main challenge is the design of heat exchangers, as the engineering system that enables the flow of energy from the sources (renewable and non-renewable) to the TSM, disregarded in recent comprehensive reviews on thermal energy storage [6, 7]. Namely, this design has a significant impact on the charging and discharging times, if using renewable energy sources, given their limited time-window throughout the day.

The standard approach in the design of heat exchangers is to optimize the thermal and hydrodynamic energy flows. It uses an iterative process based on previous work, and typical working conditions, such as the amount of fouling and pressure drop in the system, testing a significant number of trial-and-error designs until the values for the heat transfer performance, hydrodynamic effects and longevity are within pre-established requirements. A common trait in this standard approach is the lack of evaluation criteria grounded on the underlying physical processes. Constructal design distinguishes from the standard approach in providing the evaluation criteria in such a way. Therefore, the purpose of this chapter is to synthesize and present an evolutionary design approach (not optimization) using tools based on constructal theory. Therefore, the novelty is to include the architecture of thermal energy storage systems at the design stage [8] and investigate the best way to introduce the freedom to morph to overcome the shortcomings on charging and discharging periods due to prescribed, rigid and fixed designs when subjected to daily and seasonal changes.

## 2. Design tools in constructal theory

In 1996, Adrian Bejan [9], professor at Duke University, proposed a Constructal Theory to explain the evolution of configurations in nature stating,

> *"for a finite-size flow system (not infinitesimal, one particle, or sub particle) to persist in time (to live) it must evolve with freedom such that it provides easier and greater access to what flows."*

In practice, when using the constructal theory in engineering, one finds the best direction for the flow structures emerging from of what facilitates movement, designated as *constructal design*. And this design establishes a relation between what flows (energy, fluids, people, etc.) and the geometry of the flow architecture, in such a way that it becomes a global property of the engineering system. This property, the *Svelteness*, is a tool in constructal design.

In thermal energy storage, there are several length and time scales competing in the unfolding heat transfer processes, characterized by mass, momentum and energy balances of the system. However, not all the terms correspond to the dominant scales setting the overall result of charging and discharging of energy. A *scale analysis* is the second tool in constructal design explored in this section, allowing a proper definition of the relevant scales, and their implication to the heat exchanger design.

### 2.1 Svelteness of flow configuration

The *Svelteness* (Sv) is this global geometric property of the flow space, which guides the engineering practice in the assessment of the flow design performance. This property corresponds to the relation between two length scales of the flow system configuration: an external ($L_e$); and an internal length ($L_i$), usually associated with the volume as $V^{1/3}$.

$$\mathrm{Sv} = \frac{L_e}{L_i} \tag{3}$$

According to Bejan and Lorente [10], the evolutionary direction is that of *vascularization*, implying an increase of the Svelteness. Therefore, as an example applied to TES, what is the best constructal design solution for a simple tube inside a tank?

Should the design be a PCM in the inner tube and having the thermal fluid flowing through the tank (see [11]), or the opposite (see [12])?

In this example, the inner tube and tank are cylindrical with a length of $L$, and diameter of $d$ and $D$, respectively. Therefore, the Svelteness in both cases has $L$ as its external length scale. However, the volume associated to the thermal fluid depends on the situation. When the PCM is inside the inner tube, the volume where the thermal fluid flows is $V = (\pi/4)(D^2 - d^2)L$, and in the opposite case, $V = (\pi/4)d^2L$. If $\mathrm{Sv}^i$ corresponded to the case where the energy flows inward to the PCM on the inner tube, and $\mathrm{Sv}^o$ when it flows outward to the PCM in the tank, the relation between Sveltenesses would be

$$\frac{\mathrm{Sv}^o}{\mathrm{Sv}^i} = \left[\left(\frac{D}{d}\right)^2 - 1\right]^{1/3} \tag{4}$$

If a larger Svelteness points in the evolutionary design of the outward energy flow from the thermal fluid inside the inner tube, $\mathrm{Sv}^o/\mathrm{Sv}^i > 1$, thus $D^2/d^2 > 2$, and $D^2/d^2 < 2$ otherwise. Consider the case of choosing the best design to favor the flow of energy inward to a PCM inside a tube, as in the work of Ghoneim [11]. What should be the value of the void fraction $\phi$, which is the ratio between the thermal fluid volume ($V_{tf}$) and the storage tank volume ($V_{st}$)? First, the volume of thermal fluid is the difference between the storage tank volume and the volume of all tubes containing the PCM energy storage material, $V_{tf} = V_t - n_t V_t$, with $n_t$ as the number of tubes and $V_t = (\pi/4)d^2L$ as the volume of each tube. Therefore, one defines the defines the void fraction as

$$\phi = 1 - \frac{n_t V_t}{V_{st}} \tag{5}$$

Considering $D$ as the external diameter containing the tube diameter $d$, and the thermal fluid circulating in the tank, the storage tank volume should equal the total volume of all the tubes with the "necessary" thermal fluid volume, $V_{st} = n_t(\pi/4)D^2L$ thus, using this reasoning in the void fraction implies that

$$\frac{D^2}{d^2} = \frac{1}{1 - \phi} \tag{6}$$

Considering the previous constructal analysis using the Svelteness, storing energy with PCM material inside the tubes is only worthy when $D^2/d^2 < 2$. Therefore, when applied to Eq. (6), it points to the need of void fraction values of $\phi < 0.5$. In fact, all the LHS systems investigated using the configuration of Ghoneim [11], choose $\phi = 0.3 < 0.5$. Constructal theory corroborates this option, but indicates that $\phi$ could assume higher values, eventually leading to the insertion of more tubes with PCM, allowing the storage of more energy.

On the other hand, recent works as that of Agyenim et al. [12], point toward having the PCM in the tank, instead of inside the inner tubes, and in these cases with a single tube through which circulates the thermal fluid, $D^2/d^2 = 7.35 > 2$.

## 2.2 Method of scale analysis

*Scale analysis* or *scaling* is a problem solving method useful to obtain essencial and expedite information of several energetic processes [13]. It is not the same as a the dimensional analysis performed in fluid mechanics, but to assess the importance of the order of magnitude of the parameters involved in heat transfer processes, and extract the relevant scales from their governing equations. For more details on the principles of scale analysis, see Bejan [13] (pp. 17–20). Here, one uses an example to illustrate the method.

Consider the example above of an LHS system with the PCM inside a tube and the thermal fluid circulating around it. If there is a sudden change in the thermal fluid temperature ($T_{tf}$ ), how long will it take for that perturbation to reach the PCM material at the central axis of the tube? Assuming heat transfer by diffusion in cylindrical coordinates, and that changes in the thermal diffusivity ($\alpha$) are negligible within that time scale, the energy equation for the thermal energy storage process is

$$\frac{\partial T}{\partial t} = \frac{\alpha}{r}\frac{\partial^2 T}{\partial r^2} \tag{7}$$

Scaling means using the symbol $\sim$ to establish the order of magnitude of a differential term with the main parameters of the flow configuration. Therefore,

$$\frac{\partial T}{\partial t} \sim \frac{\Delta T}{\tau} \tag{8}$$

$$\frac{\alpha}{r}\frac{\partial^2 T}{\partial r^2} \sim \alpha\frac{\Delta T}{(d/2)^2} \tag{9}$$

where $\tau$ corresponds to the time scale under evaluation, $\Delta T$ to the temperature difference between the tube's boundary and the center, and $d$ is the tube's diameter. In scaling terms, Eq. (7) becomes,

$$\frac{\Delta T}{\tau} \sim \alpha\frac{\Delta T}{(d/2)^2} \tag{10}$$

which solved for the time scale results in

$$\tau \sim \frac{(d/2)^2}{\alpha} \tag{11}$$

Bejan [13] contains the synthesis for all the rules in a scale analysis. However, the example above is enough to explain the procedure applied later in section 3. The following section exemplifies the application of constructal theory as an evolutionary design method to develop heat exchangers in sensible and latent heat storage engineering systems.

## 3. Constructal theory in thermal energy storage heat exchangers

One of the essential elements in a constructal theory analysis is the freedom to morph of flowing configurations. Therefore, once we identify what is the flow under analysis, one can better understand what its freedom to morph means. On the other hand, the heat exchanger in thermal energy storage corresponds to the structure obtained after morphing through which energy flows from a source, usually the thermal fluid, to the storage material (e.g. a solid or a phase-change material, PCM). Depending on the storage material, the heat transfer mechanisms vary, and, accordingly, the energy storage scales. For example, if the material is solid, the heat transfer mechanism is diffusion and the mode is the one on the left of **Figure 1**. But if one uses a PCM, the energy storage story follows the second mode on the right, involving natural convection, a melting process, a solid–liquid interface moving boundary, and all these elements lead to additional complexity of the heat exchanger, affecting the energy storage scales.

### 3.1 Heat exchangers in sensible heat storage

Consider an underground volume of solid with a network of channels through which a thermal fluid transports energy for storage purposes. The storage mechanism is heat diffusion from the channels outer area to the volume of solid. What should be the structure of the channels network? Combelles et al. [14] explored this TES system with tree-shaped configurations of 2D channels made of parallel plates the length $L_1$ and $D_1$ of width within an area of $2L_1 \times 2L_1$, and 3D pipes configurated within a solid of $2L_1 \times 2L_1 \times L_1$ of volume, as depicted in **Figure 2**.

The first step is to characterize the architecture of each configuration type (2D or 3D) in terms of their Svelteness, as the global property of the system, which relates the external length scale given by the total length of the flow network, $L_{total} = \sum_{i=1}^{n} L_i$, depending on its complexity ($n$ ), and the internal flow length scale varying with the 2D or 3D nature of the flow. If the flow network is bi-dimensional, the internal length scale corresponds to $A_f^{1/2}$ with $A_f = \sum_{i=1}^{n} 2^{i-1} L_i D_i$ , while in the three-dimensional configuration, this scale is $V_f^{1/3}$ with $V_f = \sum_{i=1}^{n} 2^{i-1} \frac{\pi}{4} D_i^2 L_i$ . When one increases the complexity of the flow network, Lorente et al. [15] show the relation between the length and diameter of one branch ($i$) and the next ramification ($i+1$) follows the Hess-Murray rule. Thus,

$$\frac{L_{i+1}}{L_i} = \frac{D_{i+1}}{D_i} = 2^{-1/2} \tag{12}$$

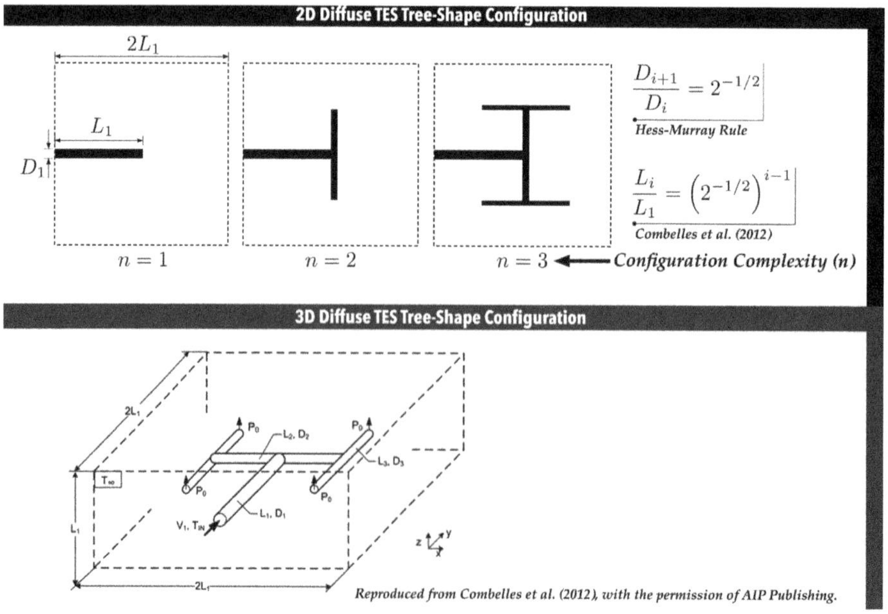


**Figure 2.**
*2D and 3D diffusive TES tree-shape configurations.*

which developed to depend on the length and diameter of the first branch $(L_1, D_1)$, simplify to

$$L_i = L_1\left(2^{-1/2}\right)^{i-1} \tag{13}$$

$$D_i = D_1\left(2^{-1/2}\right)^{i-1} \tag{14}$$

The Svelteness for both general configurations, considering the relations in Eqs. (13) and (14), depends on two major features of the configuration: its complexity $(\psi(n))$; and the geometrical relation between the length and diameter of the first channel $(L_1/D_1)$.

$$\mathrm{Sv}_k = \psi_k(n)\left(\frac{L_1}{D_1}\right)^{q_k} \text{ with } k = \{2D, 3D\} \tag{15}$$

and

$$\begin{cases} \psi_{2D}(n) = \dfrac{1-2^{-n/2}}{1-2^{-1/2}}n^{-1/2} & \wedge\ q_{2D} = 1/2 \\ \psi_{3D}(n) = 2\left(\frac{3+2\sqrt{2}}{1-2^{-n}\sqrt{2^n}}\right)^{1/3}\left(1-2^{-n/2}\right) & \wedge\ q_{3D} = 2/3 \end{cases}$$

If the Svelteness indicates the evolution of the flow configuration, one should connect the flow architecture complexity, $\psi_k(n)$, and the geometry of the initial channel, with the scales associated to the storage of energy.

The amount of energy stored depends on the material, but in this TES system, the relevant scale is the storage time and the evolution of the temperature in the conductive solid. Considering the solid is, initially, at $T_0$, and the inflowing thermal fluid is at $T_{in}$, in time, the average temperature in the solid $(T_{avg})$ evolves toward $T_{in}$. Therefore, Combelles et al. [14] analyzes the evolution of this diffusive thermal storage configuration with a dimensionless thermal potential as

$$\theta_{avg} = \frac{T_{avg} - T_0}{T_{in} - T_0} \tag{16}$$

This analysis focuses on the timescales of energy storage and the corresponding effect of the configuration complexity (number of bifurcations, $n$). Considering the fluid, there are two essential timescales:

- the timescale of fluid traveling the $n$ channels;

$$t_f = \sum_{i=1}^{n}\frac{L_i}{V_i} = \psi_f(n)\left(\frac{L_1}{V_1}\right) \sim \frac{L_1}{V_1} \tag{17}$$

since $\psi_f(n) = \frac{2^{-3n/2}-1}{2^{3/2}-1}$ varies between 1 and 1.55, which means $\psi_f(n) \sim 1\ \forall_{n\geq 1}$;

- and the timescale of thermal diffusion accross the channel;

$$t_c \sim \frac{D_1^2}{\alpha_f} \tag{18}$$

Combelles et al. [14] argue that in the case where thermal diffusion in the channel›s boundary layer is a slower process than the fluid traveling through the channels, $t_f < t_c$, and the temperature of the fluid at exit is practically unchanged. It is a relevant result to establish a stable boundary condition.

The third timescale, and the longer, corresponds to the time it takes to store energy in the solid volume, $t_s$, meaning the timescale to heat the entire volume by thermal diffusion, expressed as

$$t_s \sim \frac{L_1^2}{\alpha_s} \tag{19}$$

From the numerical simulations, the evolution of the dimensionless thermal potential $\theta_{avg}(t)$ is given by

$$\theta_{avg}(t) = 1 - \exp\left(-C\frac{t}{\tau}\right) \tag{20}$$

where $C$ is a scale parameter, and $\tau$ is the TES response time given by

$$\tau = \frac{m_s c_{p,s}}{\dot{m}_f c_{p,f}} \tag{21}$$

with $m_s = \rho_s 4L_1^3$ as the solid mass (and $\rho_s$ its density), $c_{p,s} = \frac{k_s}{\rho_s \alpha_s}$ is the solid specific heat, and $\dot{m}_f c_{p,f} = \left(\frac{f}{\alpha_f}\right)\frac{\pi}{4}k^2 D_1^2 V_1$ is the thermal capacity rate of the fluid. Considering Eq. (15), and introducing it in the TES response time results in

$$\tau = \frac{16}{\pi}\left(\frac{Sv_k}{\psi_k(n)}\right)^{2/q_k}\frac{\tilde{\alpha}}{\tilde{k}}t_f \tag{22}$$

with $\tilde{\alpha} = \alpha_f/\alpha_s$ and $\tilde{k} = k_f/k_s$. The results from the numerical simulations in both 2D and 3D configurations reported in Combelles et al [14] evidence the decrease of the diffuse TES system response time with a higher complexity of the flow network, which is consistent with the relation obtained in Eq. (22) for a fixed $Sv_k$ as considered in their simulations.

**Figure 3** shows the results for the complexity degree scale ($\psi_k(n)$) normalized by the value obtained for $n = 4$: $\tilde{\psi}_k(n) = \psi_k(n)/\psi_k(4)$. In a 2D configuration, one obtains a maximum of the scale associated to the complexity degree at $n = 4$ (which

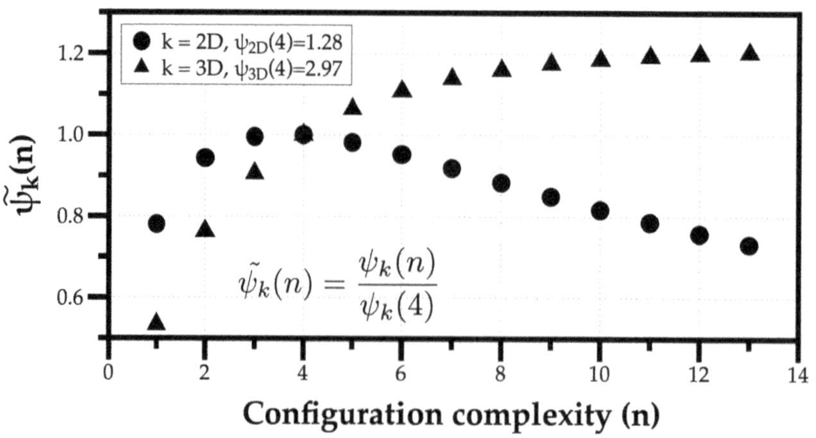


**Figure 3.**
*Evolution of the scale associated to the complexity degree of the flow network in relation to its maximum, $\tilde{\psi}_k(n)$.*

is the maximum complexity investigated in Combelles et al. [14]). In the 3D configuration, $\tilde{\psi}_{3D}(n)$ shows a monotonic behavior, although for $n > 10$, the increase of one level of complexity generates a variation of less than 1% in diminishing returns. The maximum complexity explored by Combelles et al. [14] was $n = 4$ and adding one level of complexity would produce an increment of only 6 4% compared to 10.6% between $n = 3$ and 4.

In absolute terms, the constructal design of heat exchangers in diffusive TES systems suggests the choice of a 3D configuration, rather than a 2D, for a faster energy storage, since it leads to $\psi_{3D}$ complexity scale factors of 1.6 to 2.3 times higher than the 2D scale, allowing shorter charging times. However, in applications where a 2D configuration is more appropriate, the level of complexity shoud not go beyond 4 dendritic bifurcations.

### 3.2 Heat exchangers in latent heat storage

Energy storage systems using the latent heat of a certain phase-change material (PCM) rely on the heat transfer mechanisms of diffusion and natural convection. Initially, the PCM is in its solid state and it stores heat by diffusion close to the channel containing the thermal fluid, or fin, until it reaches the fusion (or melting) temperature ($T_m$), creating a melting frontline. Thereafter, a solid–liquid interfacial boundary develops and the melting history consists in two distinct periods: invasion; and consolidation.

The invasion period corresponds to the time interval until the solid–liquid interface reaches a distance equivalent to the process characteristic length. The consolidation period corresponds to the remaining time until all the PCM in the LHS system is in its liquid state. These periods do not, necessarily, correspond to the timescales associated to the diffusive and convective heat transfer processes. The charging and discharging times of LHS depend on the heat exchanger design and the dominant heat transfer mechanisms through which energy flows from its source to the PCM.

There are several design configurations investigated with a constructal approach for the heat exchangers using phase-change to store energy in PCM. **Figure 4** presents three configurations reported in the literature. The configurations with a vertical pipe [15] is the less prone to morphing. The helical pipe [16] in a cylindrical PCM enclosure is fixed, but the ability to vary the number of turns and the diameter of each turn increases the system's freedom to morph. Finally, an advancing heat source line invading the PCM material aims at the theoretical design with the greatest freedom to morph [17]. One of the novelties in constructal design of engineering systems is determining the Svelteness as expression of its architecture, and the system's freedom to morph, considering an evolutionary path toward

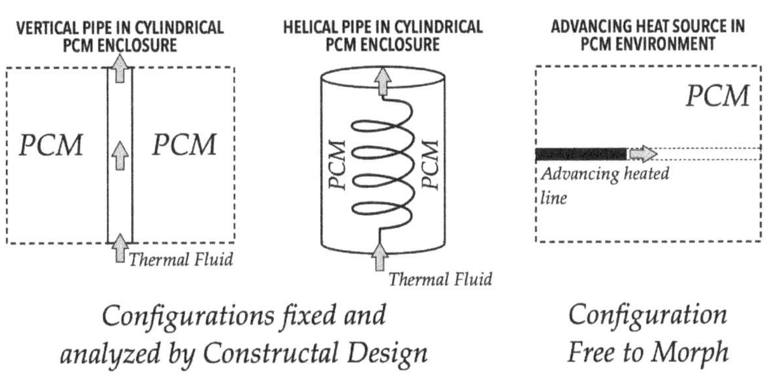


**Figure 4.**
*Heat exchanger configurations investigated with a constructal theory approach.*

vascularization, i.e. an increase of its Svelteness. The constructal analysis of all designs also implies the investigation of length and timescales associated with heat transfer mechanisms, and the possible effect of the Svelteness in these scales.

Lorente et al. [15] performed a scale analysis to analyze the latent thermal energy storage where energy flows from the thermal fluid circulating inside a central vertical pipe and the surrounding PCM. The dominant heat transfer mechanism is natural convection. The Svelteness in this case would have the height of the enclosure $(L_e = H)$ as external length scale, and an internal length scale based on the volume occupied by the thermal fluid as $L_i = \left(\frac{\pi}{4}d^2H\right)^{1/3}$, with $d$ as the diameter of the vertical pipe The final outcome relating the Svelteness with the heat exchanger geometry leads to

$$H/d = (\pi/4)^{1/3}\,\mathrm{Sv}^{3/2} \tag{23}$$

The maximum energy one can store in this first TES configuration is $Q_{max} = \frac{\pi}{4}\left(D^2 - d^2\right)H\rho h_{sl}$, with $\rho$ as the PCM density and $h_{sl}$ as its latent heat of fusion. dom According to Lorente et al. [15], the time-scale of the melting history $(\tau_m)$ considers natural convection as the dominant heat transfer mechanism where friction - inates buoyancy forces, thus, resulting in

$$\tau_m = \frac{Q_{max}}{\pi d k \Delta T C\,\mathrm{Ra}_H^{1/4}} \tag{24}$$

The relation between the maximum amount of energy stored and this time-scale, including Eq. (23) in (24) leads to $Q_{max}/\tau_m \sim \mathrm{Sv}^{-3/2}$. Considering that evolution in constructal theory occurs toward the vascularization of flow architectures, implying the increase of their Svelteness, in this case, it leads to decreasing $Q_{max}/\tau_m$ , instead of increasing as desired. This is an interesting result from the constructal design point of view because it indicates that natural convection generated by the vertical pipe alone is not the best heat transfer mechanism for a faster energy storage in LHS. However, one should point that their simplified scale analysis, and numerical simulations, which considers an annular moving melting front, seems unrealistic when confronted with later experimental works such as Zhang et al. [18], with time-scales of one order of magnitude lower than those predicted in Lorente et al [15] – $O(\tau_m) \sim 10^2$ h. Nonetheless, this result points to the need of a better solution to facilitate the flow access of energy between a surface heated by a thermal fluid and the PCM. The works of Ogoh and Groulx [19], and Kamkari and Shokouhmand [20], are examples where fins around the main energy source facilitate its flow to the PCM for storage during charging, promoting heat transfer by diffusion and mitigating natural convection.

Considering the case of Ogoh and Groulx [19], **Figure 5** depicts the adding of disc-shape fins to the original vertical pipe in cylindrical PCM enclosure configuration, where each disc-shape fin corresponds to a *construct* (represented on the right).

The total height of the cylinder ($H$), in terms of constructs corresponds to $H = n(h + h_f)$, with $n$ as the number of constructs, $h$ the total height of the PCM inside a construct, and $h_f$ as the fin thickness. Therefore, the interval between annular fins becomes

$$h = \frac{H}{n} - h_f \tag{25}$$

Considering the Svelteness (Sv) for this construct as the ratio between the external characteristic length based on the upper and bottom areas of the disc-shape fin, $L_e = \left[\frac{\pi}{2}\left(D^2 - d^2\right)\right]^{1/2}$, and the internal characteristic length given by the annular

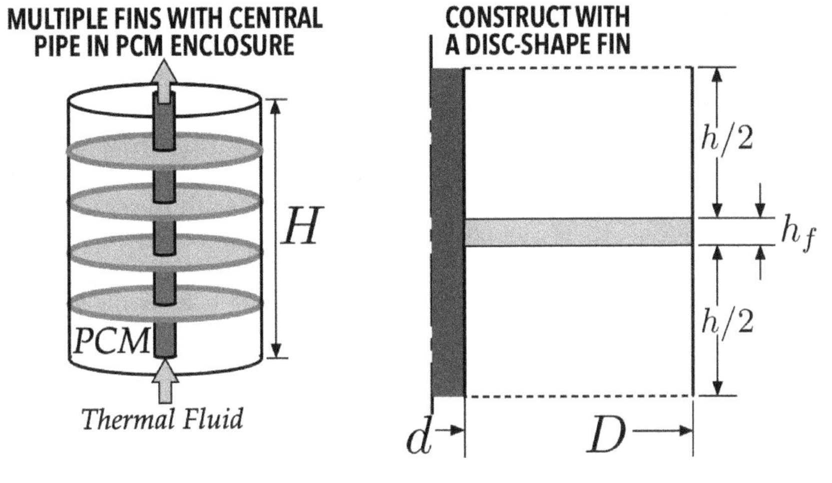


**Figure 5.**
*Disc-shape fin constructs applied to a vertical pipe in cylindrical PCM enclosure.*

disc volume, $L_i = \left[\frac{\pi}{4}(D^2 - d^2)h_f\right]^{1/3}$, considering Eq. (25), one can express the ratio between the space between fins ($h$) and the external diameter of the enclosure ($D$) as

$$\frac{h}{D} = \frac{H}{nD} - \frac{\sqrt{2\pi}}{\mathrm{Sv}^3} \tag{26}$$

Ogoh and Groulx [19] argue for a neglecting effect of convection between annular fins, thus, the energy stored by phase-changing the PCM from solid to its liquid state occur by conduction. In this sense, the most important scale characterizing the melting front ($\delta$) departs from the annular disc fin. The balance between the conduction heat flux ($q''$) supplied to the melting front and the rate of melting can be described as

$$q''dt = \rho h_{sl} d\delta \tag{27}$$

Assuming a linear temperature distribution across this layer where the heat transfer occurs by diffusion, one could quantify the heat flux as $q'' = k\frac{\Delta T}{\delta}$, which applied in Eq. (27) results in

$$\delta = \sqrt{a\tau} \tag{28}$$

with $a = \frac{2k\Delta T}{\rho h_{sl}}$, where $\Delta T = T_w - T_m$, with $T_w$ as the temperature of the fin wall, $T_m$ the fusion temperature of the PCM, and $\tau$ corresponds to the timescale of energy storage. To understand the evolution of the configuration based on this scale, one could argue that the charging finishes when $h \sim 2\delta$, thus, replacing this scale in Eq. (26), and solving it as a function of the Svelteness, results in

$$\mathrm{Sv} \sim \sqrt{2\pi}\left(\frac{nD}{H - 2n\sqrt{a\tau}}\right) \tag{29}$$

Therefore, since $\mathrm{Sv} > 0$, it implies $H - 2n\sqrt{a\tau} > 0$, resulting in an upper theoretical limit for this timescale as

$$\tau < \left(\frac{H}{2n}\right)^2 \frac{1}{a} = \tau_{max} \tag{30}$$

**Figure 6** shows the results for this limit considering the properties reported in [19].

The constructal design analysis of this heat exchanger indicates diminishing returns of less than 10% for a number of fins above $n > 18$, which is coherent with the numerical results presented by Ogoh and Groulx [19]. Applying the timescale defined by Eq. (30) to the experimental conditions in the work of Kamkari and Shokouhmand [20], which explore the effect of no fins with the cases of 1 and 3 fins, for the later case, $\tau_{max}$ is roughly 1.3× the value measured for the total melting process. In the case of the experiments with 1 fin, the results for $\tau_{max}$ are significantly larger, evidencing the role of natural convection in delaying the heat transfer to the PCM.

The helical coil illustrated in **Figure 4** is an alternative to the vertical pipe, and a geometry where the freedom to morph is larger because of the ability to change the helix diameter, number of turns and pitch angle. Alailami et al. [16] explored the morphing ability of the system in its design stage to optimize the storage of energy analyzing two scales. The timescale of heat penetrating the storage material from the boundaries of the helical coil to the cylinder diameter, $\tau_c = D^2/\alpha$; and the temperature difference, $T(t) - T_{tf}$, scaled by the initial condition, $T(0) - T_{tf}$, with $T_{tf}$ as the temperature of the thermal fluid. When Alailami et al. [16] simulated the evolution of the scaled average temperature – $T^*_{avg} = \frac{T_{avg}(t^*) - T_{tf}}{T(0) - T_{tf}}$ – its value decreased with the scaled time – $t^* = t/\tau_c$ – meaning the average temperature of the energy storage material approaches the temperature of the thermal fluid.

Afterward, focusing the analysis on $t^* = 0.3$, and varying the helical coil diameter – $D_h = \zeta D$ – and pitch height – $H_h = \varepsilon H$ – obtained as function of the cylinder diameter ($D$) and height ($H$), the authors reached an optimum diameter and pitch length of the helical coil, corresponding to $\zeta = 0.6$, and $\varepsilon = 0.3$, respectively.

Without using the work of Alailami et al. [16], Joseph et al. [21] performed an experiment of this configuration to store energy in a PCM. The authors used unoptimized values for the helical coil ($\zeta \approx 0.7, \varepsilon \approx 0.2$), from the Alailami et al. [16] point of view. However, the PCM configuration showed promising results storing 20% more energy than its equivalent mass in water. Also, the charging process in the PCM facility was slower and the authors attribute this result to the unoptimized heat exchanger design, justifying the need of more research on this topic.

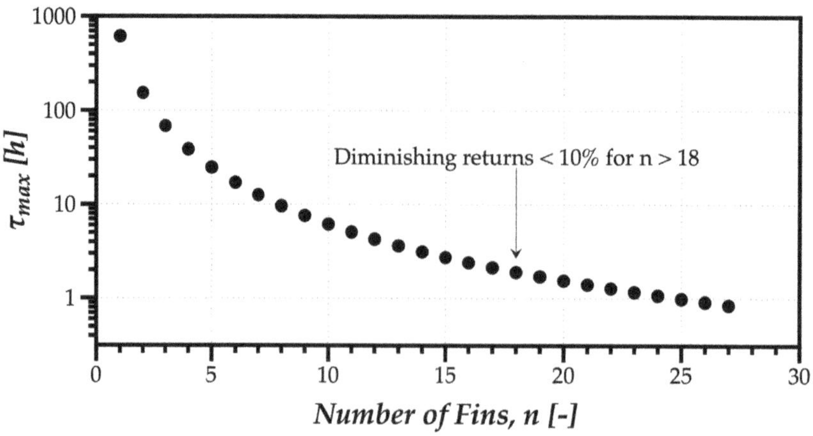


**Figure 6.**
*Variation of limit timescale for the energy storaged in a PCM through annular fins distributed around a vertical pipe inside a cylindrical enclosure.*

The last geometry in **Figure 4** is theoretical and corresponds to the greatest freedom to morph through an advancing heated line that can bifurcate at some point. The analysis performed by Bejan et al. [17] focus on the invasion (line advances until the storage boundaries) and consolidation (all PCM melts) stages, and introduces a tree invasion pattern with a complexity level up to $n = 2$ branching events. The results of this theoretical analysis point to the acceleration of the charging times. However, the conversion of this approach into a practical application is still a challenge, since there is no technology capable of this kind of morphing inside a PCM solid environment.

A final comment concerns the possible contribution of constructal design to optimize the total cost of heat exchanger design methods, which is not a direct correlation. As shown by Azad and Amidpour [22], since the constructal design provides an evolutionary perspective on the optimum geometric features of heat exchangers, it can lead to a substantial reduction of the total cost, compared to more standard design approaches. Namely, in the aforementioned work, the authors used constructal theory to optimize shell and tube heat exchangers, and the new approach allowed to reduce this cost by 50%. However, the application of a similar reasoning in the development of heat exchangers for thermal energy storage is in need of more research.

## 4. Conclusions

Thermal energy storage is one of the preeminent options to face the energy challenges of this century, providing a high energy saving potential and effective utilization. However, in these systems, the architecture of the heat exchangers through which energy flows, during charge and discharge, is of paramount importance. While most approaches optimize heat exchanger designs, the one presented in this chapter, based on constructal theory, follows an evolutionary design, meaning that the configuration explored at the design stage is dynamic and free to morph. It is not pre-defined, rigid, or still, but considers how it should evolve toward the greater access of the energy currents that flow through it.

Thermal energy storage systems follow two thermodynamic processes using the sensible heat of the energy storage material, or, besides the sensible heat, also the latent heat, as in Phase-Change Material (PCM). After introducing the general considerations on these systems, this chapter presents two design tools in constructal theory: the Svelteness, as a global property of any flow system, which tends to increase and evolve toward vascularization; and the scale analysis, as an expedite problem solving tool that allows obtaining relevant information of the several energetic processes involved.

Using the design tools presented, this chapter reviews and further explores the constructal theory approach in the development of heat exchangers for sensible and latent thermal energy storage configurations. The analysis evidences the explanatory potential of the constructal approach, increasing the sensibility of the engineer to the advantages of including the freedom to morph at the design stage of heat exchangers in thermal energy storage.

## Acknowledgements

The author would like to acknowledge project UIDB/50022/2020 and UIDP/50022/2020 of ADAI for the financial support for this publication.

## Nomenclatures and Abbreviations

| | |
|---|---|
| $c_p$ | Specific heat [J·kg$^{-1}$ K$^{-1}$] |
| $d, D$ | Diameter [m] |
| $H, h$ | Height [m] |
| $h_{sl}$ | Latent heat of fusion [J/kg] |
| $k$ | Thermal conductivity [W·m$^{-1}$ K$^{-1}$] |
| $L$ | Length [m] |
| $m$ | Mass [kg] |
| $\dot{m}$ | Mass flow rate [kg/s] |
| $n$ | Number of tubes or Complexity degree [–] |
| $Q_{st}$ | Energy stored [J] |
| $r$ | radial coordinate [m] |
| Ra | Rayleigh number [–] |
| Sv | Svelteness [–] |
| $T$ | Temperature [K] |
| $t$ | time [s] |
| $T_f$ | Final temperature [K] |
| $T_i$ | Initial temperature [K] |
| $T_m$ | Melting temperature [K] |
| $V$ | Volume [m$^3$] |

*Greek Symbols*

| | |
|---|---|
| $\alpha$ | Thermal diffusivity [m$^2$/s] |
| $\delta$ | Length [m] |
| $\Delta T$ | Temperature difference [K] |
| $\varepsilon$ | Scale factor [–] |
| $\zeta$ | Scale factor [–] |
| $\theta$ | Normalized temperature difference [–] |
| $\rho$ | Density [kg/m$^3$] |
| $\tau$ | Timescale [s] |
| $\phi$ | Void fraction |

*Subscripts*

| | |
|---|---|
| *avg* | average |
| *c* | cylinder |
| *e* | external |
| *f* | fluid, fin |
| *h* | helical |
| *i* | internal, inner |
| *m* | melting |
| *max* | maximum |
| *o* | outward |
| *s* | solid |
| *st* | stored |
| *t* | tube |
| *tf* | thermal fluid |
| 2D | Two-dimensional |
| 3D | Three-dimensional |
| HE | Heat Exchanger |
| LHS | Latent Heat Storage |

| | |
|---|---|
| PCM | Phase-Change Material |
| SHS | Sensible Heat Storage |
| TES | Thermal Energy Storage |
| TSM | Thermal Storage Material |

## Author details

Miguel Rosa Oliveira Panão
ADAI, LAETA, Department of Mechanical Engineering, University of Coimbra, Coimbra, Portugal

*Address all correspondence to: miguel.panao@dem.uc.pt

## References

[1] Zsiborács H, Baranyai NH, Vincze A, Zentkó L, Birkner Z, Máté K, Pintér G. Intermittent renewable energy sources: The role of energy storage in the european power system of 2040. Electronics.

[2] Wu S, Yan T, Kuai Z, Pan W. Thermal conductivity enhancement on phase change materials for thermal energy storage: A review. Energy Storage Materials.

[3] Soares N, Costa JJ, Gaspar AR, Santos P. Review of passive PCM latent heat thermal energy storage systems towards buildings energy efficiency. Energy and buildings.

[4] Huang BK, Toksoy M, Cengel YA. Transient response of latent heat storage in greenhouse solar system. Solar en-ergy. 1986 Jan 1;37(4):279–92.

[5] Kaygusuz K. Experimental and theoretical investigation of latent heat storage for water based solar heating systems. Energy conversion and management. 1995 May 1;36(5):315–23.

[6] Sarbu I, Sebarchievici C. A comprehensive review of thermal energy storage. Sustainability.

[7] Alva G, Lin Y, Fang G. An overview of thermal energy storage systems. Energy.

[8] Clemente MR, PanÃ£o MR. Introducing flow architecture in the design and optimization of mold inserts cooling systems. International Journal of Thermal Sciences.

[9] Bejan A. Freedom and Evolution: Hierarchy in Nature, Society and Science. Springer Nature; 2019 Dec 6.

[10] Bejan A, Lorente S. Design with constructal theory. Wiley; 2008.

[11] Ghoneim AA. Comparison of theoretical models of phase-change and sensible heat storage for air and water-based solar heating systems. Solar Energy. 1989 Jan 1;42(3):209–20.

[12] Agyenim F, Eames P, Smyth M. A comparison of heat transfer enhancement in a medium temperature thermal energy storage heat exchanger using fins. Solar Energy. 2009 Sep 1;83 (9):1509–20.

[13] Bejan A, Convection heat transfer. 4th Ed., John Wiley & Sons; 2013.

[14] Combelles L, Lorente S, Anderson R, Bejan A. Tree-shaped fluid flow and heat storage in a conducting solid. Journal of Applied Physics. 2012 Jan 1;111(1):014902.

[15] Lorente S, Bejan A, Niu JL. Phase change heat storage in an enclosure with vertical pipe in the center. International Journal of Heat and Mass Transfer. 2014 May 1;72:329–35.

[16] Alalaimi M, Lorente S, Bejan A. Thermal coupling between a helical pipe and a conducting volume. International Journal of Heat and Mass Transfer. 2015 Apr 1;83:762–7.

[17] Bejan A, Ziaei S, Lorente S. The S curve of energy storage by melting. Journal of Applied Physics. 2014 Sep 21; 116(11):114902.

[18] Zhang P, Meng ZN, Zhu H, Wang YL, Peng SP. Melting heat transfer characteristics of a composite phase change material fabricated by paraffin and metal foam. Applied Energy. 2017 Jan 1;185:1971–83.

[19] Ogoh W, Groulx D. Effects of the number and distribution of fins on the storage characteristics of a cylindrical latent heat energy storage system: a

numerical study. Heat and Mass Transfer. 2012 Oct 1;48(10):1825–35.

[20] Kamkari B, Shokouhmand H. Experimental investigation of phase change material melting in rectangular enclosures with horizontal partial fins. International Journal of Heat and Mass Transfer. 2014 Nov 1;78:839–51.

[21] Joseph A, Kabbara M, Groulx D, Allred P, White MA. Characterization and realtime testing of phasechange materials for solar thermal energy storage. International Journal of Energy Research. 2016 Jan;40(1):61–70.

[22] Azad AV, Amidpour M. Economic optimization of shell and tube heat exchanger based on constructal theory. Energy. 2011 Feb 1;36(2):1087–96.

# 11

# Heat Exchanger Design and Optimization

*Shahin Kharaji*

**Abstract**

A heat exchanger is a unit operation used to transfer heat between two or more fluids at different temperatures. There are many different types of heat exchangers that are categorized based on different criteria, such as construction, flow arrangement, heat transfer mechanism, etc. Heat exchangers are optimized based on their applications. The most common criteria for optimization of heat exchangers are the minimum initial cost, minimum operation cost, maximum effectiveness, minimum pressure drop, minimum heat transfer area, minimum weight, or material. Using the data modeling, the optimization of a heat exchanger can be transformed into a constrained optimization problem and then solved by modern optimization algorithms. In this chapter, the thermal design and optimization of shell and tube heat exchangers are presented.

**Keywords:** log-mean temperature difference (LMTD), effectiveness-number of transfer units (ε-NTU), genetic algorithm (GA), particle swarm optimization (PSO)

## 1. Introduction

Heat exchangers are systems used to transfer heat between fluids with different temperatures. These devices have vast applications in many areas, such as refrigeration, heating, and air conditioning systems, power plants, chemical processes, food industry, automobile radiators, and waste heat recovery units. Heat exchangers can be classified according to different criteria such as construction, flow arrangement, heat transfer mechanism, etc [1]. The heat exchanger design can be divided into two main categories, thermal and hydraulic design and mechanical design. In ther-mal and hydraulic design, the focus is on calculating an adequate surface area transfer a certain amount of heat, pressure dope, pumping power work, etc. The goal of the mechanical design is to design the mechanical integrity of the exchanger, as well as designing various pressure and non-pressure components. In this chapter, the thermal and hydraulic design of heat exchangers is presented. To achieve better performance of heat exchangers, they optimize based on their application. Heat exchanger optimization can be performed using different optimization algorithms. Since most heat exchanger optimization problems are nonlinear, using traditional methods such as linear and dynamic programming and steepest descent may not lead to the desired solution and may even fail. Also, most traditional methods need gradient information to solve an optimization problem. On the other hand, advanced optimization algorithms are developed, which are gradient-free. Several advanced optimization methods, such as genetic algorithm, non-nominated sorting genetic algorithm, bio-geography-based optimization, particle swarm optimization, Jaya algorithm, and teaching-learning-based optimization, can be more efficient in solving an optimization problem. However, each of these methods has its advantages and disadvantages, which are discussed in the optimization section. In this chapter, genetic algorithm and particle swarm optimization are discussed in detail due to the vast applications that arise from their acceptable accuracy, as well as short computational time [2].

## 2. Basic equation of heat transfer

In most heat transfer problems, hot and cold fluids are divided by a solid wall. In this case, the mechanism of heat transfer from hot fluid to the cold fluid can be categorized into three steps:

- Heat transfer from the hot fluid to the wall by convection.
- Heat transfer through the wall by conduction.
- Heat transfer from the wall to the cold fluid by convection.

**Figure 1** shows a schematic of heat transfer between two fluids. As it can be seen, thermal resistance ($R$) is present at each stage of the transfer. Thermal resistance is a thermal (physical) property that indicates the resistance of each material to heat transfer due to temperature differences that can be calculated from [3]:

$$\begin{aligned} R &= \frac{L}{KA}\ , \ for\ conduction \\ R &= \frac{1}{hA}\ , \ for\ convection \end{aligned} \tag{1}$$

Where $L$ is the thickness of the wall, $A$ is the cross-sectional area in which heat transfer occurs, and $K$ and $h$ are conduction and convection heat transfer coefficient, respectively.

Heat transfer in each stage can be calculated as follows [4]:

$$Q = \frac{T_1 - T_2}{R_{c,H}} = \frac{T_2 - T_3}{R_{f,H}} = \frac{T_3 - T_4}{R_w} = \frac{T_4 - T_5}{R_{f,C}} = \frac{T_5 - T_6}{R_{c,C}} \tag{2}$$

Where:

$R_{c,H}$ = thermal resistance for convection in the hot side.

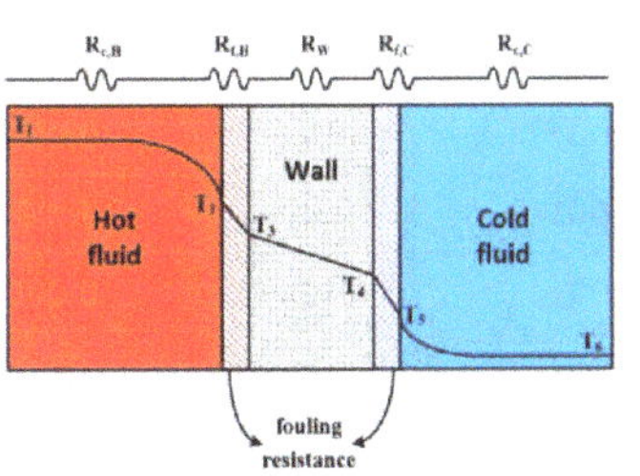


**Figure 1.**
*A schematic of heat transfer in heat exchangers.*

$R_{f,H}$ = fouling resistance of hot side.
$R_w$ = wall resistance.
$R_{f,C}$ = fouling resistance of cold side.
$R_{c,C}$ = thermal resistance for convection in cold side.

**Table 1** shows the fouling resistance of the most common fluids used in heat exchangers. The overall heat transfer coefficient can be obtained from Eq. (2) as follows [4]:

$$Q = \frac{T_1 - T_6}{R_{c,H} + R_{f,H} + R_w + R_{f,C} + R_{c,C}} \tag{3}$$

or:

$$Q = \frac{T_1 - T_6}{\frac{1}{h_H A_H} + \frac{r_{f,H}}{A_H} + \frac{r_w}{A_H} + \frac{r_{fC}}{A_C} + \frac{1}{h_C A_C}} \tag{4}$$

| Gas and vapors | Fouling Factor $\left(\frac{hr-ft^2-°F}{Btu}\right)$ | Liquids | | | | Fouling Factor $\left(\frac{hr-ft^2-°F}{Btu}\right)$ |
|---|---|---|---|---|---|---|
| **Industrial** | | **Industrial liquids** | | | | |
| Manufactured Gas | 0.01 | Industrial organic heat transfer media | | | | 0.001 |
| Engine Exhaust Gas | 0.01 | Refrigerating liquids | | | | 0.001 |
| Steam (non-oil bearing) | 0.0005 | Molten heat transfer salts | | | | 0.0005 |
| Exhaust Steam (oil bearing) | 0.001 | Hydraulic fluid | | | | 0.001 |
| Refrigerant Vapors (oil bearing) | 0.002 | **Industrial oils** | | | | |
| Compressed Air | 0.002 | Fuel oil | | | | 0.005 |
| Industrial Organic Heat Transfer Media | 0.001 | Engine lube oil | | | | 0.001 |
| **Chemical Processing** | | Transformer oil | | | | 0.001 |
| Acid Gas | 0.001 | Quench oil | | | | 0.004 |
| Solvent Vapors | 0.001 | Vegetable oils | | | | 0.003 |
| Stable Overhead Product | 0.001 | **water** | | | | |
| **Petroleum Processing** | | Temperature of heating medium | 240°F≤ | | 240–400°F | |
| Atmospheric Tower Overhead Vapors | 0.001 | Temperature of water | 125°F≤ | | 125°F> | |
| Light Naphthas | 0.001 | velocity | 3 ft.≤ | 3 ft.> | 3 ft. ≤ | 3 ft. > |
| Vacuum Overhead Vapors | 0.002 | Seawater | 0.0005 | 0.0005 | 0.001 | 0.001 |
| Natural Gas | 0.001 | Brackish water | 0.002 | 0.001 | 0.003 | 0.002 |
| Overhead Products | 0.001 | Distilled water | 0.0005 | 0.0005 | 0.0005 | 0.0005 |
| Coke Unit Overhead Vapors | 0.002 | Boiler blowdown | 0.002 | 0.002 | 0.002 | 0.002 |

**Table 1.**
*Fouling factors for different types of fluid [5, 6].*

Where:
$h_H$ and $h_C$ = convection heat transfer coefficient of the hot and cold sides, respectively.
$A_H$ and $A_C$ = and surface area of wall in the hot and cold side, respectively. The $r_w$ can be calculated for flat wall and cylindrical walls using Eqs. (5) and (6), respectively.

$$r_w = \frac{d_w}{KA}, \; for\ flat\ wall \tag{5}$$

$$r_w = \frac{\ln\left(\frac{r_o}{r_i}\right)}{2\pi LK}, \; for\ cylindrical\ wall \tag{6}$$

Where $d_w$ is the thickness of the wall, and $r_o$ and $r_i$ are the outside and inside diameter of the wall, respectively.
Total thermal resistance can be expressed as [7]:

$$R_t = \frac{1}{h_H A_H} + \frac{r_{f,H}}{A_H} + \frac{r_w}{A_H} + \frac{r_{f,C}}{A_C} + \frac{1}{h_C A_C} \tag{7}$$

The rate of heat transfer ($Q$) can be determined from

$$Q = UA\Delta T \tag{8}$$

Where $U$ is the overall heat transfer coefficient.

$$U = \frac{1}{\frac{1}{h_H \frac{A_H}{A_{ref}}} + \frac{r_{f,H}}{\frac{A_H}{A_{ref}}} + \frac{r_w}{\frac{A_H}{A_{ref}}} + \frac{r_{f,C}}{\frac{A_C}{A_{ref}}} + \frac{1}{h_C \frac{A_C}{A_{ref}}}} \tag{9}$$

Where $A_{ref}$ is a reference area. If the heat transfer is carried out over a pipe, the inside and outside surface areas of the pipe are not equal. Hence, the $A_{ref}$ must be determined (The outer surface of the pipes is usually selected).

## 3. Thermal design of heat exchangers

The thermal design of heat exchangers can be performed by several methods. The most commonly used methods are log-mean temperature difference (LMTD) and effectiveness-number of transfer units ($\varepsilon$-NTU) [8]. The LMTD was used to calculate heat transfer when the inlet and outlet temperatures of fluids are specified. When more than one inlet and/or outlet temperature of the heat exchanger is unknown, LMTD may be calculated by trial and errors solution. In this case, the ε-NTU method is commonly used [3].

### 3.1 The log-mean temperature difference (LMTD) method

As mentioned earlier, by determining the temperature difference between hot and cold fluids, the amount of heat transfer can be calculated from Eq. (3). **Figure 2** shows temperature changes of hot and cold fluids along with a heat exchanger with different types of flow configuration. As it can be seen the temperature difference between hot and cold fluids can vary along with the heat exchanger. Terminal temperatures of hot and cold fluids ($T_{H,out}$ anf $T_{C,out}$) are very effective factors in a heat exchanger design. If $T_{C,out}$ is lower than $T_{H,out}$ for countercurrent flow,

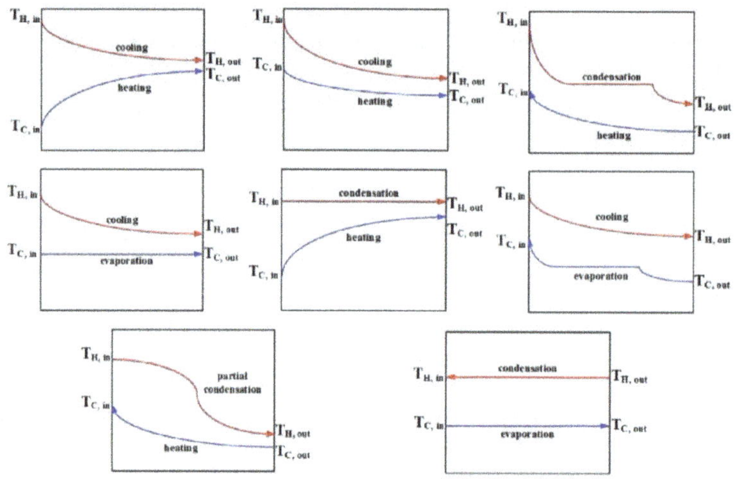


**Figure 2.**
*Temperature changes of hot and cold fluids along with a heat exchanger with different types of flow configuration.*

temperature approach occurs. In contrast, if $T_{C,out}$ is higher than the $T_{H,out}$ for countercurrent flow, temperature cross happens [9]. But if $T_{C,out}$ is equal to $T_{H,out}$, temperature meet takes place. Based on the second law of thermodynamics, temperature cross can never take place for heat exchangers with co-current flow configuration [10]. In 1981, Wales [11] proposed a parameter G, which can be used to determine the temperature conditions in the heat exchangers. Eq. (10) defines the $G$ parameters that can be between −1 and 1.

$$\begin{aligned} &G = \frac{T_{H,out} - T_{C,out}}{T_{H,in} - T_{C,in}}, \\ &G > 0 \rightarrow \ temperature \ \ approach \\ &G = 0 \rightarrow temperature \ \ meet \\ &G < 0 \rightarrow temperature \ \ cross \end{aligned} \tag{10}$$

Since the temperature difference between hot and cold streams varies along with the heat exchanger, the basic question is which temperature difference should be considered to calculate the heat transfer rate. To answer this question, consider **Figure 3** which shows heat transfer between two parallel and co-current fluids. Based on **Figure 3**, the heat transfer for the specified heat transfer area can be written in the form:

$$dQ = U \ \ \Delta T \ \ dA \tag{11}$$

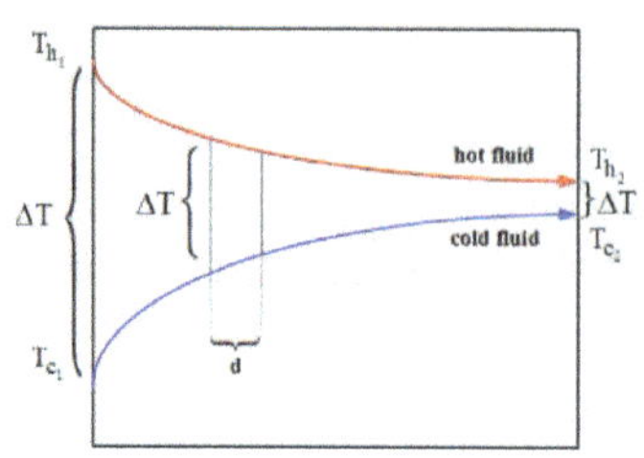


**Figure 3.**
*Heat transfer between two parallel and co-current fluids [3].*

It can be said that the amount of heat transferred is reduced from the hot fluid and added to the cold fluid. Therefore [3]:

$$dQ = -C_{p,H}\ dT_H \tag{12}$$

$$dQ = C_{p,C}\ dT_C \tag{13}$$

Where $C_{p,H}$ and $C_{p,C}$ are specific heat capacities of hot and cold fluids, respectively. The temperature difference can be written as below:

$$d(\Delta T) = dT_H - dT_C \tag{14}$$

By the combination of Eqs. (12) and (13) with Eq. (14):

$$d(\Delta T) = -\frac{dQ}{C_{p,H}} - \frac{dQ}{C_{p,C}} \tag{15}$$

By defining $\frac{1}{M} = \frac{1}{C_{p,C}} + \frac{1}{C_{p,H}}$, Eq. (15) can be written as follow:

$$dQ = M\ d(\Delta T) \tag{16}$$

Assuming the $M$ is constant along with the heat exchanger:

$$\int_0^Q dQ = -M \int_{\Delta T_{in}}^{\Delta T_{out}} d(\Delta T) \rightarrow Q = -M(\Delta T_{out} - \Delta T_{in}) \tag{17}$$

On the other hand, by placing Eq. (16) in Eq. (11):

$$-M\frac{d(\Delta T)}{\Delta T} = UdA = A\frac{UdA}{A} \tag{18}$$

$$A\int_0^A \frac{UdA}{A} = -M \int_{\Delta T_{in}}^{\Delta T_{out}} \frac{d(\Delta T)}{\Delta T} \rightarrow U_m A = -M\ \ln\left(\frac{\Delta T_{out}}{\Delta T_{in}}\right) \tag{19}$$

Where $U_m$ is the mean overall heat transfer coefficient which can be defined as below:

$$U_m = \int_0^A \frac{U\ dA}{A} \tag{20}$$

The heat transfer rate can be written as flow:

$$Q = U_m A \Delta T_{LMTD} \tag{21}$$

Where $\Delta T_{LMTD}$ is the logarithmic mean temperature difference (LMTD) which can be defined as below:

$$\Delta T_{LMTD} = \frac{\Delta T_{out} - \Delta T_{in}}{\ln\left(\frac{\Delta T_{out}}{\Delta T_{in}}\right)} \tag{22}$$

The simplicity of the LMTD method has led to its use in the design of many heat exchangers by introducing a correction factor, $F$, according to Eq. (23) [12]. The F is generally expressed in terms of two non-dimensional parameters, thermal effectiveness ($P$), and heat capacity ratio ($R$). The $P$ and $R$ are defined as Eqs. (24) and (25), respectively [13]. **Figure 4** shows the correction factor for common shell and tube heat exchangers.

$$Q = UAF\Delta T_{LMTD} \ , \quad 0 < F < 1 \tag{23}$$

$$P = \frac{T_{C,out} - T_{C,in}}{T_{H,in} - T_{C,in}} \tag{24}$$

$$R_{cr} = \frac{T_{H,in} - T_{H,out}}{T_{C,out} - T_{C,in}} \tag{25}$$

### 3.2 Effectiveness-number of transfer units (ε-NTU)

When more than one of the inlet and outlet temperatures of the heat exchanger is unknown, LMTD may be calculated by trial and errors solution. Another approach to calculating the rate of heat transfer is the effectiveness number of transfer units ($\varepsilon$-NTU) method. The $\varepsilon$-NTU can be expressed according to Eq. (26) where $Cp_{min}$ is the minimum value between the heat capacity of cold fluid ($Cp,_C$) and hot fluid ($Cp,_H$). The effectiveness ($\varepsilon$) can be defined as the ratio of the actual heat transfer rate ($q$) and the maximum possible heat transfer rate ($q_{max}$) according to Eq. (27).

$$NTU = \frac{UA_m}{Cp_{\min}} \tag{26}$$

$$\varepsilon = \frac{q}{q_{\max}} \tag{27}$$

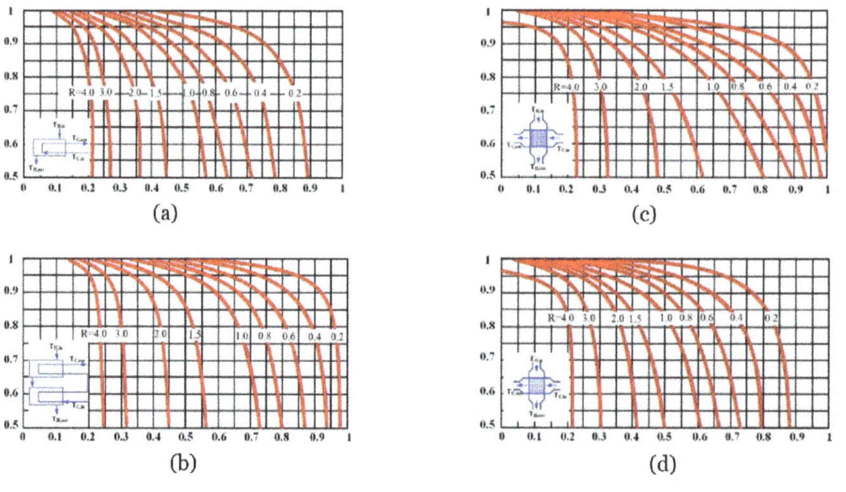

**Figure 4.**
*Correction factor for common shell and tube heat exchangers [3]. (a) One-shell pass and 2, 4, 6, etc. (any multiple of 2), tube pass. (b) Two-shell pass and 4, 8, 12, etc. (any multiple of 4), tube pass. (c) Single-pass cross-flow with both fluids unmixed. (d) Single-pass cross-flow with one fluid unmixed and other unmixed.*

Where:

$$q = Cp_{,H}(T_{H,in} - T_{H,out}) = Cp_{,C}(T_{C,out} - T_{C,in}) \tag{28}$$

$$q_{\max} = C_{p\min}(T_{H,in} - T_{C,in}) \tag{29}$$

The heat transfer rate using the $\varepsilon$-NTU method can express as [14]:

$$Q = \varepsilon C_{P\min}(T_{H,in} - T_{C,in}) \tag{30}$$

**Tables 2** and **3** show the effectiveness and NTU relations for heat exchangers, respectively. It should be noted that $C_r$ is the capacity and it can be defined as follows:

$$C_r = \frac{C_{P\min}}{C_{P\max}} \tag{31}$$

| Heat exchanger type | Effectiveness relation |
|---|---|
| Double pipe:<br>• Parallel-flow<br>• Counter flow with $C_r < 1$<br>• Counter flow with $C_r = 1$ | $\varepsilon = \frac{1-\exp[-NTU(1+C_r)]}{1+C_r}$<br>$\varepsilon = \frac{1-\exp[-NTU(1-C_r)]}{1-C_r\exp[-NTU(1-C_r)]}$<br>$\varepsilon = \frac{NTU}{1+NTU}$ |
| Sell and tube:<br>• On-pass and 2, 4, ... tube | $\varepsilon = 2\left\{1 + C_r + \sqrt{1 + C_2}\frac{1+\exp\left[-NTU\sqrt{1+C_2}\right]}{1-\exp\left[-NTU\sqrt{1+C_2}\right]}\right\}^{-1}$ |
| Cross flow (single-pass):<br>• $C_{max}$ and $C_{min}$ unmixed<br>• $C_{max}$ mixed and $C_{min}$ unmixed<br>• $C_{max}$ unmixed and $C_{min}$ mixed | $\varepsilon = 1 - \exp\left[\frac{1}{C_r}(NTU)^{0.22}\left\{\exp\left[-C_r(NTU)^{0.78}\right] - 1\right\}\right]$<br>$\varepsilon = \frac{1}{C_r}\{1 - \exp[-C_r(1 - \exp[NTU])]\}$<br>$\varepsilon = 1 - \exp\left\{-\frac{1}{C_r}(1 - \exp[-C_r NTU])\right\}$ |
| All heat exchangers with $C_r = 0$ | $\varepsilon = 1 - \exp(-NTU)$ |

**Table 2.**
*Effectiveness relation for heat exchangers [15].*

| Heat exchanger type | NTU relation |
|---|---|
| Double pipe:<br>• Parallel-flow<br>• Counter flow with $C_r < 1$<br>• Counter flow with $C_r = 1$ | $NTU = -\frac{\ln[1-\varepsilon(1+C_r)]}{1+C_r}$<br>$NTU = \frac{1}{C_r-1}\ln\left(\frac{\varepsilon-1}{\varepsilon C_r-1}\right)$<br>$NTU = \frac{\varepsilon}{1-\varepsilon}$ |
| Sell and tube:<br>• On-pass and 2, 4, ... tube | $NTU = -\frac{1}{\sqrt{1+C^2}}\ln\left[\frac{\frac{2}{\varepsilon}-1-C_r-\sqrt{1+C^2}}{\frac{2}{\varepsilon}-1-C_r+\sqrt{1+C^2}}\right]$ |
| Cross flow (single-pass):<br>• $C_{max}$ mixed and $C_{min}$ unmixed<br>• $C_{max}$ unmixed and $C_{min}$ mixed | $NTU = -\ln\left[1 + \frac{\ln(1-\varepsilon C_r)}{C_r}\right]$<br>$NTU = -\frac{\ln[C_r\ln(1-\varepsilon)+1]}{C_r}$ |
| All heat exchangers with $C_r = 0$ | $NTU = -\ln[1 - \varepsilon]$ |

**Table 3.**
*NTU relation for heat exchangers [15].*

## 4. Thermal and hydraulic design of shell and tube heat exchanger

Heat exchangers can be classified according to different criteria such as construction, flow arrangement, heat transfer mechanism, etc [1]. Shell and tube heat exchangers are some of the most convenient heat exchangers due to their versatility, wide operating range, and simplicity [16]. Hence, this chapter focuses on the design of this type of heat exchanger. In the design of shell and tube heat exchangers, a lot of consideration including the number of shells and tubes, tube pitch and layout, tube passes, baffles, etc., should be taken into account. In this case, there are some methods such as Kern and Bell-Delaware to design a heat shell and tube exchanger design. Since Kern's method offers the simplest route, this chapter is focused on this method.

### 4.1 Kern's method

Kern's method is based on experimental data for typical heat exchangers. In this method, it is assumed the shell flow is ideal, and leakage and bypass are negligible. Based on this flow model, only a single stream flows in the shell that is driven by baffles. This can lead to a very simple and rapid calculation of shell-side coefficients as well as pressure drop [17]. **Figure 5** shows a schematic of a shell and tube heat exchanger.

### 4.2 Number of tubes

The number of tubes ($N_t$) can be calculated as follows:

$$N_t = \frac{4\,\dot{m}_t}{\rho_t \nu_t \pi d_i^2} \tag{32}$$

Where $\dot{m}_t$ is the flow rate of fluid inside the tube, $\rho_t$ is the density of the fluid inside the tube, $\nu_t$ is the velocity of the fluid inside the tube, $A_t$ is the cross-sectional area of the tube, and $d_i$ is the tube inside diameter.

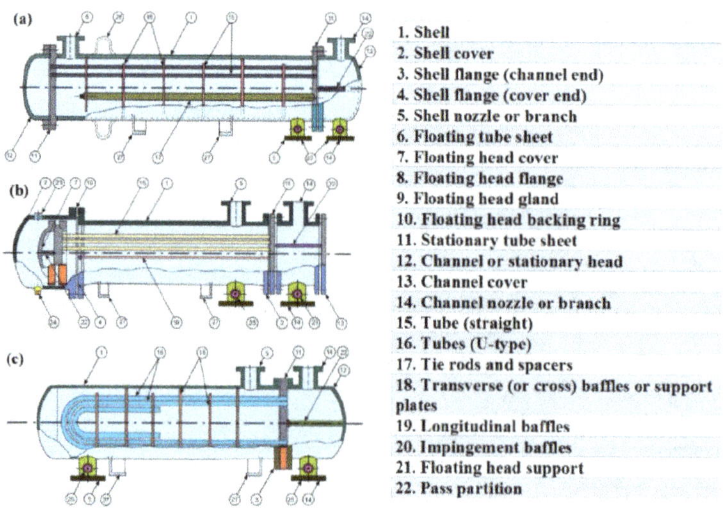


**Figure 5.**
*Schematic of a shell and tube heat exchanger a) fixed-tube b) floating-head c) removable U-tube [18].*

### 4.3 Tube-side heat transfer coefficient

The heat transfer coefficient for the tube side ($h_t$) is calculated as follows:

$$h_t = Nu_t \frac{k_t}{d_i} \tag{33}$$

Where $Nu_t$ is the Nusselt number for the tube-side fluid and $k_t$ is the thermal conductivity of the tube-side fluid. The $Nu_t$ is a function of Reynolds number (*Re*) and Prandtl number (*Pr*). *Re* and *Pr* can be obtained by the following:

$$\mathrm{Re}_t = \frac{\rho_t v_t d_t}{\mu_t} \tag{34}$$

$$\mathrm{Pr}_t = \frac{C_p \mu_t}{K} \tag{35}$$

Where $\mu_t$ is the dynamic viscosity of the tube-side fluid, $K$ is the heat conductivity coefficient, and $C_p$ is the heat capacity of the tube-side fluid. The $Nu_t$ can be calculated according to the type of flow as follows:

$$Nu_t = \frac{(f_t/2)\,\mathrm{Re}_t \mathrm{Pr}_t}{1.07 + 12.7(f_t/2)^{1/2}\left(\mathrm{Pr}_t^{2/3} - 1\right)} ;for : 10^4 < \mathrm{Re} < 5 \times 10^6 \,\&\; 0.5 < \mathrm{Pr} < 200 \tag{36}$$

$$Nu_t = 1.86\left(\frac{\mathrm{Re}_t \mathrm{Pr}_t d_i}{L}\right)^{1/3} \; ; \; for: \; \left(\frac{\mathrm{Re}_t \mathrm{Pr}_t d_i}{L}\right)^{1/3} > 2 \;\&\; 0.48 < \mathrm{Pr} < 16700 \tag{37}$$

Where $L$ is the length of the tube and $f_t$ is the friction factor of the tube side, which can be calculated from

$$f_t = (1.58 \ln \mathrm{Re}_t - 3.28)^{-2} \tag{38}$$

The convection heat transfer coefficient in the tube is obtained based on the value of the $\mathrm{Re}_t$ from [19]:

$$h_t = \frac{k_t}{d_i}\left[3.657 + \frac{0.0677\left(\mathrm{Re}_t \mathrm{Pr}_t \frac{d_i}{L}\right)^{1.3}}{1 + 0.1\mathrm{Pr}_t\left(\mathrm{Re}_t + \frac{d_i}{L}\right)^{0.3}}\right] \; ; \; for \;\; \mathrm{Re}_t < 2300 \tag{39}$$

$$h_t = \frac{k_t}{d_i}\left[\frac{\frac{\lambda}{8}(\mathrm{Re}_t - 1000)\mathrm{Pr}_t}{1 + 12.7\sqrt{\frac{\lambda}{8}}(\mathrm{Pr}_t^{0.67} - 1)} \quad 1 + \left(\frac{d_i}{L}\right)^{0.67}\right] \; ; \; for \;\; 2300 < \mathrm{Re}_t < 10000 \tag{40}$$

$$h_t = \frac{k_t}{d_i} 0.027\,\mathrm{Re}_t^{0.8} \mathrm{Pr}_t^{\frac{1}{3}} \left(\frac{\mu_t}{\mu_{w,t}}\right)^{0.14} \; ; \; for \;\; \mathrm{Re}_t > 10000 \tag{41}$$

Where $\mu_{w,t}$ is the dynamic viscosity of the tube-side fluid at the wall temperature and $\lambda$ is the Darcy friction coefficient which can be defined as [19]:

$$\lambda = (1.82 \log_{10} \log_{10} \mathrm{Re}_t - 1.64)^2 \tag{42}$$

The tube-side pressure drop is calculated by the following:

$$\Delta P_t = \left(4f_t \frac{LN_p}{d_i} + 4N_p\right) \frac{\rho_t \mu_t^2}{2} \tag{43}$$

Where $N_p$ is the tube passes.

### 4.4 Shell diameter

Inside sell diameter ($D_s$) is calculated as follows:

$$D_s = \sqrt{\frac{4AN_t}{(CTP)\pi}} \tag{44}$$

Where $A$ is the projected area of the tube layout expressed as an area corresponding to one tube and can be obtained from Eq. (45). Also, $P_t$ is tube pitch and $CL$ is the tube layout constant. **Figure 6** depicts two common tube layouts, square pitch and triangular pitch. The $CTP$ is the tube count calculation constant that accounts for the incomplete coverage of the shell diameter by the tubes, due to necessary clearances between the shell and the outer tube circle and tube omissions due to tube pass lanes for multitude pass design [15]. Eq. (46) shows the $CTP$ for different tube passes.

$$\begin{aligned} &A = P_t^2(CL) \\ &CL = 1 \rightarrow for\ square-pitch\ layout \\ &CL = 0.866 \rightarrow for\ triangular-pitch\ layout \end{aligned} \tag{45}$$

$$\begin{aligned} &CTP = 0.93 \rightarrow for\ one-tube\ pass \\ &CTP = 0.9 \rightarrow for\ two-tube\ pass \\ &CTP = 0.85 \rightarrow for\ three-tube\ pass \end{aligned} \tag{46}$$

Combining Eq. (44) with Eq. (45) as well as defining tube pitch ratio as $P_r$, one gets:

$$D_s = \sqrt{\frac{4(P_r d_o)^2(CL)N_t}{(CTP)\pi}} \tag{47}$$

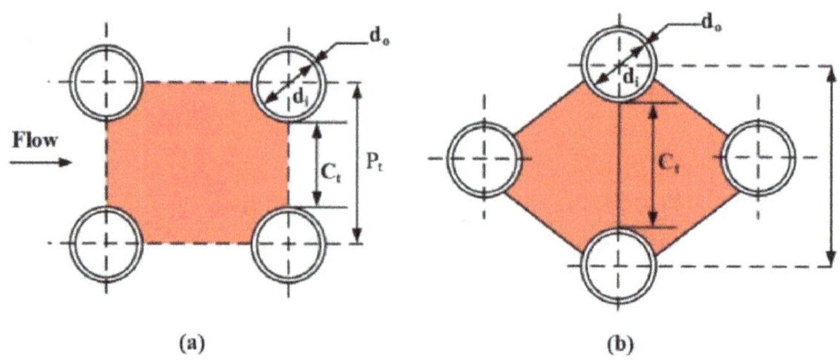


**Figure 6.**
*(a) Square-pitch (b) triangular-pitch layout [20].*

$$P_r = \frac{P_t}{d_o} \tag{48}$$

Where $d_o$ is tube outside the diameter. Eq. (47) can be written as follows:

$$D_s = \sqrt{\frac{4(P_r^2)(CL)A_o d_o}{(CTP)\pi^2 L}} \tag{49}$$

Where $A_o$ is the outside heat transfer surface area based on the outside diameter of the tube and can be calculated from:

$$A_o = \pi d_o N_t L \tag{50}$$

The shell side flow direction is partially along the tube length and partially across to tube length or heat exchanger axis. The inside shell diameter can be obtained based on the cross-flow direction and the equivalent diameter ($D_e$) is calculated along the long axes of the shell. The equivalent diameter is given as follows:

$$D_e = \frac{4 \times free - flow \;\; area}{wetted \;\; perimeter} \tag{51}$$

From **Figure 6** the equivalent diameter for the square pitch and triangular pitch layouts are as below:

$$D_e = \frac{4\left(P_t^2 - \frac{\pi d_o^2}{4}\right)}{\pi d_o} \; ; \; for \;\; square - pitch \;\; tube \tag{52}$$

$$D_e = \frac{4\left(\frac{P_t^2\sqrt{3}}{4} - \frac{\pi d_o^2}{8}\right)}{\frac{\pi d_o}{2}} \; ; \; for \;\; triangular - pitch \;\; tube \tag{53}$$

Reynolds number for the shell-side ($Re_s$) can be obtained as follows:

$$\mathrm{Re}_s = \left(\frac{\dot{m}_s}{A_s}\right)\frac{D_e}{\mu_s} \tag{54}$$

Where $\dot{m}_s$ is the flow rate of shell-side fluid, $\mu_s$ is the viscosity of the shell-side fluid, and $A_s$ is the cross-flow area at the shell diameter which can be obtained as below:

$$A_s = \frac{D_s}{P_t}(B \times C_t) \tag{55}$$

Where $B$ is the baffle spacing and $C_t$ is the clearance between adjacent tubes. According to **Figure 6** $C_t$ is expressed as follows:

$$C_t = P_t - d_o \tag{56}$$

The shell-side mass flow rate ($G_s$) is found with:

$$G_s = \frac{\dot{m}_s}{A_s} \tag{57}$$

In Kern's method, the heat transfer coefficient for the shell-side ($h_s$) is estimated from the following:

$$h_s = \frac{0.36k_s}{D_e} \mathrm{Re}_s^{0.55} \mathrm{Pr}_s^{1/3}$$
$$for \ 2 \times 10^3 < \mathrm{Re}_s = \frac{G_s D_e}{\mu_s} < 1 \times 10^6 \quad (58)$$

Where $k_s$ is the thermal conductivity of the shell-side fluid. The tube-side pressure drop is calculated by the following:

$$\Delta P = \frac{f_s G_s^2 (N_b + 1) D_s}{2\rho_s D_e \left(\frac{\mu_b}{\mu_{w,s}}\right)^{0.14}} \quad (59)$$

Where $N_b$ is the number of baffles, $\rho_s$ is the density of the shell-side fluid, $\mu_b$ is the viscosity of the shell-side fluid at bulk temperature, and $\mu_{w,s}$ is the viscosity of the tube-side fluid at wall temperature. The $f_s$ is the friction factor for the shell and can be obtained as follows:

$$f_s = \exp\left[0.576 - 0.19 \ln\left(\mathrm{Re}_s\right)\right] \ ; \ for \ 400, \ \mathrm{Re}_s < 1 \times 10^6 \quad (60)$$

The wall temperature can be calculated as follows:

$$T_w = \frac{1}{2}\left(\frac{T_{H,in} + T_{H,out}}{2} + \frac{T_{C,in} + T_{C,out}}{2}\right) \quad (61)$$

According to Eq. (21), the heat transfer surface area ($A$) of the shell and tube heat exchanger is obtained by the following:

$$A = \frac{Q}{U_m F \Delta T_{LMTD}} \quad (62)$$

The required length of the heat exchanger can be calculated based on the heat transfer surface area as follows:

$$L = \frac{A}{\pi d_o N_t} \quad (63)$$

## 5. Optimization of heat exchangers

The applications of heat exchangers are very different. Therefore, they are optimized based on their application. The most common criteria for optimizing heat exchangers are the minimum initial cost, minimum operating cost, maximum effectiveness, minimum pressure drop, minimum heat transfer area, minimum weight or material, etc. These criteria can be optimized individually or in combination. It is clear from the above that the optimal design of heat exchangers is based on many geometrical and operational parameters with high complexity. So it is difficult to design a cheap and effective heat exchanger. The optimization techniques are usually applied to ensure the best performance as well as lower the cost of the heat exchanger. The optimization is carried out using different techniques. Traditional techniques such as linear and dynamic programming as well as steepest descent usually fail to solve nonlinear large-scale problems. The need for gradient information is another drawback of traditional techniques. Therefore, it is not possible to solve non-differentiable functions using these methods. To overcome these difficulties, advanced optimization algorithms are developed which are gradient-free. Several advanced optimization methods, such as genetic algorithm (GA) [21], non-nominated sorting GA (NSGA-II) [22], bio-geography-based optimization (BBO) [23], particle swarm optimization (PSO) [24], Jaya algorithm, and teaching–learning-based optimization (TLBO) [25], had been used for the optimization of heat exchangers by many researchers each of which has its advantages and disadvantages. Using GA, it is possible to solve all optimization problems, which can be described with the chromosome encoding and solves problems with multiple solutions. But in order to use GA, it is necessary to set a number of specific algorithmic parameterssuchasjumpprobability,selection operator, cross probability. NSGA-II has an explicit diversity preservation mech-anism and elitism prevents an already found Pareto optimal solution from being removed, but crowded comparison can limit the convergence and it needs the tuning of algorithmic-specific parameters including mutation probability, cross-over probability, etc. The optimization using BBO is also effective and it inhibits the degradation of the solutions, but poor exploiting the solutions is the main drawback of this method. The PSO is a heuristic and derivative-free technique that has the character of memory but it needs the tuning of algorithmic specific parameters and plurality of the population is not enough to achieve the global optimal solution. Similarly, TLBO and Jaya need the tuning of their own algorithmic-specific parameters [26].

Generally, an optimization design starts by selecting criteria (quantitatively) to minimize or maximize, which is called an objective function. In an optimization design, the requirements of a particular design such as required heat transfer, allowable pressure drop, limitations on height, width and/or length of the exchanger are called constraints. Several design variables such as operating mass flow rates and operating temperatures can also participate in an optimization design [15]. The single target optimization can be expressed as [27]:

$$\begin{aligned} &\min\ f(x) \\ &g_i(x) \geq 0,\ j = 1,\ 2,\ \ldots,\ J \\ &h_k(x) \geq 0,\ k = 1,\ 2,\ \ldots,\ K \end{aligned} \tag{64}$$

Where $f(x)$ is the objective function, $g_i(x) \geq 0$ is the inequality constraint, and $h_i(x) \geq 0$ is the equality constraint. Multi-objective combination optimization can be indicated as [27]:

$$\begin{aligned} &\min\ f(x) = \left( f_1(x),\ f_2(x),\ \ldots,\ f_m(x)\right) \\ &g_i(x) \geq 0,\ j = 1,\ 2,\ \ldots,\ J \\ &h_k(x) \geq 0,\ k = 1,\ 2,\ \ldots,\ K \end{aligned} \tag{65}$$

Using the data modeling, the optimization of a heat exchanger can be transformed into a constrained optimization problem and then solved by modern optimization algorithms. In this chapter, the focus is on GA and PSO because many researchers mentioned that these algorithms lead to remarkable savings in computational time and have an advantage over other methods in obtaining multiple solutions of the same quality. So it gives more flexibility to the designer [2].

## 5.1 Genetic algorithm

GA is a search heuristic that is inspired by Charles Darwin's theory of survival of the fittest, which explains inferior creatures pass away and superior creatures remain [28]. In GA, sets of design variables are codified by sequences with fixed or variable lengths, similar to chromosomes or individuals in biological systems. Each chromosome is formed of several design variables, which are known as genes. In repetitive processes such as GA, each repetitive stage is a generation and a collection of solutions associated with each generation is a population. Generally, the initial population is generated randomly [29]. In GA statistical methods are used to achieve optimum points. In the process of natural selection, populations are selected based on their fitness. A new population is formed using genetic operations containing selection, crossover, mutation, etc. This cycle continues until a certain result is achieved or the stop criterion is satisfied [30]. **Figure** 7 shows the flowchart of GA and the steps of binary GA are discussed below [32].

**Step 1: Initialization of population**

The initial population of GA includes binary numbers generated randomly which are chromosomes or GA strings, consisting of bits called genes. Actually, the initial population is the probable solution to the optimization problem. The number of gens ($n_g$) assigned to represent a variable in the chromosome depends on the precision ϵ and the range of the variable $[x_{min}, x_{max}]$, and is given by

$$n_g = \log_2\left(\frac{x_{\min} - x_{\max}}{\in}\right) \tag{66}$$

**Step 2: Fitness evaluation**

The fitness value of each GA string is examined by first determining the decoded values of the variables $D$, and next the corresponding real values are obtained as follows:

$$x = x_{\min} + \frac{x_{\max} - x_{\min}}{2^{n_g} - 1} \tag{67}$$

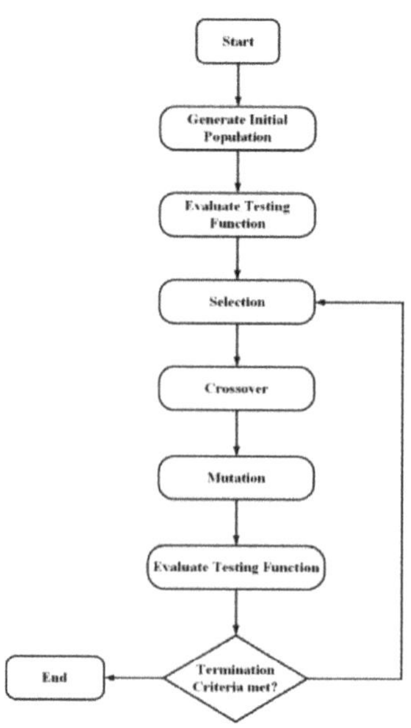


**Figure 7.**
*Flow chart of GA [31].*

The fitness function values are then computed knowing the real values of design variables.

**Step 3: Reproduction/selection**

In this step, chromosomes with better fitness values to participate in the cross-over are selected. Several selection modes such as roulette wheel selection or pro-portionate selection, rank-based selection, and tournament selection can be used in this step [33]. In proportionate selection, the probability of a chromosome to be selected is directly proportional to its fitness value. Hence, the chromosome having a better fitness value has a higher chance of selection for reproduction. This may result in premature convergence of the solution because there is a chance of losing diversity. The tournament selection is faster compared with the other two selection methods. In this method, $n$ chromosomes are randomly picked from the population of solutions, where n represents the tournament size. The chromosome having the best fitness value is copied to the mating pool and all the n GA strings are returned to the population. This process is repeated for obtaining all the individuals of the mating pool.

**Step 4: Crossover**

The genes are exchanged between two-parent chromosomes in the crossover step, which leads to a new set of solutions, called children. The crossover operation represents the selection pressure or exploitation of fit chromosomes for even better solutions. The crossover probability ($P_c$) specifies the number of individuals taking part in the crossover operation, and this control parameter value is optimally chosen as nearly equal to 1.0. Several schemes of crossover such as single-point crossover, two-point crossover, multipoint crossover, and uniform crossover can be used in this step. A comparison of these methods is given in the literature [34].

**Step 5: Mutation**

Mutation means the change of a bit from 0 to 1 and from 1 to 0 in the solution chromosome, which is used for the exploration of new solutions. It helps to come out of the local basin and search for a global solution. The mutation probability $P_m$ specifies the number of mutations and is commonly kept very low. Because if its value is high, the qualified solutions may be lost. The range of $P_m$ is given as

$$\frac{0.1}{l} \leq P_m \leq \frac{1}{l} \tag{68}$$

Where $l$ represents the length of the GA string. Steps 2, 3, 4, and 5 are repeated until the termination criterion (maximum number of generations or desired precision of solution) is met.

## 5.2 Particle swarm optimization

The PSO is inspired by the way fish and birds swarm search for food [35]. In this method, each particle represents one solution to a problem and they aim to find optimum points in a search space. This method is also based on the behavior of birds that they use to find their orientation. Based on this direction, the collective location of the swarm and the best individual location of particles per time are calculated and a new search orientation is composed of these two orientations and the previous orientation. In a search space of the D dimension, the best individual location of a particle and the best location of the overall particle are defined as Eqs. (69) and (70), respectively.

$$\vec{P}_1 = C_1\left(\vec{P}_{i1},\ \vec{P}_{i2},\ \ldots,\ \vec{P}_{iD}\right) \tag{69}$$

$$\vec{g}_1 = C_1\left(\vec{g}_1,\ \vec{g}_2,\ \dots,\ \vec{g}_D\right) \tag{70}$$

The best location in the vicinity of each particle is given as below:

$$\vec{n} = \left(\vec{n}_{i1},\ \vec{n}_{i2},\ \dots,\ \vec{n}_{iD}\right) \tag{71}$$

Displacement of particles after determination their velocity is as follows:

$$\vec{x}(t) = \vec{x}(t-1) + \vec{v}(t) \tag{72}$$

$$\vec{v}(t) = \vec{v}(t-1) + \vec{F}(t-1) \tag{73}$$

The best individual location of the particle and the best collective location of particles as two springs connected to the particle are used to model the force entered in the particle. The first spring is directed to the best individual experience and the second spring is directed to the best swarm experience. Eq. (74) shows the force entered in the particle.

$$\vec{F}_{i-1} = C_1\left(\vec{P}_{i-1} - \vec{x}_{i-1}\right) + C_2\left(\vec{g}_{i-1} - \vec{x}_{i-1}\right) \tag{74}$$

Where $C_1$ and $C_2$ are acceleration coefficients. The particle velocity at dimension $d$ ( $v_{id}$) and the next repetition can be obtained as follows [35]:

$$\begin{aligned} v_{id}(t) &= \omega v_{id}(t-1) + C_1 rand_1(P_{id}(t-1) - x_{id}(t-1)) \\ &\quad + C_2 rand_2(n_{id}(t-1) - x_{id}(t-1)) \end{aligned} \tag{75}$$

This shows the velocity of particle i at the star topology or global best. The $rand_1$ and $rand_2$ are random numbers that have a constant distribution in the range 0–1. **Figure 8** shows the flowchart of PSO and the steps of PSO are discussed below [32].

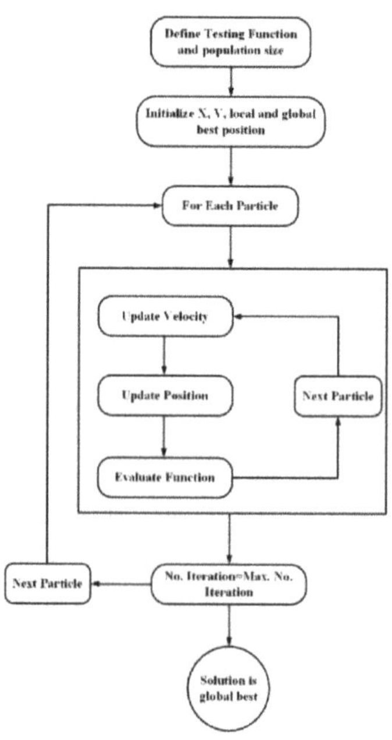


**Figure 8.**
*Flow chart of PSO [31].*

**Step 1: Initialization**

The swarm of potential solutions is generated with random positions and velocities. The $ith$ particle in D-dimensional space may be denoted as $X_i = (x_{i1}, x_{i2}, \ldots, x_{id})$ and $i = 1, 2, \ldots, N$, where $N$ denotes the size of the swarm.

**Step 2: Fitness evaluation**

The corresponding fitness values of the particles are evaluated.

**Step 3: Determination of personal and global best**

The best individual location of a particle ($\vec{P}_1$) is sorted, the particle having the best fitness value is determined for the current generation, and the best location ($\vec{g}_1$) is updated.

**Step 4: Velocity and position update**

The velocity and position of the ith particle are updated based on Eq. (75). Here, an example of the design and optimization of a shell and tube heat exchanger is presented. This example was used by Karimi et al. (2021) [36]. Their aim was to minimize the total annual cost ($C_{tot}$) for a shell and tube heat exchanger based on optimization algorithms. The total annual cost is the sum of the initial cost for the construction ($C_i$) of the heat exchanger and the cost of power consumption in the shell and tube heat exchanger ($C_{od}$). Hence, the total annual cost was consid-ered as an objective function that should be minimized using GA and PSO. Process input data and physical properties for this case study are presented in **Table 4**. Also, bounds for design parameters are listed in **Table 5**. The objective function can be written as follows:

$$C_{tot} = C_i + C_{od} \tag{76}$$

The results show that the use of PSO has been led to lower Ctot, which means that the minimization of cost function was performed better using this algorithm. Also, the use of PSO resulted in lower Δp and A as well as higher U (**Table 6**).

| | $\dot{m}$ $\left(\frac{kg}{s}\right)$ | $T_{in}$ (°C) | $T_{out}$ (°C) | $\rho$ $\left(\frac{kg}{m^3}\right)$ | $C_p$ $\left(\frac{J}{kg}\right)$ | $\mu$ (Pas) | $k$ $\left(\frac{W}{m.K}\right)$ | $R_f$ $\left(\frac{m^2K}{W}\right)$ | $\mu_w$ (Pas) |
|---|---|---|---|---|---|---|---|---|---|
| **Shell side: methanol** | 27.8 | 95 | 40 | 750 | 2840 | 0.00034 | 0.19 | 0.00033 | 0.00038 |
| **Tube side: sea water** | 68.9 | 25 | 40 | 995 | 4200 | 0.0008 | 0.59 | 0.0002 | 0.00052 |

**Table 4.**
*Process input data and physical properties for three case studies [36].*

| Parameters | Lower value | Upper value |
|---|---|---|
| **Tubes outside diameters(m)** | 0.015 | 0.051 |
| **Shell diameters(m)** | 0.1 | 1.5 |
| **Central baffle spacing(m)** | 0.05 | 0.5 |

**Table 5.**
*Bounds for design parameters [36].*

| | PSO | GA |
|---|---|---|
| $L\ (m)$ | 2.6871 | 3.9089 |
| $d_o\ (m)$ | 0.015063 | 0.015 |
| $B\ (m)$ | 0.49967 | 0.49989 |
| $D_s\ (m)$ | 0.81143 | 0.74105 |
| $N_t\ (m)$ | 1238 | 1365.5 |
| $v_t\ (m/s)$ | 0.83349 | 0.893 |
| $Re_t$ | 27,909 | 13,386 |
| $Pr_t$ | 5.69 | 5.69 |
| $h_t\ (W/m^2K)$ | 3740.8 | 4639.7 |
| $f_t$ | 0.0212 | 0.0073 |
| $\Delta P_t\ (Pa)$ | 4730 | 5191.3 |
| $D_e\ (m)$ | 0.0141 | 0.0106 |
| $v_s\ (m/s)$ | 0.498 | 0.499 |
| $Re_s$ | 15,489 | 11,716 |
| $Pr_s$ | 5.1 | 5.1 |
| $h_s\ (W/m^2K)$ | 9075.7 | 1648.9 |
| $f_s$ | 0.313 | 0.353 |
| $\Delta P_s\ (Pa)$ | 21,355 | 18,033 |
| $U\ (W/m^2K)$ | 900.98 | 686.71 |
| $A\ (m^2)$ | 198.78 | 252.58 |
| $C_i\ (\$)$ | 44,116 | 50,737 |
| $C_o\ (\$)$ | 2561.5 | 1085.2 |
| $C_{od}\ (\$)$ | 2340 | 6685 |
| $C_{tot}\ (\$)$ | 46,456 | 57422.51 |

**Table 6.**
*Optimal parameter of heat exchanger using GA and PSO algorithms [36].*

## 6. Conclusion

This chapter has discussed the thermal design and optimization of shell and tube heat exchangers. The basic equations of heat transfer were investigated and log-mean temperature difference (LMTD) and effectiveness-number of transfer units ($\varepsilon$-NTU) were presented. The thermal design was focused on Kern's method. In this method, it is assumed the shell flow is ideal, and leakage and bypass are negligible. Based on this flow model, only a single stream flows in the shell that is driven by baffles. This can lead to a very simple and rapid calculation of shell-side coefficients and pressure drop. The optimization of heat exchangers is presented based on the genetic algorithm (GA) and particle swarm optimization (PSO) due to the recommendation of these methods by man researchers because of quick convergence and obtaining multiple solutions.

## Conflict of interest

The authors declare no conflict of interest.

## Nomenclature

| | |
|---|---|
| $A$ | *Total heat transfer area* |
| $A_s$ | *Cross-flow area at the shell diameter* |
| $B$ | *Baffle spacing* |
| $Cp$ | *Heat capacity* |
| $C_i$ | *Capital investment cost* |
| $C_o$ | *Total operating cost* |
| $C_{od}$ | *Total discounted operating cost* |
| $C_r$ | *Capacity ratio* |
| $C_t$ | *Clearance between adjacent tubes* |
| $C_{tot}$ | *Total annual cost* |
| $CL$ | *Tube layout constant* |
| $CTP$ | *Tube count calculation constant* |
| $d_w$ | *Wall thickness* |
| $D_s$ | *Shell diameter* |
| $F_s$ | *Friction factor for shell* |
| $G_s$ | *Mass flow rate of shell-side fluid* |
| $h$ | *Convection heat transfer coefficient* |
| $k$ | *Thermal conductivity* |
| $K$ | *Heat conductivity coefficient* |
| $L$ | *Tube length* |
| $LMTD$ | *Logarithmic mean temperature difference* |
| $\dot{m}_s$ | *Flow rate of shell-side fluid* |
| $N_t$ | *Number of tubes* |
| NTU | Number of heat transfer unit |
| $P$ | *Thermal effectiveness* |
| $P_r$ | *Tube pitch ratio* |
| $P_t$ | *Tube pitch constant* |
| $Pr$ | *Prandtl number* |
| $q$ | *Actual heat transfer rate* |
| $q_{max}$ | *Maximum possible heat transfer rate* |
| $Q$ | *Heat transfer rate* |
| $r_i$ | *Inside diameter* |
| $r_o$ | *Outside diameter* |
| $R$ | *Thermal resistance* |
| $Re$ | *Reynolds number* |
| $T$ | *Temperature* |
| $U$ | *Overall heat transfer coefficient* |
| $U_m$ | *Mean overall heat transfer coefficient* |

### Greek symbols

| | |
|---|---|
| $\Delta$ | Difference |
| $\varepsilon$ | *Effectiveness* |
| $\rho$ | *Density* |
| $\mu$ | *Dynamic viscosity* |
| $\nu$ | ***Velocity*** |

***Subscripts***

| | |
|---|---|
| *b* | *Bulk* |
| *C* | *Cold* |
| *c* | *Convection* |
| *H* | *Cold* |
| *i* | *Inside* |
| *max* | *Maximum* |
| *min* | *Minimum* |
| *o* | *Outside* |
| *s* | *Shell side* |
| *t* | *Tube side* |

## Author details

Shahin Kharaji
Research and Develop Department, FAPKCO Engineering Group, Shiraz, Iran

*Address all correspondence to: shahinkharaji@gmail.com

## References

[1] Ghalandari, M., et al., Applications of intelligent methods in various types of heat exchangers: a review. Journal of Thermal Analysis and Calorimetry, 2021. 145(4): p. 1837-1848

[2] Lahiri SK, Khalfe NM,Wadhwa SK. Particle swarm optimization technique for the optimal design of shell and tube heat exchangers. Chemical Product and Process Modeling. 2012;7(1), Article 14.

[3] Ezgi CN. Basic Design Methods of Heat Exchanger. In: Heat Exchangers-Design, Experiment and Simulation. Rijeka: IntechOpen; 2017

[4] Chen Q, Ren J. Generalized thermal resistance for convective heat transfer and its relation to entransy dissipation. Chinese Science Bulletin. 2008;**53**(23): 3753-3761

[5] Marner W, Suitor J. Survey of Gas-Side Fouling in Industrial Heat-Transfer Equipment. Final Report. Pasadena, CA (USA): Jet Propulsion Lab; 1983

[6] Whitaker S. Fundamental Principles of Heat Transfer. Oxford, United kingdom: Pergamon press, Elsevier; 2013

[7] Çengel YA, Ghajar AJ. Heat and Mass Transfer: Fundamentals [and] Applications. New York City, United States: McGraw-Hill Education; 2020

[8] Borjigin S et al. Coupling ε-NTU method for thermal design of heat exchanger in cabinet cooling system. Applied Thermal Engineering. 2019;**159**: 113972

[9] Smith R. Chemical Process Design And Integration. New York City, United States: John Wiley & Sons; 2005

[10] Vengateson U. Design of multiple shell and tube heat exchangers in series: E shell and F shell. Chemical Engineering Research and Design. 2010; **88**(5–6):725-736

[11] Wales, R. and W. RE, Mean Temperature Difference in Heat Exchangers. New York, United state Chemical Engineering Journal; 1981

[12] Fakheri A. Log mean temperature correction factor: an alternative representation. In: Proceedings of the International Mechanical Engineering Congress and Exposition. New Orleans, Louisiana: Citeseer; 2002

[13] Gulyani BB, Jain A, Kumar S. Optimal synthesis of multipass heat exchanger without resorting to correction factor. International Journal of Mechanical, Aerospace, Industrial, Mechatronic and Manufacturing Engineering. 2011;**5**(5):898-904

[14] Iu I et al. Applying the effectiveness-NTU method to elemental heat exchanger models. ASHRAE Transactions. 2007;**113**(1):504-513

[15] Shah RK, Sekulic DP. Fundamentals of Heat Exchanger Design. New York City, United States: John Wiley & Sons; 2003

[16] Silaipillayarputhur K, Khurshid H. The design of shell and tube heat exchangers–A review. International Journal of Mechanical and Production Engineering Research and Development. 2019;**9**(1):87-102

[17] Alperen MA, Kayabaşi E, Kurt H. Detailed comparison of the methods used in the heat transfer coefficient and pressure loss calculation of shell side of shell and tube heat exchangers with the experimental results. Energy Sources, Part A: Recovery, Utilization, and Environmental Effects. 2019:1-20

[18] Standard I. Specification for Shell and Tube Type Heat Exchangers. New Delhi: BIS 2007; 2007

[19] Jalilirad S, Cheraghali MH, Ashtiani HAD. Optimal design of shell-

and-tube heat exchanger based on particle swarm optimization technique. Journal of Computational Applied Mechanics. 2015;**46**(1):21-29

[20] Anescu, G., A DMSACO Approach to Economic Heat Exchanger Design. Politechnica University of Bucharest in Romania.

[21] Guo J, Cheng L, Xu M. Optimization design of shell-and-tube heat exchanger by entropy generation minimization and genetic algorithm. Applied Thermal Engineering. 2009;**29**(14–15):2954-2960

[22] Yin Q et al. Optimization design and economic analyses of heat recovery exchangers on rotary kilns. Applied Energy. 2016;**180**:743-756

[23] Hadidi A, Nazari A. Design and economic optimization of shell-and-tube heat exchangers using biogeography-based (BBO) algorithm. Applied Thermal Engineering. 2013;**51**(1–2):1263-1272

[24] Patel V, Rao R. Design optimization of shell-and-tube heat exchanger using particle swarm optimization technique. Applied Thermal Engineering. 2010;**30**(11–12):1417-1425

[25] Rao RV, Patel V. Multi-objective optimization of heat exchangers using a modified teaching-learning-based optimization algorithm. Applied Mathematical Modelling. 2013;**37**(3): 1147-1162

[26] Rao RV et al. Design optimization of heat exchangers with advanced optimization techniques: A review. Archives of Computational Methods in Engineering. 2020;**27**(2):517-548

[27] Yao J. A review of industrial heat exchange optimization. In: IOP Conference Series: Earth and Environmental Science. JIOP Publishing Ltd in Bristol, England; 2018

[28] Haupt RL, Haupt SE. Practical Genetic Algorithms. New York City, United States: John Wiley & Sons; 2004

[29] Najafi H, Najafi B, Hoseinpoori P. Energy and cost optimization of a plate and fin heat exchanger using genetic algorithm. Applied Thermal Engineering. 2011;**31**(10):1839-1847

[30] Konak A, Coit DW, Smith AE. Multi-objective optimization using genetic algorithms: A tutorial. Reliability Engineering & System Safety. 2006; **91**(9):992-1007

[31] Sadeghzadeh H, Ehyaei M, Rosen M. Techno-economic optimization of a shell and tube heat exchanger by genetic and particle swarm algorithms. Energy Conversion and Management. 2015;**93**: 84-91

[32] Swayamsiddha S. Bio-inspired algorithms: principles, implementation, and applications to wireless communication. In: Nature-Inspired Computation and Swarm Intelligence. Cambridge, Massachusetts, United States: Academic Press; 2020. pp. 49-63

[33] Blickle T, Thiele L. A Comparison of Selection Schemes used in Genetic Algorithms , TIK-Report No. 11. Computer Engineering and Communication Networks Lab (TIK) Zurich, Switzerland; 1995

[34] Spears WM, Anand V. A study of crossover operators in genetic programming. In: International Symposium on Methodologies for Intelligent Systems. Berlin, Heidelberg: Springer; 1991

[35] Shi Y. Particle swarm optimization: developments, applications and resources. In: Proceedings of the 2001 Congress on Evolutionary Computation (IEEE Cat. No. 01TH8546). Seoul, Korea (South); 2001

[36] Karimi, H., H. Ahmadi Danesh Ashtiani, and C. Aghanajafi, Optimization Techniques for Design a shell and tube heat exchanger from economic view. AUT Journal of Mechanical Engineering, 2021. **5**(2): 9-9

# 12

# Heat Exchangers for Electronic Equipment Cooling

*Abdelhanine Benallou*

**Abstract**

Recent developments in the electronic equipment market have been very demanding on two important design parameters: the size of the equipment and the efficiency of the cooling system. Indeed, the race for more applications handling in reduced sizes in the case of smartphones requires the use of important amounts of energy in tiny volumes. Similar constraints are encountered in the design of the new generation of vehicles (electric cars, hybrid vehicles, high-speed trains, airplanes), which impose the use of highly integrated electronic structures, resulting in significant power densifications (up to several hundred Watts/cm$^2$). In some CPU boards, the power generated per unit chip area is in the order of 500 kW/m$^2$. Cooling of such boards requires low volume and lightweight heat exchangers to transfer tremendous amounts of heat. The same situation is encountered for most newly developed demotics' equipment. This chapter reviews available state-of-the-art technologies for electronic equipment cooling, including *passive* and *active techniques*, as well as *one and two-phase heat exchange*. Directions for the design of the different heat exchangers will also be given.

**Keywords:** passive cooling, active cooling, spray cooling, refrigerated cooling, heat pipes, cold plates, vapor chambers, microchannel heat sinks, thermal resistance, carrying capacity, thermal program, effective thermal conductivity

## 1. Introduction

Operation of active electronic components (transistors, integrated circuits, microprocessors, CPUs, GPUs, etc.) generates significant amounts of heat, around 500 kW/m$^2$, quite comparable to the flux densities encountered at the nose of a space shuttle entering the atmosphere.

If this heat is not extracted, it leads to important increases in the temperature of these components. Such increases ineluctably induce degradation of the bonding wires, delamination of solders, and/or the appearance of leakage currents. Moreover, temperature gradients between electronic chips and their soles generate cyclical thermomechanical stresses throughout the lifetime of the chips, which results in thermal fatigue, leading ultimately to component failure.

Thus, one of the problems with the operation of active electronic components is energy dissipation; How can an electronic system get rid of the heat its operation generates? Usually, for each electronic component, manufacturers specify a maximum operating temperature (the maximum junction operating temperature,$T_{op}$).

If they are not properly cooled when operating, electronic components will certainly reach $T_{op}$ and probably go over it. They then can completely lose their properties, which could lead to an alteration of the operation of the circuit boards or the systems they compose. This alteration can manifest through a complete breakdown of the systems or through a shortening of the mean time between failures (MTBF), leading to a premature aging.

Consequently, design techniques of electronic systems always consider a built-in cooling system, often called a *heatsink* to allow the component to get rid of the generated energy by dissipating it toward its surroundings. Heat sinks are actually special *heat exchangers*; they ensure the evacuation of the heat generated by the operation of electronic components.

This chapter reviews state-of-the-art technologies available for electronic equipment cooling. These technologies include *passive* and *active*, as well as *one and two-phase heat exchange*. Directions for the design of the different heat exchangers are also be given.

## 2. Passive heatsink technologies

Passive heat sinks are heat exchangers in the form of finned radiators, generally made of aluminum. They are called *passive* because they do not have any moving mechanical component (fan) designed to force airflow. Air moves through them only due to density difference: Hot air goes upward as it is replaced by cooler air. It is for this reason that passive sinks transfer heat to air essentially by natural convection[1]. This type of heat sinks is, by far, preferred for cooling electronic systems [2]; they are cost-effective, simple to find and assemble, and they generate no power consumption or noise.

The most widely used passive heat exchangers are *finned natural convection heat sinks*.

### 2.1 Finned heat exchangers

In their simplest version, these heat sinks are constituted of finned surfaces. **Figure 1** shows examples of *finned heat exchangers* used as heat sinks on electronic components. They are usually constituted of materials having a high thermal conductivity, such as aluminum or copper. But other new materials, like ceramics, can also be found.

The sink is usually mounted on the electronic component, which generates heat, thus increasing its heat transfer area[2] with ambient air. **Figure 2** shows a transistor mounted on a finned heatsink.

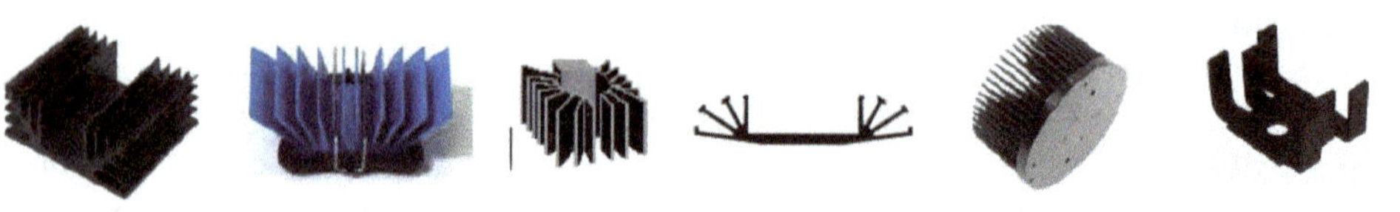

**Figure 1.**
*Examples of finned heat sinks. Sources [3, 4].*

---

[1] For a better understanding of natural convection see Ref. [1].

[2] The heat transfer area is the surface thru which energy is transferred from the electronic component to the environment.

**Figure 2.**
*A heatsink fixed on a transistor. Source [5].*

It can be seen from this figure that this type of mounting permits to increase the heat transfer area; the heat generated by the transistor is transferred to ambient, not only through the external surface of the transistor but also through the surfaces of all the fins.

In such situations, the heat transfer process depends on several parameters such as the temperature difference between air and the electronic component, the total area of the fins and their position (vertical/horizontal), the fins spacing, etc.

## 2.2 Design fundamentals

### *2.2.1 Thermal resistance*

Despite this complexity of the heat transfer process, design techniques usually use a simple model to represent the flow of energy through a heatsink. This simple model is based on a similarity between the way that heat moves from one medium to another and the way that electric current flows from one potential to another. We know that if a point at a potential $V_0$ (**Figure 3a**) is separated by an electrical resistance $R$ from a second point (at $V_1$), and then an electric current $i$ flows between these two points, such that:

$$i = \frac{V_0 - V_1}{R} \tag{1}$$

Similarly, it can be shown that the power, $Q$, of heat dissipation between a point at temperature $T_c$ and a point at temperature $T_a$ (**Figure 3b**) is given by:

$$Q = \frac{T_c - T_a}{R_{th}} \tag{2}$$

Where $R_{th}$ is called the *thermal resistance* between these points [6].

$R_{th}$ is expressed differently for conduction and convection as follows:

- For a conduction heat transfer:

$$R_{th}^{cond} = \frac{L}{kA} \tag{3}$$

**Figure 3.**
*Thermal resistance model for heat flow. (a) Electric current* i *flows between* $V_o$ *and* $V_1$. *(b) Heat flux* Q *flows between* $T_c$ *and* $T_a$.

Where $L$ is the material thickness, $k$ is the thermal conductivity, and $A$ is the heat transfer area.

- For a convection heat transfer:

$$R_{th}^{conv} = \frac{1}{hA} \quad (4)$$

Where $h$ is the convective heat transfer coefficient and $A$ is the heat transfer area.

*2.2.2 Electronic component without a heatsink*

Consider an electronic component with junction temperature $T_j$, in an ambient environment (air for example) at $T_a$.

As shown in **Figure 4**, the heat generated by the component is transferred by conduction from the junction to the external surface of the case. Then, heat is conveyed by convection and radiation to the ambient environment.

For normal operating temperatures, conduction and convection are the prevailing modes. The total heat transfer resistance is therefore given by the following:

$$R_{JA} = R_{JC} + R_{CA} \quad (5)$$

Where:

$R_{JA}$ is the junction to ambient resistance.

$R_{JC}$ is the conduction resistance.

$R_{CA}$ is a convection resistance.

Note that $R_{JC}$ is generally quite low, but $R_{CA}$ is usually high enough to limit heat transfer from the component to environment.

*2.2.3 Electronic component with a heatsink*

**Figure 5** shows the electronic component of **Figure 4** to which a finned heatsink was added in order to take advantage of its large contact area with ambient, thus permitting a better spread of the heat generated by the component.

In this case, as well, heat is transferred from the component to the ambient essentially by conduction and convection. Energy transfers are therefore represented by a series of thermal resistances:

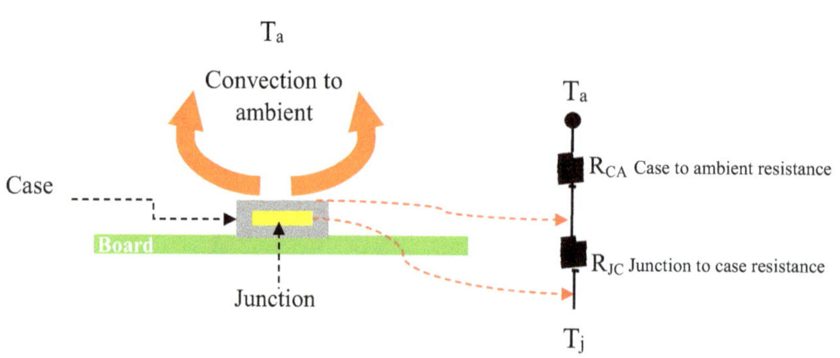


**Figure 4.**
*Heat resistance of a sink.*

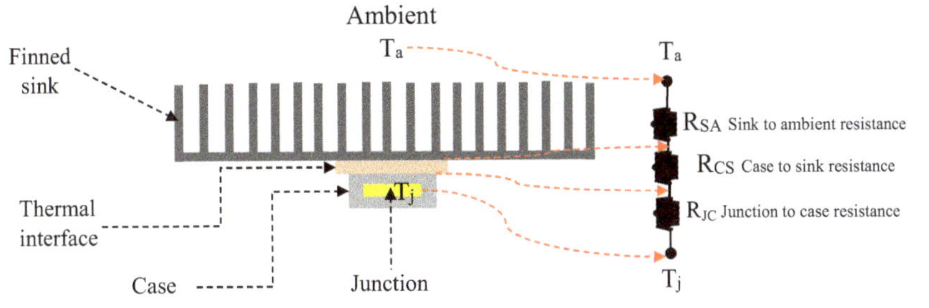


**Figure 5.**
*Total resistance of a heatsink.*

$R_{JC}$, *the junction to case thermal resistance*; a conductive resistance which depends on the thickness, $e_{JC}$ (between the junction and the thermal interface), on the area $S_{JC}$, and on the thermal conductivity $k_{JC}$. $R_{JC}$ is given by the following:

$$R_{JC} = \frac{e_{JC}}{k_{JC} S_{JC}} \tag{6}$$

$R_{CS}$, the case to heatsink thermal resistance; a conductive resistance that takes into consideration the thickness and conductivity of the case and interface material. $R_{CS}$ is given by the following:

$$R_{CS} = \frac{e_{CS}}{k_{CS} S_{CS}} \tag{7}$$

$R_{SA}$, *convection resistance* between the fins and air. $R_{SA}$ is a convective resistance given by the following:

$$R_{SA} = \frac{1}{h S_F} \tag{8}$$

Where $S_F$ is the fins surface area in contact with ambiance and $h$ is the convection heat transfer coefficient between the fins and the ambient environment (see Appendix 1).

The sum $R_{JC} + R_{CS} + R_{SA}$ represents the *heat exchanger thermal resistance* between the junction and ambient. *The thermal resistance of the heat exchanger* is then given by the following:

$$R_{th} = R_{JC} + R_{CS} + R_{SA} = \frac{e_{JC}}{k_{JC} S_{JC}} + \frac{e_{CS}}{k_{CS} S_{CS}} + \frac{1}{h S_F} \tag{9}$$

Thus, the *heat exchanger thermal resistance* is a parameter, which depends on the materials constituting the heatsink, the casing, and the thermal interface material. It also depends on the surface area in contact with air, the configuration of the fins, their number, the position of the heatsink (horizontal/vertical), air temperature, etc. Its determination is somewhat complicated [6] and often necessitates the running of experiments.

Hopefully, when integrated systems are offered, values of the thermal resistance, $R_{th}$, are given by manufacturers' data. In the case where the heat exchanger is mounted from separate pieces, the heat exchanger thermal resistance can be computed from Eq. (9). Electronic component suppliers' data sheets give the parameters necessary for the computation of $R_{CS}$. Similarly, characteristics of gaskets

| N° | Source | Heatsink | Thermal resistance $R$ (°C/W) |
|---|---|---|---|
| 1 | [7] |  | $R_{th}$ = 2.4°C/W |
| 2 | [8] |  | $R_{th}$ = 17°C/W |
| 3 | [9] |  | $R_{th}$ = 28°C/W |
| 4 | [10] |  | $R_{th}$ = 55°C/W |
| 5 | [11] | 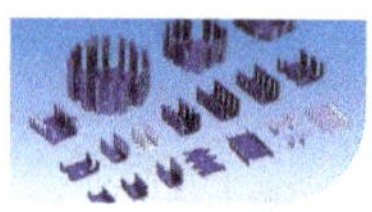 | $R_{th}$: 7–80°C/W |
| 6 | [12] | 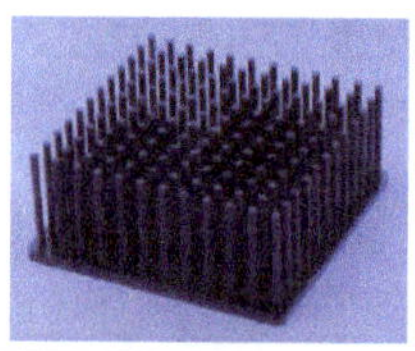 | $R_{th}$: 2–56°C/W |
| 7 | [13] |  | $R_{th}$ = 11°C/W |
| 8 | [14] |  | $R_{th}$ = 4.75°C/W |
| 9 | [15] |  | $R_{th}$ = 63°C/W |
| 10 | [16] |  | $R_{th}$ = 14°C/W |

**Table 1.**
*Values of* $R_{th}$ *for different heat sinks.*

(eventually), gap fillers, and other interface materials will permit the calculation of $R_{CS}$. Finally, $R_{SA}$ can either be taken from heatsink data sheets (**Table 1** shows values of $R_{SA}$ for different heatsink makes), or calculated from Eq. (8) and Appendix 1.

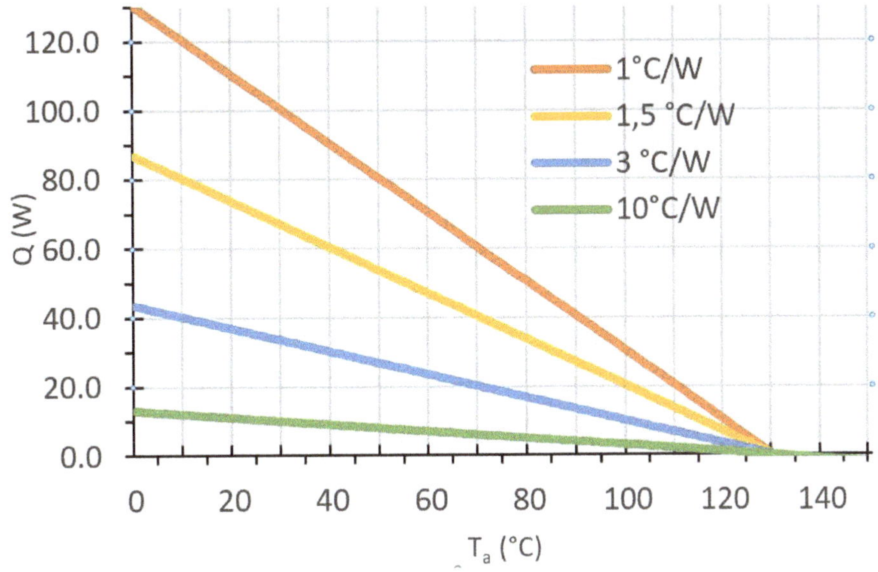


**Figure 6.**
*Dissipation power as a function of ambient air temperature.*

Remarks:

i. Sometimes, some manufacturers give two values of the thermal resistance of the heatsink. This can be the case in two situations:

- If the heatsink is such that it can be mounted vertically or horizontally, the lower value of thermal resistance corresponds to the *vertical mounting* of the fins, which does not get in the way of air movement from the low to the high parts of the fins. Heat transfer is therefore more efficient. In contrast by opposing air movement, horizontal assembly of the fins leads to a 20% of efficiency.

- If the gasket and the filling material resistances are not included, the second value corresponds to the resistance of these mounting accessories. In this case, one should add the two values to obtain the thermal resistance, $R_{CS} + R_{SA}$.

ii. In this chapter, $R_{th}$, refers to the total resistance of the *heat exchanger*, a s shown in Eq. (9).

iii. It should be noticed that the smaller the *heat exchanger thermal resistance* is, the higher is the *dissipation power* $Q$, and the better is the heat exchanger and the cooling of the electronic equipment. **Figure 6** shows the evolution of the dissipation power as a function of the surroundings temperature for a component with $T_{op}$ = 130°C, using different heat exchangers: $R_{\text{th}}$ = 1°C/W to $R_{\text{th}}$ = 10°C/W

## 3. Passive heat exchanger design

The simplest design technique uses the thermal resistance model of the heat exchanger. It supposes that the coolant (air for example) temperature, $T_a$, i s constant in contact with the heatsink. Using this model, the power of heat dissipated by a heatsink is written as follows:

$$Q = \frac{T_J - T_a}{R_{th}} \tag{10}$$

Where:
$Q$ is the heat flux from the component to ambient air (Watts), $T_J$ and $T_a$ are, respectively, junction and air temperatures (°C), and $R_{\mathrm{th}}$ is the thermal resistance of the heat exchanger (°C/W). Three design problems can generally be encountered:

- *A sizing problem:* Given an electronic component or system with a maximum power generation P, how can one choose, among commercially available products, the heatsink that would ensure dissipation of the heat generated during operation?
- *A verification problem:* You have an electronic component/system with a heatsink mounted on. Is this heatsink correctly sized; that is, will it fit your electronic component or system?
- *Positioning circuit boards on a rack for optimal heat exchange:* How can a number of circuit boards are placed in a rack so that cooling is optimal?

### 3.1 Heat exchanger sizing

In this case, it is desired to find a commercially available heatsink, capable of dispersing the heat generated by the operation, in surroundings of temperature $T_a$, of an electronic component with a maximum power generation $Q_{max}$ and a maximum junction operating temperature, $T_{op}$.

The sizing procedure is carried out in two steps.

STEP 1: From component, thermal interface, and heatsink datasheets select those which satisfy the space constraint. Get the values of $Q_{max}$, $T_{op}$, $R_{JC}$, $R_{CS}$, and $R_{SA}$.

STEP 2: Write Eq. (10) for $Q_{max}$ and determine the thermal resistance of the heatsink which will insure operating temperature not exceed $T_{op}$.

Written for $Q_{max}$ and $T_{op}$ Eq. (10) gives the following:

$$Q_{max} = \frac{T_{op} - T_a}{R_{th}} \tag{11}$$

Rearranging, we obtain:

$$R_{th} = \frac{T_{op} - T_a}{Q_{max}} \tag{12}$$

*Illustration 1*

*An electronic chip is required to operate at a surrounding temperature of 45°C. This chip is mounted with a casing having a thermal resistance $R_{JC}$, and with thermal interface materials having a global resistance $R_{CS}$. The data sheet of the chip shows that it can generate up to $Q_{max}$ at $T_{op}$.*

*Supposing there are no space or geometric constraints, what are the heat sinks that can be used among those of **Table 1**.*

*Data:*
$R_{JC} = 0.3°C/W$ $R_{CS} = 0.4°C/W$.
$Q_{max}$ =*10 W* $T_{op}$ = *85°C.*
*Solution*
*Using Eq. (12), the thermal resistance at maximum power is:* $R_{th} = \frac{85-45}{10} = 4°C/W$
*The adequate heatsink resistance is obtained using Eq. (9) as follows:*

$$R_{SA} = R_{th} - (R_{JC} + R_{CS})$$

$$R_{JC} + R_{CS} = 0.7\ {}^{\circ C}\!/\!_{W} \Longrightarrow R_{SA} = 3.3\ {}^{\circ C}\!/\!_{W}$$

*This means that only heat sinks with thermal resistances lower or equal to 3.3°C/W will be able to give heat exchangers with* $R_{th} \leq 4°C/W$, *thus ensuring safe operation of this chip.*
*This is the case for references 1 and 6 only.*

### 3.2 Heat exchanger verification

In this case, it is desired to verify if a given heatsink will permit the safe operation of an electronic component at a known surroundings temperature.
The verification procedure is carried out in four steps.
STEP 1: From data sheets (component, thermal interface, and heatsink) get the values of $Q_{max}, T_{op}, R_{JC}, R_{CS}$, and $R_{SA}$.
STEP 2: Calculate $R_{th}$ from Eq. (12): $R^*_{th}$.
STEP 3: Calculate $R^*_{SA}$ from Eq. (13) as follows:

$$R^*_{SA} = R^*_{th} - (R_{JC} + R_{CS}) \tag{13}$$

STEP 4: If $R_{SA} \leq R^*_{SA}$ then the heatsink will do the job.
*Illustration 2*
*You bought an electronic component with a heatsink mounted on, thus constituting a heat exchanger ready to use. The heatsink is the one shown under N° 10 in **Table 1**.*
*The data sheet of the electronic component gives the following information:*
$Q_{max}$ = *10 W and* $T_{op}$ = *115°C.*
*Will this system operate safely at a surrounding temperature of 27°C?*
*Solution*
*STEP 1:* $Q_{max}$ = *10 W,* $T_{op}$ = *115°C and*
*N° 10 in **Table 1** gives:* $R_{th}$ = *14°C/W.*
*STEP 2: Eq. (11) gives:* $R^*_{th} = \frac{115-27}{10} = 8, 8° \frac{C}{W}$.
*STEP 3:* $R_{th} > R^*_{th}$: *the mounted heatsink will not do the job.*

### 3.3 PCBs optimal assembly

A typical passive heat exchange problem is encountered when one needs to assemble a set of circuit boards on a rack.
**Figure** 7 shows a number of boards of the same size, $H$, placed on a rack of length $L$ and width $w$. The spacing between the different circuit boards is assumed to be constant, $\delta$, whereas each board releases a constant flux density, $q$ ($W/m^2$). The cooling of the circuit boards is ensured by natural convection.
In such a configuration, the temperature of the boards is not uniform, rather it increases with height, reaching its maximum value at $H$. The spacing $\delta$ needs to be such that heat evacuation is optimum.

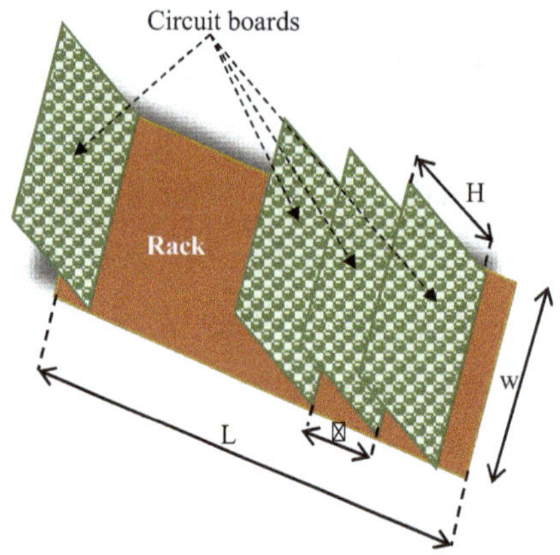


**Figure 7.**
*Assembling a set of circuit boards on a rack.*

### *3.3.1 Optimum spacing of PCBs*

A frequent question is; *how should the boards be placed on the rack for optimum heat evacuation?* In other words, *what is the optimum spacing, $\delta^*$?*
The optimal spacing is given by [17]:

$$\delta^* = 2,12\frac{w}{Ra_q^{0,2}} \tag{14}$$

$Ra_q$ being the Rayleigh number for, $q$, given by:

$$Ra_q = 2,12\frac{w^4 C_p \rho^2 g \beta q}{\mu k^2} \tag{15}$$

Where

- $C_p$, $\rho$, $\beta$, $\mu$ and $k$ are respectively the specific heat, the density, the expansion factor, the viscosity, and the thermal conductivity of air[3].

- $g$ gravity constant; $g = 9.81\ \mathrm{ms}^{-2}$

### *3.3.2 Optimal number of boards*

The optimum number of boards of thickness $e$, to be assembled on a rack of length $L$, is then given by:

$$n^* = \frac{L}{\delta^* + e} \tag{16}$$

Other issues of importance in rack design such as the determination of heat transfer coefficients between electronic boards and air, and the calculation of the heat flux evacuated by natural convection are discussed in Ref. [1].

---

[3] See Data Bank in Ref. [18].

## 4. Active heatsink technologies

During the last decade, unprecedented technological developments imposed tremendous flows of information (data or multimedia), implying an *ever-growing need for fast data transfers* (Bluetooth, Wi-Fi) and for *larger computing and storage capacities*. In addition, the advent of 5G and the *Internet of Things* means that the number of connected devices is constantly increasing [19]. Moreover, *onboard electronics* and *mobile telephony* created new ergonomics and space (volume) constraints. From a systems engineering perspective, this implies that *larger power circuits have to be integrated in smaller volumes*.

Under such conditions, the amounts of energy generated by electronic circuits are so great that *no passive heatsink of a reasonable size will be able to do the cooling job*. This is because passive sinks dissipate heat mainly through natural convection, but natural convection is no longer sufficient to extract the heat generated by the operation of these power components. Thus, the use of active cooling systems, with or without refrigeration or phase change is necessary [20, 21] for the safe operation of power electronics.

Active cooling systems are heat exchangers where the flow of the heat transfer medium (air or liquid) is forced by a fan or a pump. Several technologies exist. They can be sorted into two categories:

i. single-phase heat sinks, and

ii. two-phase heat sinks.

### 4.1 Single phase

Single-phase heat sinks are heat exchangers in which the cooling fluid (liquid or gas) *does not undergo any phase change*, meaning that if the cooling fluid is a liquid, then it remains liquid throughout the cooling process.

Single-phase heat exchangers used for electronic systems cooling include the following:

- Forced convection heat sinks
- Cold plate heat sinks
- Microchannel coolers
- Refrigerated coolers

By far, forced convection heat sinks are the simplest.

#### *4.1.1 Forced convection heat sinks*

Forced convection heat sinks are typically heat exchangers formed of finned surfaces similar to those presented in **Table 1**, where cooling air is forced through the fins using *fans*.

**Figure 8** shows such a heat exchanger fixed on a CPU board. The fan is mounted on top of the heatsink to draw air through the fin surfaces.

Depending on the type of fins configuration, the fan can also be mounted in the middle of the finned heatsink as shown in **Figure 9**.

**Figure 8.**
*A fan on a heatsink mounted on a CPU board. Source [22].*

**Figure 9.**
*Fan in the middle of a finned heatsink. Source [23].*

**Figure 10.**
*Miniaturized forced convection heatsink module. Source [24].*

Very often, the assembly fins-fan is miniaturized in a cooling module similar to the one presented in **Figure 10**.

Forced convection heat exchangers become necessary when $Q_{max}$ exceeds 70 W/cm$^2$ [25]: Using a fan to force air through the finned heatsink increases significantly heat transfer to air; *convective heat transfer coefficients* can reach values as high as 3000 Wm$^{-2}$°C$^{-1}$ [26]. For comparison, this coefficient is between 25 and 500 Wm$^{-2}$°C$^{-1}$ for natural convection heat sinks [1].

We should underline, however, that including a fan in an electronic system will require an additional power supply. This will certainly impact the final size of the system.

### *4.1.2 Forced convection heatsink design*

An important parameter in the design of a forced convection heatsink is the flow rate of the cooling fluid (e.g., air) going through the fins of the sink [2]. As a matter of fact, the flow rate generated by the fan actually impacts the heat transfer coefficient: A higher flow rate will mean a greater cooling fluid velocity, leading to a better convection heat transfer [1]. The flow rate also impacts the pressure drop across the heatsink given that for a given fin configuration, higher flow rates generate higher pressure drops[4].

Consequently, the proposed design procedure will search for *a balance between a reduced thermal resistance and an acceptable pressure drop*. It can be organized in the following seven-step procedure.

*STEP 1: Collect heatsink pressure drop data.*

Flow over a fanned heatsink generates a pressure drop that depends on several parameters including the number of fins, the distance between two fins, the value of the flow rate, etc. [26]. Suppliers of heat sinks will generally be able to provide, for the different models they propose, plots of the pressure drop versus the cooling fluid (air) flow rate. These plots are referred to as *heatsink working diagrams* where the flow rate is generally expressed in cubic feet per minute (CFM: ft$^3$/mn).

**Figure 11** shows an example of such plots.

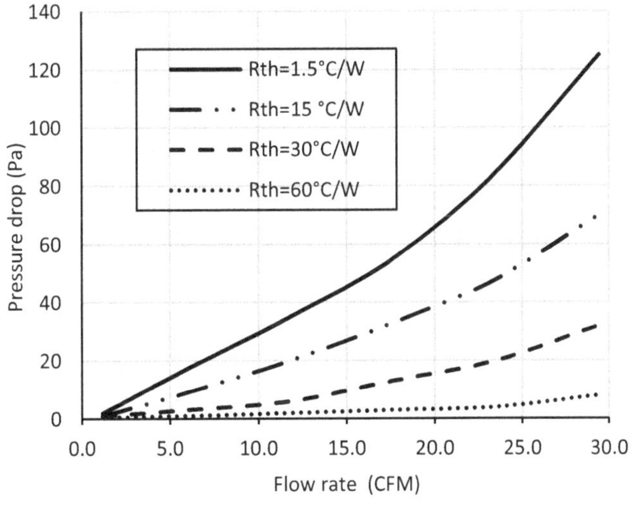


**Figure 11.**
*Heatsink working diagram.*

[4] For laminar flow, the pressure drop is proportional to the square of the flow velocity [26].

*STEP 2: Get the fan curves.*

Similarly, fan suppliers provide plots that give, for each model proposed, the flow rate achievable by the fan under a given pressure drop. These plots are often referred to as *performance curves of the fans*.

**Figure 12** presents examples of performance curves for four fans. *STEP 3: Select heatsink candidates based on their working diagrams.* The selection criterion here is the individual thermal resistance, $R^s_{th}$. *STEP 4: Define the allowable coolant temperature rise.*

The *coolant temperature rise* is defined as follows:

$$\Delta T = T_o - T_i \tag{17}$$

where $T_i$ and $T_o$ are, respectively, the inlet and outlet temperatures of the coolant.

$T_i$ being known (generally, the coolant enters the heatsink around ambient temperature: $T_i \approx T_a$), the definition of $\Delta T$ permits to calculate$T_o$:

$$T_o = T_i + \Delta T \tag{18}$$

*STEP 5: Determine coolant flow rate for each heatsink candidate.*

For each heatsink s under evaluation, the thermal resistance, $R^s_{th}$, is known. This makes it possible to calculate the coolant flow rate by combining Eq. (9) and a heat balance on the cooling fluid:

Equation (9) gives:

$$Q_s = \frac{T_c - T_a}{R^s_{th}} \tag{19}$$

The heat balance for the cooling fluid gives the following:

$$Q_s = \dot{m}_s C_p (T_o - T_i) = F_s \rho C_p \Delta T \tag{20}$$

Where:

- $\dot{m}_s$ is the mass flow rate of the coolant fluid (air) over the heatsink $s$, kg/sec,
- $C_p$ is its sensible heat (J/kg°C),

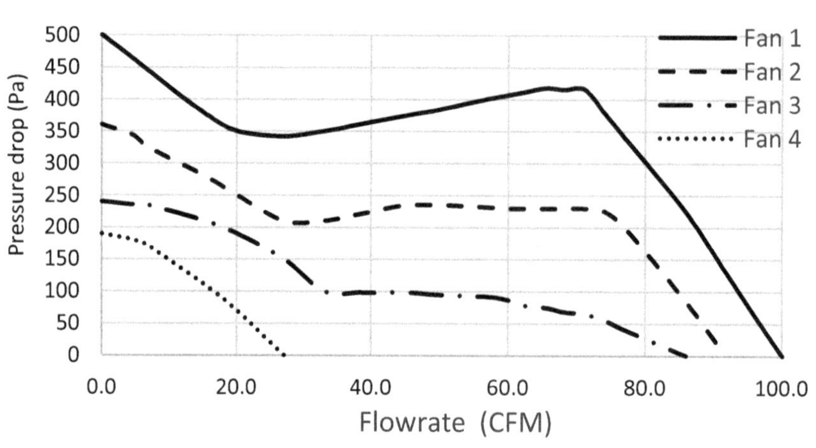


**Figure 12.**
*Performance curves of fans.*

- $\rho$ is its density (kg/m$^3$), and
- $F_s$ is the coolant volume flow rate (m$^3$/s).

Substituting for $Q_s$ in Eq. (19) and extracting the coolant volume flow rate, $F_s$:

$$F_s = \frac{T_c - T_a}{\rho C_p \, \Delta T \, R^s_{th}} \tag{21}$$

*STEP 6: Determine the pressure drop for each of the heatsink candidates.*

Injecting the values of flow rates into the working diagram gives the pressure drop which will be generated by each of the heatsink candidates. **Figure 13** shows this procedure for the first two heat sinks considered in **Figure 10** ($R_{th}$ = 1.5°C/W and $R_{th}$ = 15°C/W), which permits the determination of pressure drops $\Delta P_1$ and $\Delta P_2$ generated by when $F_1$ and $F_2$ flow over sink 1.

*STEP 7: Determine compatible fans using the performance curves.*

For each heatsink candidate, s, inject the pressure drop, $\Delta P_s$, determined in the previous step in the performance curves. This will generate the series of flows, $F_s^{fan_i}$, which will be delivered by fan i, operating against the pressure drop $\Delta P_s$.

**Figure 14** shows that, for fan 1, all performance curves cross the line $\Delta P_1$. This means that any of the fans under evaluation can be used. The choice will then be made on price and volume criteria.

### *4.1.3 Cold plate heat exchangers*

Cold plate heat sinks could be considered as a particular class of *plate heat exchangers* [27]. They are used when thermal powers released by electronic systems (smartphones, electric cars, onboard avionics systems, TGV, etc.) become so important that forced convection heat sinks are no longer sufficient [28–30]. They are constituted of plates that are fitted with pipes (see **Figure 15**) through which cooling fluid passes. They are attached to the surfaces of the electronic component or to the board to be cooled. The coolant is conveyed by means of a pump. The coolant itself is, in turn, cooled using a compact exchanger [27].

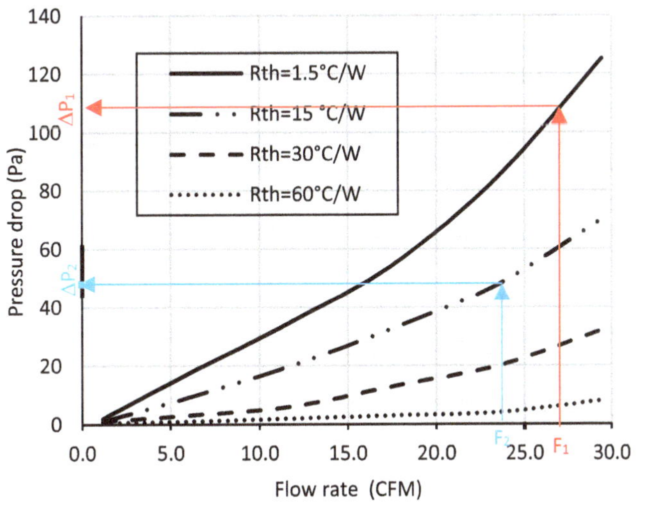


**Figure 13.**
*Determination of pressure drops for heat sinks 1 and 2.*

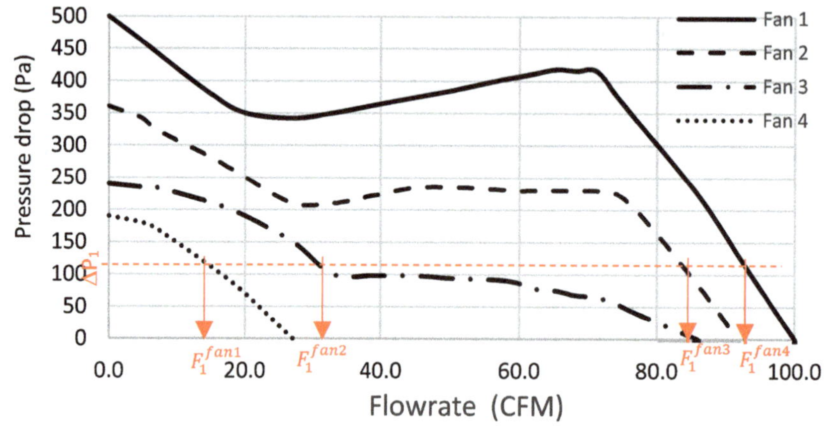


**Figure 14.**
*Determination of deliverable flows under* $\Delta P_1$.

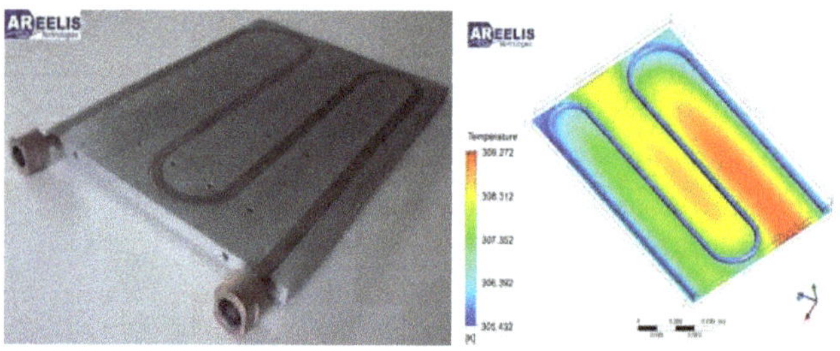


**Figure 15.**
*A cold plate heatsink and its thermal scan. Sources [28, 29].*

### *4.1.4 Microchannel heat exchangers*

Microchannel heat sinks can be considered as a particular subclass of *printed circuit heat exchangers* [27]. These are indeed very small exchangers with overall dimensions not exceeding a few millimeters.

In contrast to their very small dimensions, they allow heat transfers in the order of 800 W/cm$^2$ [30–33]. The cooling fluid (generally air) circulates in microchannels, of microscopic equivalent diameters (approximately 10–60 μm), formed by etching on metal plates or in composite materials [34]. These microchannels have heights of the order of 0.5 mm. The modules thus formed are placed under the electronic components to be cooled [35]. **Figure 16** illustrates such an assembly where the circuit board to be cooled is shown in semi-transparency, above the microchannels.

However, microchannel cooling suffers from several drawbacks: complex implementation, significant pressure drops associated with microchannel flow. Moreover, it does not quite meet the requirements of thermal management in power electronic systems [36].

### *4.1.5 Refrigerated heat exchangers*

Significant efforts have been devoted during the last 10 years to the development of a new kind of heat sinks. These are constituted of heat exchange plates associated with refrigeration cycles and a refrigerant as the heat transfer fluid [37–39]. Such refrigerated heat exchangers are used to cool power electronic

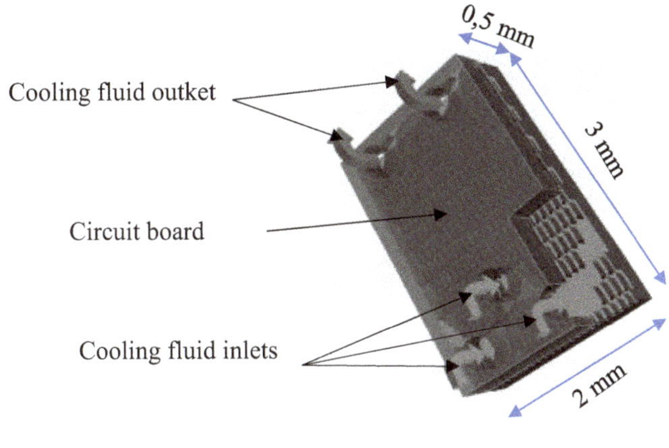


**Figure 16.**
*Overall dimensions of a micro-channel heatsink.*

systems such as laser power supplies and their optical systems. They are able to extract heat flux densities exceeding 1000 W/cm$^2$ while keeping microchips at temperatures below 65°C [39–44].

Note, however, that mounting this type of heat exchanger on electronic equipment remains difficult due to the fact that all the components of a refrigeration cycle (compressor, expansion valve, evaporator, and condenser) must be assembled in rather tiny spaces (**Figure 17**).

These systems must therefore meet an important challenge: the miniaturization required by designs favoring small sizes [21, 42, 43].

### 4.2 Two-phase heat sinks

These are phase change micro-exchangers, where the cooling fluid (generally a liquid) undergoes phase changes (from liquid to vapor and back to liquid) during the heat transfer process. Two-phase heat sinks are generally constituted of a vessel containing a heat transfer fluid; water in most cases. The lower surface of the vessel is in contact with the heat source (electronic equipment), and the upper surface is in contact with the cooling medium (generally air). **Figure 18** presents the operating principle of this system.

When in contact with the lower surface (heat source), the heat transfer fluid evaporates by absorbing the heat generated at this surface. The vapor then goes

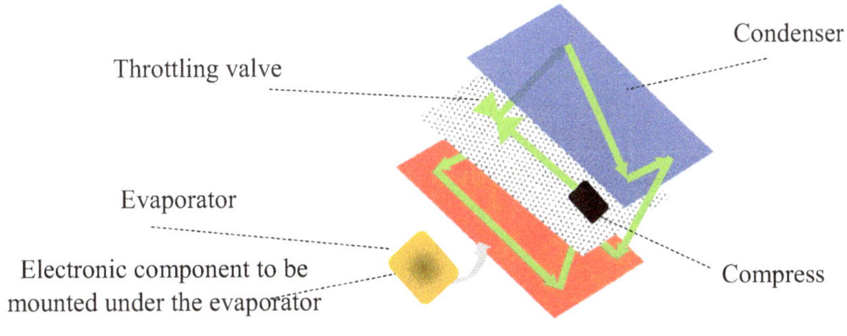


**Figure 17.**
*Exploded view of a chip mounted under a refrigerated heatsink.*

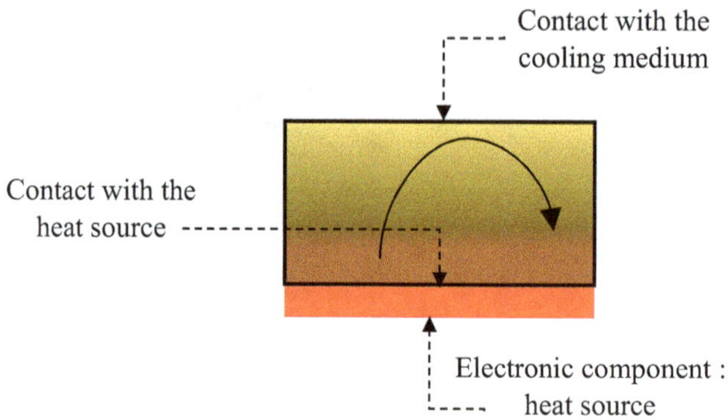


**Figure 18.**
*Operating principle of a two-phase heatsink.*

upwards, toward the upper surface where it gets cooled and condensates. Condensed liquid falls back along the walls of the vessel, down to the lower surface, where it absorbs again the heat and evaporates. This cycle is repeated as long as heat is generated by the electronic equipment.

This way, two-phase heat sinks are used to remove heat from the heat source to the ambient environment. The lower surface acts as an evaporator and the upper one plays the role of a condenser.

Note that two-phase heat exchangers involve latent heat instead of sensible heat in energy transfers, thus allowing large amounts of heat to be exchanged over small areas, which leads to great compactness.

Two-phase heat sinks include the following:

- Spray coolers
- Vapor chambers
- Heat pipes.

### *4.2.1 Spray coolers*

Spray coolers are a special class of two-phase heat exchangers since the heat transfer fluid undergoes a series of evaporations and condensations [45, 46]. The electronic component is cooled by a jet of fluid, which partially evaporates while absorbing heat (see **Figure 19**). The heat transfer fluid is sprayed directly onto the surface of the power component to be cooled [47–49].

Spraying enhances the vaporization of the fluid even at relatively low temperatures (60–75°C, under 1 Atm.). It is generally realized using a single injector [50] or multiple injectors [51].

However, like refrigerated exchangers, the realization of spray cooling is relatively complex. As shown in **Figure 20**, it requires, in addition to the sprayers, the installation of a condenser to collect the vapors generated and a pump or compressor to feed the injectors.

### *4.2.2 Vapor chambers*

Vapor chambers are the most common two-phase-heat exchangers where a heat transfer fluid (water in most cases) is placed in a sealed envelope: *the chamber* [52]. The lower surface of the chamber is mounted on the electronic component to be

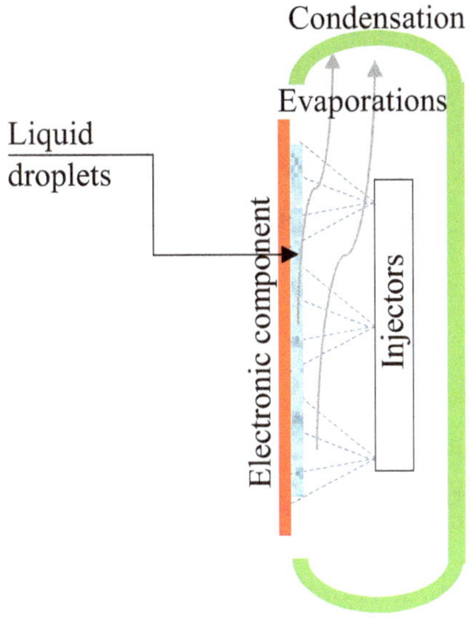


**Figure 19.**
*Spray cooling system.*

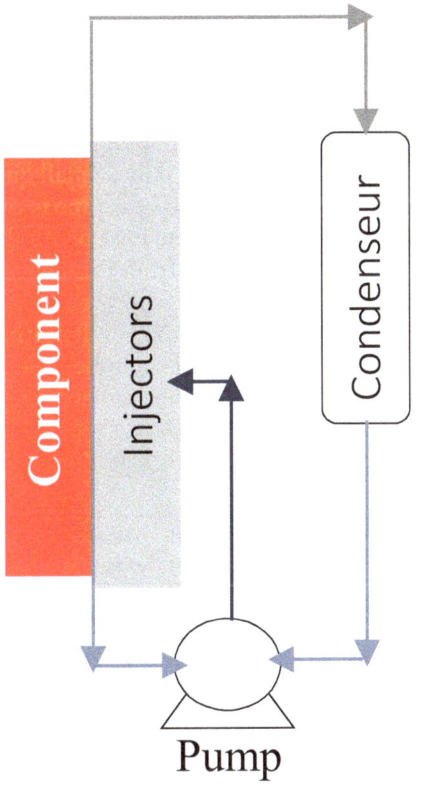


**Figure 20.**
*Spray cooling cycle.*

cooled. As described above, the liquid absorbs the heat generated by this component to evaporate. The vapors then condense on the cold surface of the chamber, which transfers the ambient environment the energy liberated by the condensation process [53, 54].

**Figure 21.**
*Vapor chamber heat exchanger. Source [55].*

**Figure 21** shows this type of heat exchanger. It should be noted that the dimen-sions are extremely small to meet the requirements of miniaturization of onboard electronics and mobile telephony. The typical thickness of such an exchanger is 2–8 mm.

Current developments focus on the ultra-miniaturization of these exchangers by introducing ultra-thin evaporation chambers: thicknesses from 0.3 to 2 mm using walls made of titanium, stainless steel, or copper alloys. **Table 2** shows the current uses of these types of exchangers, as well as their typical dimensions and the thermal powers conveyed.

### *4.2.3 Heat pipes*

A heat pipe consists of a sealed tube, containing a heat transfer fluid, without any other gas (see **Figure 22**). In most cases, only a small quantity of water suffices: About 1 cc of water for a 150 mm long, 6-mm heat pipe is typical.

In one of its zones (vaporization zone), the heat pipe tube is in contact with the hot source to be cooled. The heat recovered from this source increases the temperature of the heat transfer fluid causing it to evaporate.

The resulting vapors then accumulate in the condensation zone of the heat pipe where they condense on the internal walls of the tube, releasing their latent heat to the ambient environment (see **Figure 22**). The condensate flows in droplets on

| Domain of use | Thickness (mm) | Length (mm) | Width (mm) | Power range (W) | Construction materials |
|---|---|---|---|---|---|
| Smart phones | 0.3–0.4 | 80 | 50 | 5 | Copper alloy or stainless steel |
| Portable microcomputers | 0.5 | 220 | 60 | 15–25 | Copper alloy |
| Base telecommunication stations | 2–4 | 100 | 100 | 50–100 | Copper or copper alloy |
| Graphic boards | 2–4 | 200 | 90 | 200–350 | Copper or titanium |
| Servers | 3–4 | 150 | 80 | 200–400 | Copper |
| Game consoles | 3–5 | 150 | 150 | 150–200 | Copper |
| Power inverters | 3–8 | 450 | 450 | 500–2000 | Copper |

**Table 2.**
*Vapor chamber uses and construction materials.*

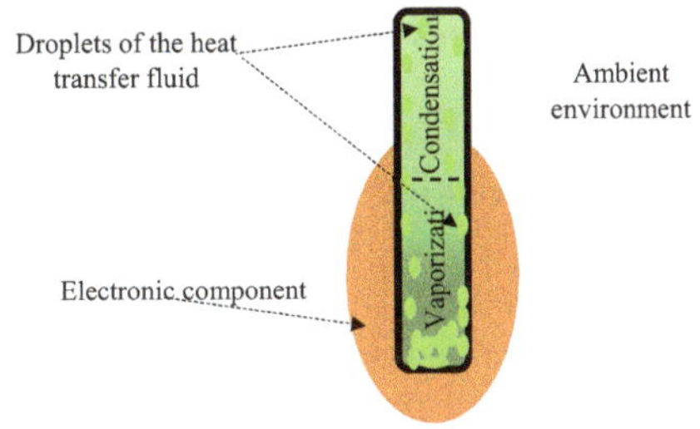


**Figure 22.**
*Heat pipe working principle.*

the walls of the tube and returns to the vaporization zone to be again submitted to the heat flow and to evaporate.

### *4.2.4 Heat pipe exchanger*

These special heat exchangers are generally formed of a number of heat pipes usually made of copper, where several fins (the fin stack) have been mounted in the condensation zone.

As shown in **Figure 23**, each pipe is bent in a U form, slightly flattened where contact is made with the mounting base plate (hot area), and filled with a heat transfer fluid. When the copper pipes are going to be directly in contact with the environment, protection against corrosion could be obtained by nickel plating.

Each heat pipe acts actually as a *heat mover* from the area where heat is generated (the base plate) to the cooling area (fin stack) where air movement carries the heat away.

Rules of thumb

1. To maximize the amount of energy received from the heat source (the electronic component), each heat pipe should be placed directly above this source. The closer and tighter the contact is, the better is the heat transfer.

2. To improve the contact between the source and the heat pipe, sections of the pipes that are in contact with the source are flattened to about 30–65% of their original diameter.

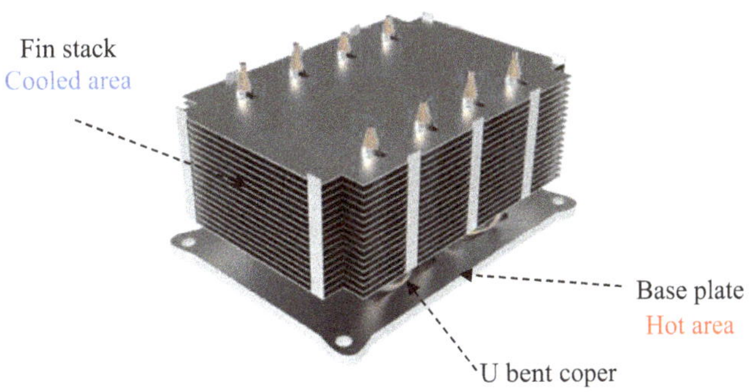


**Figure 23.**
*Fixing the fin stack to the base plate.*

3. The bend radius of the heat pipes is to be fixed with care in order not to alter the integrity of the pipes.

4. The minimum *bend radius* should be *three times the diameter of the heat pipe*.

Note that the heat pipes can be directly fixed on the electronic component to be cooled. **Figure 24** shows a heat exchanger made of six copper heat pipes that are directly fixed on a central processing unit (CPU) to convey heat to a 13 fin stacks. The advantage is a direct contact between the component and the heat pipes, thus reducing the thermal resistance of the base plate.

A more elaborated version is the U-shaped vapor chamber design where a U-shaped heat plate replaces the heat pipes, the heat plate operating in the same way as flattened pipes. **Figure 25** shows a heat exchanger formed from a vapor chamber bent to offer a large copper base to be in direct contact with the power electronic component. This design offers the advantage of a large direct contact area, which translates into a better performance: more than 20% better than the design of the pipe.

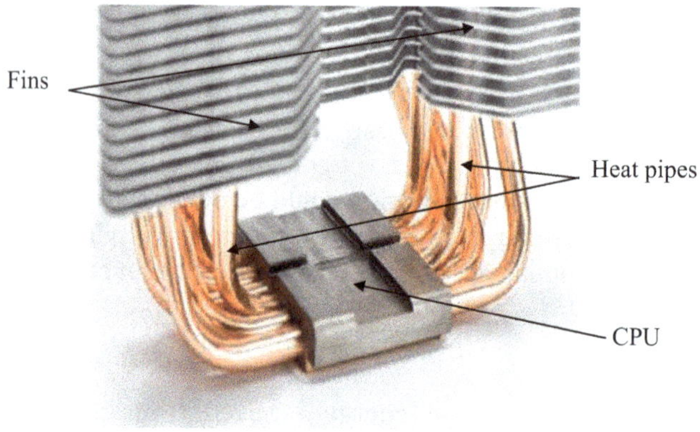


**Figure 24.**
*Heat pipe heatsink to cool a CPU.*

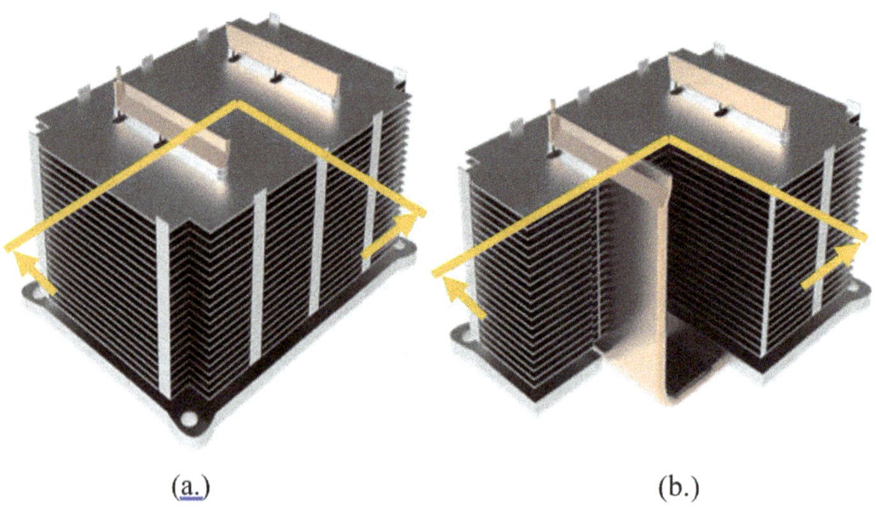


**Figure 25.**
*U bent vapor chamber with a fin stack. (a) Complete vapor chamber and (b) showing inside the chamber.*

## 4.3 Two-phase heat exchanger design

### 4.3.1 Design fundamentals

#### 4.3.1.1 Carrying capacity

The performance of a two-phase heatsink is measured by its carrying capacity, $Q$, defined below.

Definition

The *carrying capacity*, $Q$, of a two-phase heat exchanger is defined as the amount of heat the device can move out per unit time.

From its definition, $Q$ is an *energy per unit time*, which is a *power* which should be expressed in *Watts*.

For a given two-phase heat exchanger, the value of $Q$ depends on several parameters: the size of the vessel, the latent heat of the transfer liquid used, the thermal conductivity of the metal constituting the vapor chamber or the heat pipe, convections (inside and outside the vessel), etc. However, we should note that, in two-phase heat sinks, energy is essentially transferred through the vaporization-condensation process, involving latent heat.

#### 4.3.1.2 Two-phase thermal budget

Definition

The *thermal budget*, $\Delta T$, of a two-phase heat exchanger is defined as the difference between the temperature of the electronic component (heat source),$T_c$, and the temperature of the ambiance where the component will operate,$T_a$:

$$\Delta T = T_c - T_a \tag{22}$$

The carrying capacity, $Q$, is proportional to the thermal budget $\Delta T$ (see Eq. (9)).

Thus for low thermal budgets, it will be difficult to achieve high values of $Q$.In this case, two-phase heat sinks should be used, mainly when $\Delta T < 40°C$.

#### 4.3.1.3 Effective thermal conductivity

Heat exchange in two-phase sinks is actually complicated to model. It involves several heat transfer mechanisms: conduction through the vapor chamber or the heat pipe metal envelope, evaporation, condensation, convection inside and outside the vessel, etc. The exact representations of these transfers are interesting on the theoretical level of analysis. But, for sizing and design purposes, a more practical approach has been adopted by manufacturers and system designers. This approach is based on the thermal resistance model, in a way similar to that presented in Subsection 2.1 for single-phase heat sinks.

Let us recall that in conduction and in convection, thermal resistance models permit to express the power transferred $Q$ in terms of $R_{\text{th}}$ and the thermal budget, $\Delta T$, as follows [2, 47]:

$Q = \frac{\Delta T}{R_{th}}$, where, $R_{\text{th}}$ is the thermal resistance, which is expressed differently for conduction and convection:

- For a convection heat transfer:

$$R_{th}^{conv} = \frac{1}{hA} \tag{23}$$

Where $h$ is the convective heat transfer coefficient and $A$ the heat transfer area.

- For a conduction heat transfer:

$$R_{th}^{cond} = \frac{L}{kA} \quad (24)$$

Where $L$ is the material thickness, $A$ the heat transfer area, and $k$ is the *thermal conductivity*.

Similarly, two-phase heat exchanger energy transfers are represented using an *effective thermal resistance*, $R_{eff}$, such that the carrying capacity, $Q$, is given by:

$$Q = \frac{\Delta T}{R_{eff}} \quad (25)$$

With $R_{eff} = \frac{L}{A\, k_{eff}}$, where $k_{eff}$ is the *effective thermal conductivity* of the two-phase heat exchanger and $A$ is its *heat transfer area of the heat sink*.

Definition

The *effective thermal conductivity*, $k_{eff}$, of a two-phase heat exchanger is defined as follows:

$$k_{eff} = \frac{QL}{A\Delta T} \quad (26)$$

Where $Q$ is the *carrying capacity*, $A$ is its *heat transfer area*, $\Delta T$ is the thermal budget (temperature difference between evaporator and condenser sections) and $L$ is the *effective length* of the vessel which is the *distance from the midpoint of the evaporator to the midpoint of the condenser*.

It should be noted that, unlike the thermal conductivity of materials, the *two-phase heat exchanger thermal conductivity* varies with length. It is not a physical property of the material, but it is a modeling representation of heat transfer in two-phase heat sinks.

This modeling—using effective thermal conductivity, $k_{eff}$—makes it possible to represent *two-phase heatsink energy transfers* by *a simple thermal resistance model*, just the same way thermal resistance was used in single-phase heat sinks.

Effective thermal conductivities of two-phase heat exchangers range from 1500 to 50,000 W/m°C. For comparison, the thermal conductivities of the best energy conductors are around 400 W/m°C. Effective thermal conductivities of two-phase heat exchangers can reach 10–100 times the conductivity of the best thermal conductors like copper.

### *4.3.2 Design steps*

Generally, manufacturers of two-phase heat exchangers supply, among other information on their product, the following parameters:

- Operating temperature limits: generally between 20 and 250°C.
- The heat exchanger carrying capacity.
- The heat exchanger thermal resistance.

The following five-step procedure is proposed to design two-phase heat exchangers.

STEP 1: The datasheet of the electronic component to be cooled gives the power and operating temperature: $Q$ (W) and $T_{op}$ (°C).

STEP 2: Knowing ambient temperature, $T_a$, calculate the thermal budget $\Delta T$ as follows:

$$\Delta T = T_{op} - T_a \tag{27}$$

STEP 3: Calculate the *design thermal budget* $\Delta T_D$ as follows:

$$\Delta T_D = \Delta T - 5 \tag{28}$$

STEP 4: Divide the *design thermal budget*, $\Delta T_D$, by the power, $Q$, of the electronic equipment to be cooled. This division will give the maximal thermal resistance to be assured by the two-phase heatsink: $R_{Max}$

STEP 5: Select the two-phase heatsink having a thermal resistance $R < R_{Max}$.
Rules of thumb

1. Derate the thermal budget by 5°.
2. In the case of heat pipes, if bending is considered, derate the carrying power by 2.5% for every 45-degree bend.
3. When heat pipes are flattened to ensure a better contact with the base plate, the flattening should be in the range of 30–65%.
4. Flattening implies a 15–30% reduction in the carrying capacity of the heatsink.

## A. Appendix 1

Determination of heat transfer coefficients for electronic components and circuit boards under natural convection

### A.1 Electronic components under natural convection in air

$$h = 3.53\left(\frac{T_w - T_a}{H}\right)^{0.25} \tag{29}$$

Where:
$H$ is the height of the small electronic component, *expressed in meters*
$T_w$ and $T_a$ are, respectively, the wall temperature and the ambient temperature

### A.2 Circuit boards under natural convection in air

$$h = 2.44\left(\frac{T_w - T_a}{H}\right)^{0.25} \tag{30}$$

Where:
$H$ is the height of the electronic board considered, *expressed in meters*
$T_w$ and $T_a$ are, respectively, the board and the ambient temperatures.

## Author details

Abdelhanine Benallou
SIGMA TECH, Rabat, Morocco

*Address all correspondence to: pr.benallou@gmail.com

## References

[1] Benallou A. Energy Transfers by Convection (Energy Engineering Set—Volume 3). NJ, USA: John Wiley & Sons Inc.; 2019

[2] Wilcoxon R, Martin G. Fan cooling—Part 1: Determining flow rate. In: Electronics Cooling Magazine. PA, USA: Lectrix, Spring; 2021. pp. 6-9. Available from: https://www.electronics-cooling.com/ wp- content/ uploads/ 2021/ 03/ Electronics-Cooling_Spring-2021.pdf

[3] https://encryptedtbn0.gstatic.com/images?q=tbn:ANd9Gc RQ6FHjqIqAQ3svtm02NWba vJzOMuFOyviR Q&usqp=CAU

[4] https://ae01.alicdn. com/ kf/ HTB1e3UmoY_I8KJjy1Xaq6zsxp XaK /Round-D133mm-Pre-drilled-led-pin-fin- heats ink- fit- for- CREE- cob -cxb 3590-Bridgelux-V29-50.jpg_Q90.jpg_webp

[5] Nuts and Volts: https://www.nutsvolts.com/ uploads/ wygwam/ NV_0709_Post_Fig 2_1.jpg

[6] Simons RE. Estimating parallel plate-fin heat sink thermal resistance. Electronics Cooling. 2003;**9**(1):8-9

[7] Sonelec: http://www.sonelec-musique .com/ images/ radiateur_to3_ml25_simple.jpg

[8] Sonelec: http://www.sonelec-musique. com/images/radiateur_to220_ml24.jpg [9] Sonelec: http://www sonelec-musique com/ images/ radiateur_to5_co 180.jpg

[10] Sonelec: http://www.sonelec-musique . com/ images/ radiateur_to5 _ml61.jpg

[11] Mikkelsen-electronics: http://www.mikkelsen-electronics.com/images/FingerKK.jpg

[12] Myheat sinks: https://myheat sinks.com/docs/images/standard- heat- sinks/cold-forged/round-pin-heat-sink.jpg

[13] https://asset.conrad.com/media10/isa/160267/c1/-/en/183947_BB_00_FB/image.jpg?x=600&y=600

[14] Fischerelektronik: https://www.fischerelektronik. de/fileadmin/_ migrated /pics/aufsteckgr.jpg

[15] https://cdn.webshopapp.com/shops/297304/files/322145950/800x1024x2/to-5-heatsink-63c-w-9mm-x-127mm-skk56-fisher.jpg

[16] https://i.ebayimg.com/images/g/3R4AAOSwTKZfMltz/s-l640.jpg

[17] Bar-Cohen A, Rohsenow W. Thermally optimum spacing of vertical natural convection cooled parallel plates. Journal of Heat Transfer. 1984; **106**:116

[18] Benallou A. Mass Transfers and Physical Data Estimation (Energy Engineering Set—Volume 1). NJ, USA: John Wiley & Sons Inc.; 2019. pp. 179-180

[19] Amalfi RL, Enright R, Kafantaris V. Emerging 5G Networks: Potential Economic Benefits of Two-phase Thermal Management. Electronics Cooling Magazine. PA, USA: Lectrix; June 2021. pp. 19-23. Available from: https://www.electronics-cooling.com/wp-content/uploads/2021/07/Electronics-Cooling-Summer-2021.pdf

[20] Zhang H, Che F, Lin T, Zhao W. Modeling, Analysis, Design, and Tests for Electronics Packaging Beyond Moore. Woodhead Publishing, UK: Elsevier Science; 2019

[21] Phelan PE, Chiriac VA, Lee T-YT. Current and future miniature refrigeration cooling technologies for high power microelectronics. IEEE Transactions on Components and Packaging Technologies. 2002;**25**: 356-365

[22] https://upload.wikimedia.org/wikipedia/commons/thumb/2/25/AMD_heatsink_and_fan.jpg/220px-AMD_heatsink_and_fan.jpg

[23] https://2.imimg.com/data2/WI/RM/MY-615568/pin-type-heat-sink 500x500 . jpg

[24] Advanced Thermal Solutions Inc.: https://www. mouser.fr/ images/ marketingid/2019/microsites/122375246/ Quad%20Flow%20Dual%20Flow.png

[25] PRIATHERM. Design your cooling system using forced convection in just three easy steps. Available from: https:// priatherm. com/ design -your- cooling-system-using-forced-convection- in-just - three - easy - steps / [Accessed: 13 December 2018]

[26] Simons RE. Estimating parallel plate-fin heat sink pressure drop. In: Electronics Cooling Magazine. PA, USA: Lectrix; April 2016. Available from: h ttps:// www. electronics - cooling.com/ 2016/04/calculation-corner- estimating-parallel-plate-fin-heat- sink- pressure-d rop/ [Accessed: 24 July 2021]

[27] Benallou A. Série Ingénierie de l'Énergie, volume 6, Échangeurs de Chaleur: Conception et Algorithmes de Calcul. Section 1.5, ISTE Editions. London; 2021. pp. 5-16 (under press)

[28] AREELIS Technologies: https:// www. areelis. fr/ wpcontent/ uploads/ 2019/05/plaques_froides-495x400.jpg

[29] AREELIS Technologies: https:// www.areelis.fr/ wp-content/ uploads/ 2019/05/FRE-0814-01-2-495x400.png

[30] Milnes P.D. Phase separation in two phase microfluidic heat exchangers [PhD dissertation]. Stanford University; March 2011 Dissertation available online at: http:// purl.stanford.edu/ns144db9649

[31] Roland Baviere. Etude de l'Hydrodynamique et des Transferts de Chaleur dans des Microcanaux. Dynamique des Fluides [physics.flu-dyn]. PhD dissertation Diffusée par Université Joseph-Fourier - Grenoble I, 2005

[32] Meysenc L. Étude des micro-échangeurs intégrés pour le refroidissement des semi-conducteurs de puissance [PhD dissertation]; France: Institut National Polytechnique Grenoble; 1998. pp 117-123

[33] Tuckerman DB, Pease RFW. High performance heat sink for very-large-scale integrated circuits. IEEE Electron Device Letters. 1981;**2**(5):126-129

[34] Ohadi M, Choo K, Dessiatoun S, Cetegen E. Next Generation Microchannel Heat Exchangers. Nature Switzerland: Springer; 2013

[35] Colgan EG, Furman B, Gaynes M, Graham WS, LaBianca NC, Magerlein JH, et al. A practical implementation of silicon microchannel coolers for high power chips. IEEE Transactions on Components and Packaging Technologies. 2007;**30**(2): 218-225

[36] ITRS. International Technology Roadmap for Semiconductors, 2007 Edition. Washington, DC: International Roadmap Committee. Semiconductor Industry Association; 2007

[37] Marcinichen JB, Lamaison N, Thome JR. Electronic micro-evaporator cooling systems and flow control. In: Thome JR, Kim R, editors. Encyclopedia of Two-Phase Heat Transfer and Flow. Singapore: World Scientific Publishing; 2016

[38] Webb RL. Next generation devices for electronic cooling with heat rejection to air. Journal of Heat Transfer. 2005; **127**(1):2-10

[39] Burnett J. Advances in vapor compression electronics cooling.

Electronics Cooling Magazine. PA, USA: Lectrix; June 2014. pp. 28-32. Available from: http://aspensystems.com/ wp-content/ uploads/2014/07/ Electronics - Cooling-June-2014-Aspen-Article.pdf

[40] Barbosa JR Jr, Ribeiro GB, de Oliveira PA. A state-of-the-art review of compact vapor compression refrigeration systems and their applications. Heat Transfer Engineering. 2012;**33**(4–5):356-374

[41] Phelan PE, Catano J, Michna G, Gupta Y, Tyagi H, Zhou R, et al. Energy efficiency of refrigeration systems for high-heat-flux microelectronics. Journal of Thermal Science and Engineering Applications. 2010;**2**(3):1-6

[42] Shannon MA, Phillpott ML, Miller NR, Bullard CW, Beebe DJ, Jacobi AM, Hrnjak PS, Saif T, Aluru N, Sehitoglu H, Rockett A, Economy J. Integrated mesoscopic cooler circuits (IMCCS). In: Proc. ASME-AES. Vol. 39. Nashville, TN: International Mechanical Engineering Congress and Exhibition; 1999. pp. 75-82

[43] de Oliveira P A., Barbosa J R. Jr., Thermal Design of a Spray-Based Heat Sink Integrated with a Compact Vapor Compression Cooling System for Removal of High Heat Fluxes. Heat Transfer Engineering, 2015;36:14-15, 1203-1217

[44] Mancin S, Zilio C, Righetti G, Rossetto L. Mini vapor cycle system for high density electronic cooling applications. International Journal of Refrigeration. 2013;**36**(4): 1191-1202

[45] Kim J. Spray cooling heat transfer: The state of the art. International Journal of Heat and Fluid Flow. 2007; **28**(4):753-767

[46] Cheng WL, Zhang WW, Chen H, Hu L. Spray cooling and flash evaporation cooling: The current development and application. Renewable and Sustainable Energy Reviews. 2016;**55**:614-628

[47] Shicheng J, Yibin J, Liping P, Yu Z, Wenxuan Y. Investigation of spray cooling performance on different heat transfer surfaces. In: Proceedings of the International Conference on Logistics, Engineering, Management and Computer Science. Atlantis Press; 2014

[48] Thibault D. Étude du refroidissement par impact de jets à travers une paroi mince et avec un écoulement cisaillant amont: application aux aubes de turbines [PhD dissertation]. École Nationale Supérieure de Mécanique et d'Aérotechnique. Poitiers, France; 2009

[49] Kercher D, Lee JB, Brand O, Allen M, Glezer A. Microjet cooling devices for thermal management of electronics. IEEE Transactions on Components and Packaging Technologies.2006;**26**:359-366

[50] Horacek B, Kiger K, Kim J. Single nozzle spray cooling heat transfer mechanisms. International Journal of Heat and Mass Transfer. 2005;**48**(8): 1425-1438

[51] Horacek B, Kim J, Kiger K. Spray cooling using multiple nozzles: Visualization and wall heat transfer measurements. IEEE Transactions on Device and Materials Reliability. 2004; **4**(4):614-625

[52] Newton E. Vapor chamber cooling finds growing role in hot products. Embedded; 30 May 2021. Available from: https://www.embedded.com/ vapor-chamber-cooling- finds- growing-role-in-hot-products/ [Accessed: 3 August 2021]

[53] Grover GM, Cotter TP, Erickson GF. Structure of very high thermal conductance. Journal of Applied Physics. 1964;**35**:1990

[54] Glover G, Chen Y, Luo A, Chu H. Thin vapor chamber heat sink and embedded heat pipe heat sink performance evaluations. 25th Annual IEEE Semiconductor Thermal Measurement and Management Symposium, 2009, pp. 30-37. DOI: 10.1109/STHERM.2009.4810739

[55] Coolermastercorp: https://web2. coolermastercorp. com/ wp-content/ uploads /2018/01/ ultra - slim - vapor-chamber-2.png

# Permissions

All chapters in this book were first published by InTech Open; hereby published with permission under the Creative Commons Attribution License or equivalent. Every chapter published in this book has been scrutinized by our experts. Their significance has been extensively debated. The topics covered herein carry significant findings which will fuel the growth of the discipline. They may even be implemented as practical applications or may be referred to as a beginning point for another development.

The contributors of this book come from diverse backgrounds, making this book a truly international effort. This book will bring forth new frontiers with its revolutionizing research information and detailed analysis of the nascent developments around the world.

We would like to thank all the contributing authors for lending their expertise to make the book truly unique. They have played a crucial role in the development of this book. Without their invaluable contributions this book wouldn't have been possible. They have made vital efforts to compile up to date information on the varied aspects of this subject to make this book a valuable addition to the collection of many professionals and students.

This book was conceptualized with the vision of imparting up-to-date information and advanced data in this field. To ensure the same, a matchless editorial board was set up. Every individual on the board went through rigorous rounds of assessment to prove their worth. After which they invested a large part of their time researching and compiling the most relevant data for our readers.

The editorial board has been involved in producing this book since its inception. They have spent rigorous hours researching and exploring the diverse topics which have resulted in the successful publishing of this book. They have passed on their knowledge of decades through this book. To expedite this challenging task, the publisher supported the team at every step. A small team of assistant editors was also appointed to further simplify the editing procedure and attain best results for the readers.

Apart from the editorial board, the designing team has also invested a significant amount of their time in understanding the subject and creating the most relevant covers. They scrutinized every image to scout for the most suitable representation of the subject and create an appropriate cover for the book.

The publishing team has been an ardent support to the editorial, designing and production team. Their endless efforts to recruit the best for this project, has resulted in the accomplishment of this book. They are a veteran in the field of academics and their pool of knowledge is as vast as their experience in printing. Their expertise and guidance has proved useful at every step. Their uncompromising quality standards have made this book an exceptional effort. Their encouragement from time to time has been an inspiration for everyone.

The publisher and the editorial board hope that this book will prove to be a valuable piece of knowledge for researchers, students, practitioners and scholars across the globe.

# List of Contributors

**Sergio García and Alfredo Trueba**
Biofouling Group Research, University of Cantabria, Santander, Spain

**Animesh Talapatra**
MCKVIE, Howrah, West Bengal, India

**Debasis Datta**
IIEST, Howrah, West Bengal, India

**Suvanjan Bhattacharyya**
Center for Renewable Energy and Environment Development (CREED), Department of Mechanical Engineering, Birla Institute of Technology and Science, Pilani, Vidyavihar, Rajasthan, India

**Devendra K. Vishwakarma and Sanghati Roy**
Department of Mechanical Engineering, Birla Institute of Technology and Science, Pilani, Vidyavihar, Rajasthan, India

**Ranjib Biswas**
Department of Mechanical Engineering, MCKV Institute of Engineering, Howrah, West Bengal, India

**Moghimi Ardekani**
Department of Design and Engineering, Staffordshire University, Stoke-on-Trent, UK

**Rania Jradi, Ali Fguiri and Razak Jeday**
Research Laboratory "Process, Energy, Environment and Electrical Systems", National Engineering School of Gabes (ENIG), Gabes, Tunisia

**Christophe Marvillet**
CMGPCE Laboratory, French Institute of Refrigeration (IFFI), National Conservatory of Arts and Crafts (CNAM), Paris, France

**Hussain H. Al-Kayiem**
Mechanical Engineering Department, Universiti Teknologi PETRONAS, Perak, Malaysia

**Muna S. Kassim and Saud T. Taher**
Mechanical Engineering Department, Al-Mustansiriyah University, Baghdad, Iraq

**Sergey Voronin**
Research Scientist, Amathsolutions, USA

**Rakhab C. Mehta**
Department of Aeronautical Engineering, Noorul Islam Centre for Higher Education, India

**Xing Chen and Yao Sun**
School of Science, Xi'an University of Technology, Xi'an Shaanxi, China

**Qiang Huang**
State Key Laboratory of Eco-Hydraulic Engineering in Shaanxi, Xi'an University of Technology, Xi'an Shaanxi, China

**Tao Min**
School of Science, Xi'an University of Technology, Xi'an Shaanxi, China
State Key Laboratory of Eco-Hydraulic Engineering in Shaanxi, Xi'an University of Technology, Xi'an Shaanxi, China

**Bayram Kılıç**
Technical Sciences Vocational School, Burdur Mehmet Akif Ersoy University, Burdur, Turkey

**Miguel Rosa Oliveira Panão**
ADAI, LAETA, Department of Mechanical Engineering, University of Coimbra, Coimbra, Portugal

**Shahin Kharaji**
Research and Develop Department, FAPKCO Engineering Group, Shiraz, Iran

**Abdelhanine Benallou**
SIGMA TECH, Rabat, Morocco

# Index

**R**



**S**



**T**



**V**

www.ingramcontent.com/pod-product-compliance
Lightning Source LLC
LaVergne TN
LVHW080739210926
848485LV00003B/582